# BRITISH NATURAL HISTORY BOOKS

1495–1900

**By the same author**

*The Works of Charles Darwin* (1965; 2nd edition by Dawson 1977)
*The Classification of the Animal Kingdom* (English University Press, 1971)
*Charles Darwin, A Companion* (Dawson 1978)
*P. H. Gosse: A Bibliography* (with Douglas Wertheimer, Dawson 1980)

# BRITISH NATURAL HISTORY BOOKS

## 1495–1900
## A Handlist

by R. B. FREEMAN

Dawson · Archon Books

First published in 1980



Wm Dawson & Sons Ltd, Cannon House
Folkestone, Kent, England

Archon Books, The Shoe String Press, Inc
995 Sherman Avenue,
Hamden, Connecticut 06514
USA

British Library Cataloguing in Publication Data

Freeman, Richard Broke
British natural history books, 1495–1900.
1. Natural history – Bibliography
I. Title
016.5009 Z7401 80-50228

ISBN 0 7129 0971 0
ISBN 0 208 01790 0 (Archon)

Printed and bound in Great Britain
by W & J Mackay Limited, Chatham

# CONTENTS

# INTRODUCTION

This handlist pretends to no more than its title, something more than a short title list and something far less than a bibliography. It contains all relevant books, including works in many volumes as well as pamphlets and single sheets, which have come to my notice; but it excludes all serials. Reprints and preprints of matter which also appeared in a serial are included, but mere offprints are not, although the distinction is sometimes hard to make.

The term natural history is often used today in a disparaging sense to describe the activities of bird lovers and flower pressers. Its true and older meaning is the study and gathering of facts on any subject which is not of ecclesiastical, military or civil significance, or more narrowly the history of wild animals and plants. It is only within this century, and only within the last few years by most people, that it has been realised that such studies, even the duller ones of finding out what kinds of wild animals and plants there are, are essential to the conservation of the environment. Study of nature was made, and still is made, for pleasure, or, if justification was needed, to examine the attributes of God through his creation. Ecology, a word coined a little over one hundred years ago, has become the respectable term for natural history out of doors, and biology for natural history in the laboratory.

This list then excludes all studies of living things which are of direct practical application to man's life, all medicine, agriculture, horticulture and gardening, hunting, shooting and fishing. Honey bees and silkworms are not British native animals and are excluded; the medicinal leech is a native and its small literature is included. Geology and palaeontology have also been excluded, although with regret for much of their earlier study was without doubt pure natural history. However, there are a number of books in all these fields which contain important studies which must be admitted. Much comparative anatomy is contained in medical works; almost all our early knowledge of British plants comes from herbals whose intention was medical. Sporting books, particularly those on game birds and fish, contain important natural history observations and have to be included. Geology has made important contributions to the study of evolution.

Topographical books are particularly difficult, but on the whole they have been excluded. Some of the earlier ones, such as Robert Plot's *Oxfordshire* (1677) describe themselves as natural histories, but contain mostly topography with a few curiosities. The earliest of all, Boate's *Ireland's natural history* (1652) contains no natural history because the author died before writing a second volume which was to contain it; it is included only as a courtesy. Later topographies often have lists of plants and animals and may be included for this reason.

What is contained then is, firstly, all books about the fauna and flora of the British Islands, including all Ireland, the Channel Islands, and, oddly but correctly, Heligoland. Secondly, all general natural history books written by inhabitants of these islands, and thirdly translations into English of relevant books by foreign authors and a few American books which have appeared in British editions. Translations of British books into foreign languages and American printings of them are excluded. The language is usually English, although early books and some botany books of the eighteenth century are in Latin; a few are in both French and English, one is polyglot, and a few are in Gaelic languages. Place of publication is usually within the British Isles, but foreign publication is not excluded provided that the books fulfil the other requirements.

There are certain categories in which the books have proved to be intractable and I am sure that the entries in these are both imperfect and incomplete. Catalogues and lists are a good example; they are usually pamphlets and ephemeral. Full dress catalogues and learned lists are important milestones and can usually be found, but there are numerous lists for exchange or labelling purposes, especially during the great collecting days of the nineteenth century when every large town had its taxidermist and purveyor of natural history specimens. These were usually based on some standard fauna or flora. I have entered them when I have seen them, but they are unimportant. Another category is that of the commercial lists which give the prices for which natural history objects may be bought, and those of such objects as can be bid for at auction. Those for bird skins or eggs and for the more showy Lepidoptera are the most frequent, but I have seen few and have omitted almost all. There are however some, such as Alford Lloyd's lists of living material for aquaria (1858–60), or Bolton's marvellous *Portfolios* (1879–84) which contain much information which is not available elsewhere and fully deserve inclusion.

Instruction manuals form another group of ephemeral and difficult works. Those most commonly seen, and dating back to the early years of the eighteenth century, are for microscopes, but they also occur for sets of microscope and magic lantern sliders, for wall charts and even, in one case for a box of natural history objects for the use of schools. When divorced from their objects they are not always easy to distinguish. Handbooks or catalogues to museums, raree shows, menageries or botanic gardens were printed over and over again as stock changed, and few survive.

Children's books are exceptionally difficult. They range from the deliberately pedagogical to the chapbooks of the country fair, and, as most copies were given to the destructive youth for whom they were intended, the few that survive are usually in poor condition. Many of them were reprinted, often undated, over and over again. I have included these as best I can, drawing the line only at animal and plant alphabets. I have also entered a few school books such as Comenius *Orbis pictus* (1659), which though not solely natural history books, contained pictures of animals and plants for children long before there were any books intended for pleasure.

However, children's books have been assiduously collected for a long time and there is an extensive literature on them. More difficult are the little natural histories intended for beginners of all ages. These, simply written and produced at moderate cost, begin early in the nineteenth century and become a flood before its close. Their content is usually derivative, although some were written by the learned to make some money, but they are important because they often induced their readers to continue to more serious study. They are not to be found in any but the largest libraries and have until recently been neglected by collectors. The better of them went into many editions or reprints, but they are difficult to survey satisfactorily.

Plant literature contains a considerable element of lore and legend, as well as such things as sentiments of flowers and floral clocks; these have been left out as have anthropomorphic animal books. Plants are also almost the sole subject of specimen books, although a few hydroids and polyzoans are sometimes found. These have been included, but there are border line examples of what are really herbaria duplicates in which the names have been printed on the sheets. There were also available, from the middle of the nineteenth century, books with printed names and sometimes a little printed text, into which the purchaser was meant to insert the specimens. In some cases, the same books may have been available with the specimens already inserted, especially so with seaweed albums. Specimen books are, by their nature, scarce and printed sheets are mostly in the herbaria for which they were intended, but albums, usually only with some of the specimens, are common.

The poetry of nature presents another vague border line. Clearly, Erasmus Darwin's *Botanic garden* (1791, 90) must be included, and I have admitted Edward Newman's *Insect hunters* (1857), but only just. Sarah Hoare's two books (1825, 1831) I have excluded; the verse is bad and the brief notes derivative, although Priscilla Wakefield saw fit to print one of them in some late editions of her *An introduction to botany*. Darwinian verse is also excluded. Books in epistolary form have to be included; indeed Kirby and Spence *An introduction to entomology* (1815–16) is an important book, and Gilbert White's *Selborne* (1789) is the only natural history classic. So too do conversation pieces deserve a place, although they are mostly juveniles. Novels are few, but Francesca Steele's bizarre *Mr Mygale's hobby* (1889), as well as two of her others, deserve entries. The first contains a complete list of British spiders.

Evolution books are an important element in the study of nature, but I have included, in general, only those whose authors were natural historians or scientists, and excluded those whose authors had religious or political backgrounds. Robert Chambers was a publisher, nevertheless it was felt essential to include his *Vestiges of the natural history of creation* (1844); Samuel Butler was a novelist and man of letters but his evolutionary essays demand a place. Henry Drummond's *Ascent of man* (1894), which appeared in a binding which was a close mimic of that on the current printings of *The origin of species*, is excluded.

Entries in the main list are alphabetical by authors, and, within any one author, in date order. Anonymous works are entered alphabetically in the same sequence, excluding first articles, unless the author is known. In the latter cases they are entered under the author's name in brackets, with cross references under the titles. Cross references are also given to pseudonyms and to works which are signed with initials only. Authors' names are given in full, with forenames where possible, whatever may appear on the title page. Titles are given only in sufficient length to indicate the contents, in full where short, but severely reduced in the characteristically verbose ones of the eighteenth century.

The rest of each entry consists of a brief description of the first edition of each book, followed by a summary of its later printing history if any, and, in some cases brief notes. So far as format is concerned, all octavos which are smaller than demy are entered as Sm. 8vo; larger octavos are entered as royal or imperial. Most books which are signed and sewn in twelves, except quartos and folios, are entered as duodecimo, although many nineteenth century ones may be octavos impressed in sheet and a half. Where a book occurs in more than one format, and some occur in as many as four, this is stated or entered in the notes as L.P. The given pagination is usually confined to those pages which are actually numbered, and first and last blanks are ignored. In a few cases where considerable preliminary matter, or even a whole book, is unpaginated the pagination is given in brackets.

This is followed by the number of plates, if any, and indication as to whether some or all of the plates are coloured. Copies of many books occur with the plates in coloured or uncoloured state, whether the colouring is by hand or printed. This is not indicated unless there is something peculiar about it when the notes will indicate that, for example, coloured plates occur only in large paper copies, or that very few copies were coloured. The presence of text figures is indicated, without their number unless there is only one. Text figures cannot be lacking, except in rare instances where the text occurs in more than one state, unless the text is itself defective. The few books in which the text figures are coloured are noted, although, as with plate books, they may be found both coloured and plain.

The publisher's name is given, where known, in abbreviated form, but, in books which are joint or multiple ventures, only the first is given. Similarly, where books are published from more than one place only the first is given. With books published for the author, so-called privately printed books, or books printed for private distribution, the printer is given where known. The date of publication, if it is neither in brackets nor parentheses, is the date given on the title page, or, in a work of more than one volume, the range of dates. If it is in brackets, without a query mark or a *c.* for *circa*, then it is known from other sources but is not given anywhere in the book. If it is in parentheses, then it is given somewhere else in the book other than on the title page. Complex dates indicate books which came out in parts. For example [1823–]1831 [–1832] indicates that the first part, which was not dated, came out in 1823; that the whole work is dated 1831 on the title page, but that some matter did come out, without a date, in the following year, although it is an integral part of the book in final form.

This brief description of each book is followed by a summary of the later history of its printing. In most cases this has been done by listing, in date order, those editions and issues which have come to my notice. The word edition is taken in a publisher's sense and should not be considered bibliographical. In many cases, perhaps in more cases than not, what a publisher calls a second or later or new edition, is no more than a reprint, in the nineteenth century often from plates, or even a reissue of the original sheets with a new title leaf. Where a book occurs merely with a changed date on the title page but with no indication of edition I have entered the new date alone. In a few cases, mostly of minor popular books which were reprinted again and again, I have noted only 'later editions' or 'many later editions'. I have done this also in the case of one important author, Charles Darwin. In his case, all editions in his lifetime which have altered text, as well as a few in which the text has been revised by his son Francis after his death, have been entered; but the very numerous reprints have been ignored. No further information is given about later editions except where there is change in format or in places of publication.

It has not been easy to summarize the later printing history of each book, and the result is in many cases far from complete. It is well known that second and later editions are far rarer in libraries than are firsts, and that booksellers tend to despise them. Those few authors for whom adequate bibliographies or even just lists are available are easy, and the summaries for them are tolerably complete. For most only entries in subject bibliographies are available and even these are few. Such as there are are mostly confined to plate books and in these the details are often too complicated to be indicated in the brief notes given here.

The notes which, in some cases, follow the printing history are intended only to draw attention to the more obvious bibliographical peculiarities of any particular book which are not noticed in the description of the first edition. A glance at a few of them will show what they cover.

The main alphabetical list is followed by a list of brief titles in date order from the beginning until 1800. During this three hundred years, about three hundred and fifty books appeared for the first time and amongst them are the beginnings of all the branches of natural history. One group alone begins and dies, the true herbals. These contained not only descriptions of the medicinal properties of the plants, but also the rudiments of plant description, classification and ecology. The subjects split late in the seventeenth century, medicine taking the pharmacy whilst the rest becomes straight botany. Nothing is left of the herbals except superstition and folk remedies. No other subject actually dies out, but creation books, those which survey the power and wisdom of God in his creation, become few. They had their peak in the early eighteenth century, but they go on, thanks to William Paley and to the Bridgewater treatises, until much later. No subject arises entirely for the first time, but some begin late and some have diffuse beginnings.

The earliest rudiments of evolutionary ideas in English begin with Erasmus Darwin, but it must not be forgotten that the examples of divine wisdom in adaptive purpose can be equally used as examples of natural, rather than divine, means of selection. Animal distribution, another tool of the evolutionists, cannot begin properly until naturalists accompanied voyages of exploration, although there was a lot of scattered information available from much earlier. True ecology, the study of the relationships of plants and animals to each other and to their physical environment, only begins late in the last century. Its seeds lie in those early naturalists who went out into the field and looked, rather than just collected; Ray and Willughby perhaps for animals and, much earlier, William Turner and some of the herbalists, for plants.

The subject index has been constructed mainly from indications given in the titles of the books, although in some cases, where the title does not show sufficiently what the book is about, the known content has been used. In subjects which contain a large number of entries, the main heading is subdivided as follows, first the general books, then those about Europe, those about the British Isles in general, those about particular places, those about particular kinds, and lastly divisions of the main heading which are none of these, in alphabetical order. The heading 'Juveniles' is arbitrary. Many of the lesser books which may have been intended for adult beginners are just as suitable for children and vice-versa. If a book says that it is written for children, then it is entered under that heading; in other cases I have used my judgement.

Numbers in the index which are printed in italic indicate those works which, in my opinion, are the most important contributions to the subject of the entry.

ACKNOWLEDGEMENTS

The most important source of information has been the books themselves, and I am grateful to the many librarians who have given me unfailing help. Most particularly, I would like to thank Mr J. W. Scott, Librarian of University College London, whose library has been my daily resource over many years. Many of his staff have given me their advice and help, but I would like particularly to name Miss Joan Nash of the Natural Sciences Library, and Mr John Spiers whose job is to solve difficulties. Mr M. J. Rowlands and his staff at the British Museum (Natural History), the finest natural history library in the world, have also shown me many kindnesses.

However, even great libraries do not carry long runs of editions and issues and they often do not hold the lesser and more popular works which make so important a contribution to the development of natural history, or, if they do, they hold the first edition only. It is here that I am grateful to the antiquarian and secondhand booksellers who have not only allowed me to browse on their shelves but have brought unusual books to my notice. Particular thanks must go to Howard Radclyffe of Bernard Quaritch Ltd, to John Collins, once with the same firm but now with Sotheby Parke Bernet & Co.; to Charles Kirke Swann* and his brother Howard of Wheldon & Wesley Ltd; and Dr M. E. Korn. These and many others are thanked not only for the hospitality of their shelves but also for their catalogues which contain the wisdom accumulated over many years and many copies.

The last great sources are bibliographies and lists for people or subjects and catalogues of learned libraries. I have listed some of these below, but there are many omissions particularly of shorter works in learned journals. Such works are accepted as the tools of the trade and their compilers and authors are seldom thanked. Without them this list could never have been put together. Finally, although I have been given much help, the omissions and the mistakes remain my own.

*University College London* R. B. FREEMAN
*June 1980*

* Charles Kirke Swann died in July 1980.

# REFERENCES

AGASSIZ, Jean Louis Rodolphe: Hugh Edwin Strickland, *editor Bibliographia zoologiae et geologiae. A catalogue of books, tracts and memoirs on zoology and geology.* 8vo, 4 vols, London, Ray Society 1848–54.

ANKER, Jean *Bird books and bird art: an outline of the literary history and iconography of descriptive ornithology, based principally on the collection in the University Library at Copenhagen.* 8vo, Copenhagen, Levin and Munksgaard 1938.

BEWICK, Thomas: Sydney Roscoe *Thomas Bewick; a bibliography raisonné.* 8vo, London, Oxford University Press 1953.

BIBLIOTHÈQUE NATIONALE *Catalogue général des livres imprimés.* 8vo, 218 vols, Paris, Imprimerie Nationale 1896–1973. In continuation.

BRITISH MUSEUM, Department of Printed Books [British Library] *General catalogue of printed books to 1955.* Fol., 263 vols, London, Trustees of the British Museum 1959–66. *Supplements* in continuation.

BRITISH MUSEUM (NATURAL HISTORY) *Catalogue of the books, manuscripts, maps and drawings in the British Museum (Natural History).* 4to, 5 vols, London, by order of The Trustees 1903–1915; *Supplement*, 3 vols, 1922–40.

BRITTEN, James & BOULGER, George Simonds: Alfred Barton Rendle, *editor A biographical index of deceased British and Irish botanists.* Second edition. 8vo, London, Taylor and Francis 1931. For third edition see Ray Desmond.

CHICAGO, FIELD MUSEUM: John Todd Zimmer *Catalogue of the Edward E. Ayer ornithological library, Field Museum, Chicago.* 8vo, 2 vols, Chicago, Field Museum 1926. *Zoological Series Publications* Vol. 16

DARWIN, Charles Robert: Richard Broke Freeman *The works of Charles Darwin: an annotated bibliographical handlist.* Second edition. 8vo, Folkestone, Dawson 1977.

DEAN, Bashford: Charles Rochester Eastman & Eugene Willis Gudger *editors A bibliography of fishes.* 8vo, 3 vols, New York, American Museum of Natural History 1916–23.

DEBY, Julien Marc *Bibliotheca Debyana; being a catalogue of books and abstracts relating to . . . microscopy. Vol. 1. Serials, Microscope, Protozoa.* 8vo, London, privately printed 1889. All published.

DESMOND, Ray *Dictionary of British and Irish botanists and horticulturalists, including plant collectors and botanical artists.* 8vo, London, Taylor and Francis 1977. This is a third edition of Boulger and Britten q.v.

DUNTHORNE, Gordon *Flower and fruit prints of the 18th and early nineteenth centuries.* Fol., Washington, for the author 1938.

ENGELMANN, Wilhelm *Bibliotheca Historico–Naturalis: Verzeichniss der Bücher über Naturgeschichte . . . 1700–1846 erschienen sind. Vol. 1 Vermischte Schriften, Zoologie, Palaeontologie.* 8vo, Leipzig, Engelmann 1846. All published; *Supplement . . . 1846–60*, 1861; *Supplement II . . . 1861–1880*, 1886.

FREEMAN, Richard Broke Children's natural history books before Queen Victoria. *Bull. Hist. Education Soc.*, No. 17: 7–21, No. 18: 6–34, 1976.

GILBERT, Pamela *A compendium of the biographical literature on deceased entomologists.* 8vo, London, British Museum (Natural History) 1977.

GOSSE, Philip Henry: Richard Broke Freeman & Douglas Lloyd Wertheimer *Philip Henry Gosse: a bibliography.* 8vo, Folkestone, Dawson 1980.

GOSSE, Philip Henry: Peter Stageman *A bibliography of the first editions of Philip Henry Gosse, F.R.S.* 8vo, Cambridge, Golden Head Press 1955.

HAGEN, Hermann August *Bibliotheca entomologica.* 8vo, 2 vols, Leipzig, Engelmann 1862–63.

HALKETT, Samuel & LAING, John: James Kennedy, William Allan Smith, & Alfred Forbes Johnson *editors Dictionary of anonymous and pseudonymous English literature.* New edition. 4to, 9 vols, Edinburgh, Oliver and Boyd 1926–62.

HARVARD UNIVERSITY, MUSEUM OF COMPARATIVE ZOOLOGY *Catalogue of the Library of the Museum of Comparative Zoology.* 4to, 8 vols, Boston, Hall 1967.

HENREY, Blanche *British botanical and horticultural literature before 1800.* 4to, 3 vols, London, Oxford University Press 1974.

HORN, Walther & SCHENKLING, Sigmund *Index litteraturae entomologicae, Series I: Die Welt-Literature über die gesamte Entomologie bis inklusive 1863.* 8vo, 4 vols in 2, Berlin, Deutsch Entomologische Gesellschaft 1928–29; *Series II: 1864–1900*, by Wilhelm Oerksen & Ulrich Scheiding.

HORTICULTURAL SOCIETY OF LONDON *The Lindley Library; catalogue of books, pamphlets, manuscripts and drawings.* 8vo, London, The Society 1927.

HUDSON, William Henry: John R. Payne *W. H. Hudson, a bibliography.* 8vo, Folkestone, Dawson 1977.

HUNT BOTANICAL LIBRARY: John Quinby & Allan Stevenson *editors Catalogue of botanical books in the collection of Rachel McMasters Miller Hunt.* 8vo, 2 vols in 3, Pittsburg, The Library 1958–61.

IRWIN, Raymond *British bird books; an index of British ornithology, A.D. 1481 to A.D. 1948.* 8vo, London, Grafton 1951.

JACKSON, Benjamin Daydon *Guide to the literature of botany; being a classified selection of botanical works, including nearly 6,000 titles not given in Pritzel's 'Thesaurus'.* 8vo, London, Index Society 1881. *Publications* Vol. VIII.

KNIGHT, David Marcus *Natural science books in English 1600–1900.* 8vo, London, Batsford 1972.

LIBRARY OF CONGRESS *The national union catalog; pre-1956 imprints.* Fol., 644 vols, Chicago & London, Mansell 1968–198–. In continuation.

LINNAEUS, Carolus: Basil Harrington Soulsby *A catalogue of the works of Linnaeus (and publications more immediately relating thereto) preserved in the Libraries of the British Museum (Bloomsbury) and the British Museum (Natural History) (South Kensington).* Second edition. 4to, London, British Museum 1933; *An index to the authors (other than Linnaeus).* London, British Museum 1936.

LINNEAN SOCIETY OF LONDON *Catalogue of the printed books and pamphlets in the Library of the Linnean Society of London.* 8vo, London, The Society 1925.

LISNEY, Arthur Adrian *A bibliography of British Lepidoptera 1608–1799.* 8vo, London, privately printed at the Chiswick Press 1960.

McGILL UNIVERSITY, BLACKER-WOOD LIBRARY *A dictionary catalogue of the Blacker-Wood Library.* 4to, 9 vols, Boston, Hall 1966.

McGILL UNIVERSITY, OSLER LIBRARY: William

Willoughby Francis, Reginald Harrison Hill & Archibald Malloch *editors Bibliotheca Osleriana: a catalogue of books illustrating the history of medicine and science collected . . . by Sir William Osler.* 8vo, Montreal, McGill-Queen's University Press 1969.

MULLENS, William Herbert & SWANN, Harry Kirke *A bibliography of British ornithology from the earliest times to the end of 1912.* 8vo, London, Macmillan [1916–]1917.

MULLENS, William Herbert, SWANN, Harry Kirke & JOURDAIN, Francis Charles Robert *A geographical bibliography of British ornithology to 1918, arranged under counties.* 8vo, London, Witherby [1919–] 1920.

NISSEN, Claus *Schöne Fischbücher.* 8vo, Stuttgart, Hiersemann 1951.

NISSEN, Claus *Die illustrierten Vögelbücher; ihre Geschichte und Bibliographie.* 4to, Stuttgart, Hiersemann 1953.

NISSEN, Claus *Die botanische Büchillustration; ihre Geschichte und Bibliographie.* Second edition. 4to, Stuttgart, Hiersemann 1966.

NISSEN, Claus *Die zoologische Büchillustration; ihre Bibliographie und Geschichte.* 4to, 2 vols, Stuttgart, Hiersemann 1966–78[–1979].

PLESCH, Arpad *Mille et un livres botaniques; repertoire bibliographique de la bibliotheque Arpad Plesch.* 4to, Brussels, Arcade 1973.

PLESCH, Arpad [Sale catalogue]. *The magnificent botanical library of the Stiftung für Botanik, Vaduz, Liechtenstein, collected by the late Arpad Plesch.* 4to, 3 vols, London, Sotheby 1975–76.

PRITZEL, Georg August *Thesaurus literaturae botanicae omnium gentium inde a rerum botanicorum initiis ad nostra usque tempora.* Second edition. 4to, Leipzig, Brockhaus 1871–77.

PRITZEL, Georg August *Iconum botanicarum index locupletissimus. An alphabetical register . . . of phanerogamic plants and ferns, compiled from botanical and horticultural publications of the XVIIIth and XIXth centuries.* New edition. 4to, Leipzig, Nicolaische Verlag 1872 [–1877].

RAY, John: Geoffrey Langdon Keynes *John Ray, a bibliography.* 8vo, London, Faber & Faber 1951. Reprinted with additions and corrections. Amsterdam 1974.

RAY SOCIETY: Richard Curle *The Ray Society; a bibliographical history.* 8vo, The Society 1954.

READING UNIVERSITY: Nellie Barbara Eales *The Cole Lib-*

*rary of early medicine and zoology: Catalogue of books and pamphlets. Part I 1472–1800.* 4to, Reading, Alden Press for the Library 1969. All published.

ROSCOE, Sydney *John Newbery and his successors 1740–1814; a bibliography.* 8vo, Wormley, Herts, Five Owls Press (1973).

ROTHAMSTED EXPERIMENTAL STATION: Mary S. Aslin *compiler Library catalogue of printed books and pamphlets on agriculture published between 1471 and 1840.* 8vo, [Harpenden] 1940.

ROYAL SOCIETY OF LONDON *Catalogue of scientific papers (1800–1900)*. 4 series, 19 vols, London & Cambridge, Stationery Office & Cambridge University Press 1867–1925.

SIMPSON, Norman Douglas *compiler A bibliographical index of the British flora.* 8vo, Bournemouth, privately printed 1960.

SITWELL, *Sir* Sacheverell *Bart*, BUCHANAN, Handasyde & FISHER, James Maxwell McConnell *Fine bird books, 1700–1900.* London & New York, Collins & Van Nostrand 1953.

SOWERBY FAMILY: R. J. Cleevely A provisional bibliography of natural history works by the Sowerby family. *J. Soc. Biblphy Nat. Hist.*, 6: 418–481. 1974.

STAFLEU, Frans Antonie *Taxonomic literature.* 8vo, Utrecht, International Bureau for Plant Taxonomy and Nomenclature 1967. *Index Vegetabile* Vol. 52.

TORONTO PUBLIC LIBRARY: Judith St John *The Osborne collection of early children's books, 1566–1910.* 8vo, Toronto, Public Library 1958; *. . . list of recent additions to the Osborne collection . . . 1911–1963*, Toronto, Public Library 1964.

WELLCOME HISTORICAL MEDICAL LIBRARY *A catalogue of printed books in the Wellcome Historical Medical Library. I. Books printed before 1641. II. Books printed from 1641–1850*, A–E, F–L. 4to, London, The Library and Institute 1962, 1966, 1976. In progress.

WHITE, Gilbert: Edward Alfred Martin *A bibliography of Gilbert White; the naturalist and antiquarian of Selborne.* [Second edition]. 8vo, London, Halton 1934.

WOOD, Casey Albert *An introduction to the literature of vertebrate zoology; based chiefly on the titles in the Blacker Library of Zoology . . . and other libraries of McGill University, Montreal.* 4to, London, Oxford University Press 1931.

YALE UNIVERSITY: Sidney Dillon Ripley & Lynette L. Scribner *Ornithological books in the Yale University Library, including the*

*Library of William Robertson Coe.* 8vo, New Haven, Yale University Press 1961.

ZOOLOGICAL SOCIETY OF LONDON: Frederick Herschel Waterhouse *Catalogue of the Library*. Fifth edition. 8vo, London, The Society 1902.

# PART I
# ALPHABETICAL LIST

# A

A., A. *Photographs of British Algae, Cyanotype impressions.* [1843–53]. *see* Atkins, *Miss* A.

A., J. *The woodland companion.* 1802. *see* Aikin, J.

A., M. *Rudiments of conchology.* 1826. *see* Venning, *Miss* A. M.

A., W. H. D. *Wonders of the vegetable world.* 1867. *see* Adams, W. H. D.

—— *Animal life throughout the globe.* 1876. *see* Adams, W. H. D.

1. ABBOT, Charles *Flora Bedfordiensis: comprising such plants as grow wild in the county of Bedford.* 8vo, xii+351 pp, 6 col. pls, Smith, Bedford 1798.

2. ABERNETHY, John *Physiological lectures, exhibiting a general view of Mr. Hunter's physiology, and of his researches in comparative anatomy.* 8vo, 352 pp, Longman, London 1817.

*An accentuated list of British Lepidoptera.* 1858. *see* Oxford and Cambridge Universities Entomological Societies.

Acheta domestica, *pseudonym. see* Budgen, *Miss* M. L.

*Across the common after wild flowers.* 1895. *see* Cooke, M. C.

3. ADAM, William *The gem of the Peak; or, Matlock Bath and its vicinity . . . flora of the High and Low Peak.* 8vo, 256 pp, Longman, London [1838]. Later abridged editions.

4. —— *Brief remarks on the geology of Derbyshire . . . also the botany of the High and Low Peak.* 12mo, 29 pp, text figs, London 2nd ed. 1841. First ed. is in No. 3.

5. ADAMS, Arthur & ADAMS, Henry *The genera of recent Mollusca.* Roy. or demy 8vo, 3 vols, 138 col. pls, text figs, Van Voorst, London [1853–] 1858. Only the Royal copies have the plates coloured.

6. ——, BAIKIE, William Balfour & BARRON, Charles *Manual of natural history for the use of travellers.* Sm. 8vo, viii+749 pp. Van Voorst, London 1854.

7. ADAMS, Estelle Davenport *Illustrated flora textbook.* 32mo, Mack, London 1884.
8. ADAMS, George *the father Micrographia illustrata; or, the knowledge of the microscope explained.* 4to, 16+263 pp, 65 pls, for the author, London 1743–46; 2ed. 1747; 4ed.1771. 3rd ed. is unknown.
9. ADAMS, George *the son Essays on the microscope.* 4to, xxiii+721 pp. 33 pls, for the author, London 1787; 2ed. 1798; 4ed. 1798. Plates fpce, 1–31, 26A. The plates also occur in oblong folio.

Adams, Henry *see* Adams, A.

Adams, Henry B. *see* Adams, H. G.

10. ADAMS, Henry Gardiner *Favorite song birds; containing a popular description of the feathered songsters of Britain.* 8vo, xii+196pp. 12 col. pls, Orr, London 1851; 2ed. 1856; 3ed. 1871. Later editions.
11. —— *Beautiful butterflies; the British species described.* 16mo, xxvi+66 pp, 8 col. pls, text figs, Groombridge, London 1854. Later editions. Young Naturalist's Library.
12. —— *Nests and eggs of familiar British birds.* [1st series]. 16mo, iv+78 pp, 8 col. pls, Groombridge, London 1854. Later editions, some in one vol. with second series. Young Naturalist's Library.
13. —— *Beautiful shells.* 16mo, 78 pp, 8 col. pls, Groombridge, London 1855. Later editions. Young Naturalist's Library.
14. —— *The story of the seasons.* 16mo, 141 pp, Dean, London 1855; 2ed. 1855.
15. —— *The grammar and dictionary of flowers.* 16mo, 110 pp, Dean, London [1856]. Later editions.
16. —— *The sea side lesson book.* 16mo, 236 pp. Groombridge, London 1856.
17. —— *Nests and eggs of familiar British birds. Second series.* 16mo, xvi+66 pp, 8 col. pls, text figs, Groombridge, London 1857. Later editions, some in one vol. with first series. Young Naturalist's Library.
18. —— *The wild flowers, birds & insects of the months.* 8vo, vi+316 pp, 1 pl., text figs, Hogg, London [1862]; 2ed. 1868; ed. [c.1870] title *The seasons and their beauties*: ed. [1878] title *The wonders and beauties of the year.*
19. —— *Our feathered families . . . birds of song.* 8vo, 301 pp, 1 pl., text figs, Hogg, London [1863]. Later editions.
20. —— *Our feathered families . . . game and water birds.* 8vo,

345 pp, text figs, Hogg, London [1863]. Later editions.

21. —— *Our feathered families . . . birds of prey.* 8vo, 320 pp, text figs, Hogg, London [1863]. Later editions.

—— *The seasons and their beauties.* [c.1870]. *see* No. 18.

—— *The wonders and beauties of the year.* [1878]. *see* No. 18.

—— *The young naturalist. A handy volume upon the collection, preservation, and arrangement of butterflies and shells.* Sm. 8vo, vii+133+156 pp, 16 col. pls, text figs, Groombridge, London 1879. A combination volume of Nos 11 & 13, with separate subtitle leaves.

22. —— & ADAMS, Henry B. *The smaller British birds.* 8vo, iv+252 pp, 32 col. pls, Bell, London 1874 [=1873]. Later editions.

23. ADAMS, Lionel Ernest *The collector's manual of British land and freshwater shells.* 8vo, 125 pp, 9 col. pls, Bell, London 1884; 2ed. Leeds 1896.

24. [ADAMS, William Henry Davenport] *Wonders of the vegetable world*, 8vo, 127 pp, text figs, Nelson, London 1867. Preface signed W.H.D.A.

25. [——] *Animal life throughout the globe.* 8vo, xi+704 pp, pls, Nelson, London 1876. Preface signed W.H.D.A.

26. —— *The bird world described with pen and pencil.* Roy. 8vo, xii+464 pp, text figs, Nelson, London 1878; 1880.

27. ADAMSON, Charles Murray *Sundry natural history scraps, more especially about birds.* 8vo, 98 pp, privately printed, Newcastle upon Tyne 1879.

28. —— *Some more scraps about birds.* 8vo, vi+273 pp, 43 pp, Bell, Newcastle upon Tyne 1880–1881. Plates also issued alone, 1881, title *Studies of birds.*

—— *Studies of birds.* 1881. *see* No. 28.

29. —— *Another book of scraps principally relating to natural history.* Obl. fol., 56 pp, 36 pls, Reid, Newcastle upon Tyne 1882.

30. —— *Some more illustrations of wild birds.* Obl. 4to, no text, 24 pls. Mawson, London 1887. 200 copies.

31. ADLER, Hermann *Alternating generations. A biological study of oak galls and gall flies.* Sm. 8vo, xl+198 pp, 1 plain & 2 col. pls, text figs, Clarendon, Oxford 1894. Translation of paper in *Z. wiss. Zool.*, Vol. 35, by Charles R. Straton.

Admiralty Hydrographic Department *see* Broderip, W. J.

*Admiralty manual of scientific enquiry.* 1849. *see* Herschel, *Sir* John F. W. *editor.*

32. *The adventures of the great whale whose skeleton is now on exhibition.* 12mo, 16 pp, Birmingham 1834.
33. AFLALO, Frederick George *A sketch of the natural history – vertebrates of the British Islands.* Sm. 8vo, xiv+498 pp, 1 pl., 1 map, text figs, Blackwood, Edinburgh 1898.
34. —— *Types of British animals.* 8vo, xx+290pp, 8 pls, text figs, Sands, London 1899; ed. 1909. The Library for Young Naturalists.
35. —— *A walk through the Zoological Gardens.* 8vo, 232 pp, Sands, London 1900.
36. AGASSIZ, Louis Jean Rodolphe *An essay on classification.* 8vo, viii+381 pp, Longman, London 1859; reprint 1962. First printed in *Contributions to the natural history of the United States*, pt. 1, Boston 1857.
37. —— *The structure of animal life.* 8vo, viii+128 pp, text figs, Sampson Low, London 1866. Graham Lectures, Brooklyn Institute, Vol. 4. USA printed.

Aikin, Arthur *see* Aikin, J. No. 39.

38. [AIKIN, John] *The calendar of nature.* 12mo, vii+82 pp, Johnson, Warrington 1784; 2ed. 1785; 3ed. 1787; 4ed. 1794; 6ed. 1801. Many later editions.
39. —— *The natural history of the year: being an enlargement of Dr. Aikin's Calendar of nature.* 8vo, vi+195 pp, 1 pl., Johnson, London 1798; 2ed. 1799; 3ed. 1805; 4ed. 1815; 5ed. 1821. Enlarged by his son, Arthur Aikin.
40. —— *Pictorial calendar of the seasons . . . embodying the whole of Aikin's Calendar of nature.* 8vo, viii+567 pp, text figs, Bohn, London 1854; ed. 1862. Enlarged by Hewitt, *Miss* Mary, as an extension of No. 39.
41. [——] *The woodland companion; or, a brief description of British trees.* 8vo, 92 pp, 28 pls, Johnson, London 1802; 2ed. 12mo 1815; 3ed. 1820 signed. By the author of *Evenings at home*; preface signed J.A. His sister *Mrs* Anne Letitia Barbauld also contributed.
42. AIKMAN, James *A natural history of beasts, birds, & fishes, or stories of animated nature.* Sm. 8vo, xii+372 pp, 1 col. pl., text figs, Nelson, London 1847; 1852; ed. 1855 title *One hundred animals; or book of natural history*; ed. 1861 title *The animal kingdom.* First edition has 36 leaves of text figures unpaginated.
43. AINSWORTH, J. G. *The flowering plant as illustrating the*

*first principles of botany.* 8vo, x+181 pp, text figs, London 1890.

44. AINSWORTH, William Francis *An account of the caves of Ballybunion, County of Kerry. Appendix notice of the birds.* 8vo, 96 pp, Curry, Dublin 1834.

45. ALBIN, Eleazar *A natural history of English insects.* 4to, 100 pp, 100 col. pls, Innys, London 1720; 2ed. 1724 title . . . *to which are added, notes and observations by W. Derham*; Latin ed. 1731 title *Insectorum Angliae naturalis historia . . . his accedunt annotationes . . . & observationes . . . a G. Derham*; ed. 1749. Derham's notes, 26 pp, were published separately with a cancel title leaf.

—— *Insectorum Angliae naturalis historia.* 1731. *see* No. 45.

46. —— *A natural history of birds.* 4to, 3 vols, 306 col. pls, Innys & Manby, London 1731, 34, 38; ed. vols 1 & 2 1738; ed. vol. 3 1740 title *A supplement to* . . . Vol. 2 has *To which are added, notes and observations by W. Derham*; ed. 8vo 1779.

47. —— *A natural history of spiders and other curious insects.* 4to, viii+76 pp, 54 col. pls, Montagu, London 1736. *see also* Martyn, Thomas *draughtsman.*

48. —— *A natural history of English song-birds.* Sm. 8vo, 97 pp, 24 col. pls, Bettesworth & Hitch, London 1737; 1738; 2ed. 1741; ed. 12mo, Edinburgh 1754 unsigned, title *A natural history of singing birds*; 3ed. 1759; issue of 1754 ed. 1776; ed. 12mo, Edinburgh 1791 unsigned, title *History of singing birds.* Editions of 1754 & 1776 are by 'A lover of birds'. Editions of 1791 & 1825 are attributed to Buffon on title pages.

[——] *The history of esculent fish.* 1794. *see The history of esculent fish.* The plates are by the Albin family.

49. ALCOCK, Randall H. *Botanical names for English readers, containing a history of botany.* 8vo, xviii+236 pp, Reeve, London 1876; 1884.

50. ALDER, Joshua *A catalogue of the land and freshwater testaceous Mollusca found in the vicinity of Newcastle-upon-Tyne.* 4to, Newcastle upon Tyne 1830.

51. —— *Catalogue of the Mollusca of Northumberland and Durham.* 8vo, 115 pp, Currie, Newcastle upon Tyne 1848. Reprinted from *Trans. Tyneside Nat. Fld Club*, Vol. 1.

52. —— *A catalogue of the zoophytes of Northumberland and Durham.* 4to, 72 pp, 8 pls, Newcastle upon Tyne 1857;

*suppl.*, 23 pp, 4 pls, 1862. Reprinted from *Trans. Tyneside Nat. Fld Club*, Vols 3 & 5.

53. —— & HANCOCK, Albany *Monograph of the British nudibranchiate Mollusca.* Roy. 4to, 2 vols, 23 plain & 77 col. pls, text figs, Ray Society, London 1845–55, 1910. Ray. Society Publ. Nos 2, 5, 9, 15, 19, 23, 27 & 90 for 1844, 45, 46, 48, 50, 52, 55 & 1909.

54. ALDRIDGE, William *A gossip on the wild birds of Norwood and Crystal Palace district.* Sm. 8vo, 110 pp, text figs, Upper Norwood 1885.

55. ALEXANDER, P. Y. *Darwin and darwinism pure and mixed; a criticism with some suggestions.* 8vo, xii+346 pp, Bale, London 1899.

56. —— *More loose links in the darwinian armour.* 8vo, viii+169 pp, Bale, London 1900.

57. *All about animals for old and young.* Obl. 4to, 240 pp, text figs, Newnes, London [1897].

*All the British butterflies.* [1868]. *see* Newman, E.

Allan, James *see* Macdonald, G.

Allan Museum, Newcastle upon Tyne *see* Fox, G. T.

Allen's Naturalist's Library (Shaw), 1894–1897. *see*

Grant, W. R. O. *Game birds.* 1895–1897.

Kirby, W. F. *British butterflies and moths.* 1894–1897.

Lydekker, R. *British Mammalia.* 1895.

—— *Carnivora.* 1895.

Sharpe, R. B. *Birds of Great Britain.* 1894–1897.

*see also* Lloyd's Natural History, and No. 3258.

58. ALLEN, Benjamin *The natural history of the mineral waters of Great Britain. To which is added some observations on the Cicindela, or glow-worm.* 8vo, 30+104 pp, 1 pl., for the author, London 1711.

59. ALLEN, Charles Grant Blairfindie *The evolutionist at large.* 8vo, xii+215 pp, Chatto & Windus, London 1881.

60. —— *Vignettes from nature.* 8vo, viii+229 pp, Chatto & Windus, London 1881.

61. —— *The colour of flowers as illustrated in the British flora.* 8vo, vi+119 pp, text figs, Macmillan, London 1882; 1891. Nature Series.

62. —— *Flowers and their pedigrees.* 8vo, 266 pp, text figs, Longman, London 1883; 2ed. 1886.

63. —— *Science in arcady.* Sm. 8vo, xi+304 pp, Lawrence & Bullen, London 1892.

64. —— *The story of plants.* Sm. 8vo, 232 pp, text figs, Newnes, London 1895; ed. [1926]. Library of Useful Knowledge.
65. —— *Moorland idylls.* 8vo, viii+257 pp, Chatto, London 1896.
66. —— *Flashlights on nature.* 8vo, viii+312 pp, text figs, Newnes, London 1899; ed. [1911].
67. ——, WILSON, Andrew et al. *Nature studies.* 8vo, Wyman, London [1883]; 1884. Reprinted from *Knowledge.*

Allen, Grant *see* Allen, C. G. B.

68. ALLIN, Thomas *Flowering plants and ferns of the County Cork.* 8vo, 113 pp, Weston super Mare 1883.
69. ALLMAN, George James *Introductory lecture delivered to the students of the natural history class in the University of Edinburgh.* 8vo, 15 pp, Black, Edinburgh 1885.
70. —— *A monograph of the fresh-water Polyzoa.* Roy. 4to, viii+121 pp, 1 plain + 10 col. pls, text figs, Ray Society, London 1856. Ray Society Publ. No. 28 for 1856.
71. —— *The method and aim of natural history studies being an introduction to a course of lectures on natural history.* 8vo, 20 pp, Edmonston & Douglas, Edinburgh 1868.
72. —— *A monograph of the gymnoblastic or tubularian hydroids.* Roy. 4to, 2 pts, xxiv+450 pp. 1 plain + 23 col. pls, Ray Society, London 1871–1872. Ray Society Publ. Nos 46 & 47 for 1869, 70.
73. ALLOM, Elizabeth Anne *Sea weed collector.* 16mo, 102 pp, 44 specimens, Keble printed, Margate 1841; ed. London 1845.
74. —— *Sea-side pleasures.* 8vo, xii+144 pp, text figs, Aylott, London 1845.
75. ALSTON, Charles *Index plantarum praecipue officinalium . . . in Horto Medico Edinburgensi: (C.Linnaei Fundamenta botanica.)* 8vo, xxv+65 pp, Sands, Edinburgh 1740.
76. —— *Tirocinium botanicum Edinburgense.* 4to, 2 pts, 116+120 pp, Hamilton & Balfour, Edinburgh 1753; translation of pt. 1 only, 8vo, London 1754 title *A dissertation on botany.* An expansion of No. 75.

—— *A dissertation on botany.* 1754. *see* No. 76.

77. ALSTON, Edward Richard *Notes on the flora and fauna of the west of Scotland.* 8vo, xxxi pp, British Association, Glasgow 1876. Issued with the *British Association Handbook for Glasgow*, 1876.
78. ANDERSON, George & ANDERSON, Peter *Guide to the*

*Highlands and Islands of Scotland . . . descriptive . . . of natural history.* 8vo, xii+759 pp, 1 map, Murray, London 1834; ed. Edinburgh 1842; 3ed. Edinburgh 1851.

79. ANDERSON, James *Recreations in agriculture, natural history.* 8vo, 6 vols, Wallis, London 1795; ed. 1799–1802.

Anderson, Peter *see* Anderson, G.

Anderson, Thomas *see* Gray, R.

Andrewe, Laurence *see The noble lyfe & nature . . . of beastes.* [?1521].

80. *Anecdotes of animals: selected by a lady for the amusement of her children.* 16mo, viii+147 pp, pls, Harvey & Darton, London 1832.

81. *Anecdotes of animals: shewing their surprising instinctive intelligence and wonderful sagacity. Designed for youth.* 12mo, 108 pp, Harrild, London 1814.

82. *Anecdotes of birds; or, short accounts of their habits in a state of nature, collected from the best authors in natural history.* 12mo, London 1809.

*Animal intellect and its relation to matter and spirit.* 1851. *see* Childs, R. W.

83. *The animal kingdom . . . on the system of Cuvier.* 8vo, viii+143 pp, Longman, London 1844. By an experienced teacher.

*Animal mechanics, or proofs of design in the animal frame*. 1822. *see* Bell, *Sir* Charles.

84. *The animal museum; or, picture gallery of quadrupeds.* 12mo, ix+239 pp, 25 pls, Harris, London [1825]. By the author of *Conversations on astronomy*.

*The animal picture book.* [1876]. *see* Wood, J. G.

85. *Animal sagacity, exemplified by facts, showing the force of instinct in beasts, birds etc.* 12mo, 180 pp, text figs, Tyrell, Dublin 1819; ed. Grace, Dublin 1819. Later editions.

86. *Animals drawn from nature and engraved in aqua-tinta.* Obl. fol., 36 pls, with letterpress, for the author, London 1788. Plates by Charles Catton. Plate dates vary, 1787 or 1788.

87. *Animals. From the . . . pictures of Landseer . . . and others.* 4to, 20 pp, pls, Dean, London 1844.

88. *Animals: their food and instincts.* 16mo, 214 pp, pls, Ward, London [1864].

89. *Animals: their homes, haunts, habits.* 8vo, 53 pp, Cox, London [1896].

90. ANNERSLEY, George, *Earl of* Mountnorris *Short instruc-*

*tions for collecting shells.* 8vo, 8 pp, [London ?1815]; ed. Bewdley 1827.

91. ANSLOW, R. *The study of the mosses of the Wrekin.* 8vo, Wellington 1868; ed. 1880.

92. ANSTED, David Thomas *The correlation of the nature history sciences.* 8vo, 54 pp, Macmillan, London 1863. Rede Lectures, Cambridge.

93. —— & LATHAM, Robert Gordon *The Channel Islands.* 8vo, xxviii+594 pp, maps, text figs, Allen, London 1862; 2ed. 1865; 3ed. [1893].

94. ——, SMITH, Edward, BREEN, Hugh, SCOFFERN, John & LOWE, Edward Joseph *Natural history of the inanimate creation.* 8vo, xvi+302+205+403 pp, 1 pl., text figs, Houlston & Stoneman, London 1856. Reprinted from Orr, W. S. editor *Circle of the sciences.*

*Anti-Darwin*, 1884. *see* Suckling, H. J.

Aplin, Benjamin D'Oyley *see* Aplin, F. C.

95. APLIN, Frederick Charles, APLIN, Benjamin D'Oyley & APLIN, Oliver Vernon *A list of the birds of the Banbury district.* 8vo, 29 pp, Banbury 1882. Preprint from *J. Northants nat. Hist. Soc.*, Vol. 2, 1883.

96. APLIN, Oliver Vernon *The birds of Oxfordshire.* 8vo, vii+217 pp, 1 col. pl., 1 col. map, Clarendon, Oxford 1889.
—— *see* Aplin, F. C.

97. ARBUTHNOT, James *An historical account of Peterhead . . . also a natural history of the fishes found on the coast of Buchan.* 8vo, 83+159 pp, Clark, Peterhead 1815.

97a —— *Natural history of those fishes that are indigenous to, or occasionally frequent the coast of Buchan; with an account of the most useful Fucii produced upon these shores.* 8vo, ii+63 pp, Brown, Aberdeen 1815.

*The architecture of birds.* 1831. *see* Rennie, H.

98. ARENDTS, Carl *Practical aids to the study of natural history.* Roy. 8vo, viii+128 pp, text figs, Ward & Lock, London 1861. Translation of *Naturhistorische Schulatlas*, Leipzig 1858, by Gustave Louis Maurice Strauss.

99. ARGYLL, George John Douglas (Campbell) *Duke of The reign of law.* 8vo, vi+434 pp, 4 pls, Strahan, London *1867* [=1866]. Many later editions.

100. —— *On variety as an aim in nature.* 8vo, 8 pp, [London 1871].

101. —— *The unity of nature.* 8vo, xv+571 pp, 1 pl., Strahan, London 1884; 2ed. 1884.

102. —— *Organic evolution cross-examined.* 8vo, vi+201 pp, Murray, London 1898.
103. ARGENT, J. *A nomenclature of British birds . . . intended for labelling collections.* 8vo, 14 col. pls, London 1839.
104. ARMSTRONG, Jessie F. *Birds and their ways.* Sm. 8vo, 137 pp, Sonnenschein, London 1884.
105. ARNOLD, Edwin Lester Linden *Bird life in England.* Sm. 8vo, x+325 pp, Chatto & Windus, London 1887.
106. ARNOLD, Frederick Henry *Flora of Sussex.* 8vo, xxiii+118 pp, 1 map, Hamilton Adams, London 1887; 2ed. 1907.
107. ARNOLD, George Matthews *Robert Pocock, the Gravesend historian, naturalist, antiquarian, botanist and printer.* 8vo, xiii+267 pp, Sampson Low, London 1883.
108. ARNOTT, George Arnott Walker *Analytical botanical tables.* [Glasgow 1848].
109. —— *Characteristics of most of the natural orders.* [Glasgow 1862].

—— *see* Greville, R. K.

—— *see* Hooker, *Sir* William J.

*Around a cornfield in a ramble after wild flowers*, 1895, *see* Cooke, M. C.

110. ARUNDEL, Walter B. *Ackworth birds, being a list of the birds of Ackworth, Yorkshire.* Sm. 8vo, viii+105 pp, Gurney & Jackson, London 1898.

*Ashmolean Museum, A catalogue of the.* 1836. *see* Duncan, P. B.

111. ASHTON, John *Curious creatures in zoology.* 8vo, xi+348 pp, text figs, Nimmo, London 1890 [=1889]. L.P. 10 copies.

111a ASHTON, Thomas *Visits to the Museum of the Manchester Natural History Society.* 8vo, 3 pts, Manchester 1857; 2ed. 1860.

112. ASHTON-under-LYNE LINNEAN BOTANICAL SOCIETY *District flora*, Sm. 8vo, 83 pp, Ashton under Lyne 1888.

*Aspects of botany* [Kilkee, Co. Clare]. [1864] *see* Ward, N. B.

113. ASTLEY, Hubert Delaval *My birds in freedom and captivity.* 4to, xvi+254 pp, 21 plain + 1 col. pls, text figs, Dent, London 1900. L.P. 100 copies.
114. [ATKINS, *Miss* Anna] *Photographs of British Algae. Cyanotype impressions.* 4to, 3 vols, 411 pls, [Halstead 1842–53]. Preface signed A.A.
115. [ATKINS, *Miss* Sarah] *Botanical rambles, designed as an early and familiar introduction to the study of botany.* Sm.

8vo, 218 pp, Baldwin, London 1822; 1824; 1826. By the author of *The India cabinet.*

116. ATKINSON, John *A compendium of the ornithology of Great Britain.* 8vo, xii+232 pp, Hurst Robinson, London 1820; issue 1820.

117. ATKINSON, John Christopher *British birds' eggs and nests popularly described.* 8vo, viii+182 pp, 12 col. pls, 1 tab., Routledge, London 1861. Many later editions.

118. —— *Sketches in natural history, with an essay on instinct and reason.* 8vo, xii+328 pp, 1 pl., text figs, Routledge, London 1861; 1865.

119. —— *Forty years in a moorland parish.* 8vo, xv+471 pp, 2 pls, 2 maps, text figs, Macmillan, London 1891; 2ed. 1891. Later editions.

120. *The atlas of nature, being a graphic display of the most interesting subjects in the three kingdoms of nature, for study and reference.* Fol., 15 col.+70 plain pls, London 1823.

121. AUBREY, John *The natural history and antiquities of the County of Surrey.* 8vo, 5 vols, 10 pls, 1 portr., 1 map, text figs, Curll, London 1719, 18; ed. 1723.

122. —— *The natural history of Wiltshire.* 4to, xii+132 pp, Wiltshire Topographical Society, London 1847.

123. [AUSTEN, Ernest Edward] *How to collect mosquitoes (Culicidae).* 8vo, 7 pp, British Museum (Natural History), London [1898]; ed. [1899].

124. [——] *How to collect Diptera (two-winged flies).* 8vo, 16 pp, text figs, British Museum (Natural History), London [1900].

Austen, Robert Alfred Cloyne Godwin- *see* Forbes, E.

125. AUSTEN, Nathaniel Laurence *Natural history papers and memoir.* 8vo, xliv+190 pp, 7 pls, 1 portr., for private circulation, London 1877.

126. AVELING, Edward Bibbins *Botanical tables for the use of students.* Obl. 8vo, 14 pp, Hamilton Adams, London [1874]; 3ed. 1880; 5ed. [1882].

127. —— *The student's Darwin.* Sm. 8vo, xii+339+viii pp, Freethought Publishing Company, London 1881. International Library of Science and Freethought.

128. —— *An introduction to the study of botany.* 8vo, iv+363 pp, text figs, Sonnenschein, London 1891; 2ed. 1898.

*Aves Britannicae. A systematic catalogue of British birds.* 1836. *see* Manchester Natural History Society.

129. AVIS, Richard *Bird-preserving, bird-mounting, and the*

*preservation of birds' eggs.* 8vo, 48 pp, Groombridge, London 1870. Practical manuals.

130. AWDRY, Frances Emily, *Mrs* William *Early chapters in science. A first book of knowledge of natural history, botany.* Post 8vo, xviii+348 pp, text figs, Murray, London 1899.

# B

B., E. *A plain and easy account of British ferns.* 1854. *see* Bosanquet, E.

B., J. *Scenes of animal life and character.* 1858. *see* Blackburn, *Mrs* Jane.

131. B., L. E. *Botany for novices.* 8vo, vii+60 pp, Whittaker, London 1864.

132. BABBAGE, Charles *On the constants of nature. Class Mammalia.* 4to, Brussels 1853.

133. BABINGTON, Charles Cardale *Flora Bathoniensis: or, a catalogue of the plants indigenous to the vicinity of Bath.* 8vo, vi+62 pp, Collings, Bath 1834; suppl. [1839]; 2ed. 1839.

134. —— *Primitiae florae Sarnicae; or, outline of the flora of the Channel Islands.* 12mo, xvi+132 pp, Longman, London 1839.

135. —— *Manual of British botany.* 12mo, xxi+400 pp, Van Voorst, London 1843, 2ed. 1847; 3ed. 1851; 4ed. 1856; 5ed. 1862; 6ed. 1867; 7ed. 1874; 8ed. 1881; 9ed. 1904; 10ed. 1922.

136. —— *A synopsis of the British Rubi.* 8vo, 34 pp, Van Voorst. London 1846. Reprinted from *Ann. Mag. nat. Hist.*

137. —— *Flora of Cambridgeshire.* 12mo, lvi+327 pp, 1 map, Van Voorst, London 1860.

138. —— *The British Rubi.* 12mo, viii+305 pp, Van Voorst, London 1869.

139. —— *Memorials, journal and botanical correspondence of Charles Cardale Babington.* 8vo, xciv+475 pp, 2 portrs, pedigree in cover pocket, Macmillan & Bowes, Cambridge 1897.

—— *see* Balfour, J. H.

140. BABINGTON, Churchill *Catalogue of the birds of Suffolk.* 8vo, viii+281 pp, 7 pls, 1 map, Van Voorst, London 1884–86. Reprinted from *Proc. Suffolk Inst. Archaeol. nat. Hist.*, Vol. 5.

141. BACKHOUSE, James *the father A monograph of the British Heiracia.* 8vo, viii+89 pp, Simpson, York, 1856.

142. BACKHOUSE, James *the son A handbook of European birds.* Sm. 8vo, viii+334 pp, 1 pl., Gurney & Jackson, London 1890.

143. BACON, Francis, *Baron* Verulam, *Viscount* St Albans *Sylva sylvarum: or, a natural historie.* 4to, 2 pts, Lee, London 1627; 2ed. fol. 1628; 3ed. 1631; 4ed. 1635; 5ed. 1639; 6ed. 1651; 7ed. 1658; 8ed. 1664; 9ed. 1670; 10ed. 8vo 1676; 1677; 11ed. 1685.

144. BADCOCK, John *Vignettes from invisible life.* Sm. 8vo, xii+183 pp, 1 pl., text figs, Cassell, London 1883. Reprinted from *The St James's Gazette.*

145. BADENOCH, L. N. *Romance of the insect world.* Sm. 8vo, xv+298 pp, 1 pl., text figs, Macmillan, London 1893.

146. —— *True tales of the insects.* Sm. 8vo, xviii+255 pp, text figs, Chapman & Hall, London 1899.

147. BADHAM, Charles David *The question concerning the sensibility, intelligence and instinctive actions of insects.* 8vo, 54 pp, Belin, Paris [1837]. By Scarabaeus.

148. —— *Insect life.* 8vo, 171 pp, 1 pl., Blackwood, Edinburgh 1845.

149. —— *A treatise on the esculent funguses of England.* Roy. 8vo, x+138 pp, 3 plain+17 col. pls, Reeve, London 1847; 2ed. 1863.

150. —— *Prose halieutics or ancient and modern fish tattle.* 8vo, xii+552 pp, Parker, London 1854.

151. BAGNALL, James Eustace *Notes on Sutton Park: its flowering plants, ferns and mosses.* 8vo, 27 pp, Birmingham 1876.

152. —— *Handbook of mosses.* Sm. 8vo, vii+96 pp, text figs, Sonnenschein, London 1886; 2ed. 1886; 3ed. 1889; 4ed. 1896; 6ed. 1910. Young Collector Series.

153. —— *The flora of Warwickshire.* 8vo, xxxiv+519 pp, 1 map in pocket, Gurney & Jackson, London 1891. 500 copies.

Baikie, William Balfour *see* Adams, A.

154. —— & HEDDLE, Robert *Historia naturalis Orcadensis.*

*Zoology Part 1 . . . Mammalia and birds.* 8vo, x+104 pp, Paterson for the authors, Edinburgh 1848; slip *More important corrections.* 1856. All publ.

Baildon, Henry C. *see* Moore, T. 1869.

155. [BAILEY, *Miss* Katherine] *The Irish flora; comprising the phaenogamous plants and ferns.* 12mo, xi+220 pp, Longman, Dublin 1833; 1836; 2ed. 1846; 1847.

156. BAILEY, William *The angler's instructor . . . on the habits of fish.* 8vo, v+111 pp, Longman, London 1857; 2ed. [1866]; 3ed. [1879].

157. BAILLON, Henri Ernest *The natural history of plants.* Imp. 8vo, 8 vols, text figs, Reeve, London 1871–88. Translation of *Histoire des plantes*, Paris 1866–85 by Marcus Monvel Hartog.

158. BAILY, William Hellier *Rambles on the Irish coast: especially relating to its geology, natural history, and antiquities.* 12mo, vi+69 pp, Sealy Bryers & Walker, Dublin [1887].

159. BAINES, Henry *The flora of Yorkshire.* 8vo, xvi+159 pp, 2 pls, 2 maps, Longman, London 1840; suppl. viii+188 pp, 1 col. map 1854. Supplement by Baker, J. G. & Nowell, J.

160. BAIRD, William *Natural history of the British Entomostraca.* 8vo, viii+364 pp, 19 plain+17 col. pls, Ray Society, London 1850. *Ray Soc. Publ.* No. 17 for 1849.

161. —— *A cyclopaedia of the natural sciences.* 8vo, xvi+613 pp, 1 pl., 1 map, text figs, Griffin, London 1858; ed. 1860 title *A dictionary of natural history*; ed. 1863 title *The student's natural history.*

162. —— *Tabular view of the Mammalia.* 4to, no text, 4 pls, SPCK or Stanford, London [c.1862].

163. —— *Reptiles.* Roy. 8vo, 102+lv pp, 9 plain+3 col. pls, [Edinburgh 1860]. Reprinted from Richardson, *Sir* John, *editor The museum of natural history.* 1859–62.

—— *A dictionary of natural history.* 1860. *see* No. 161.

—— *The student's natural history.* 1863. *see* No. 161.

—— *see* Bell, T.

164. BAKER, Henry *The miscroscope made easy.* 8vo, xvi+311 pp, 14 pls, Dodsley, London 1742; 2ed. 1743; 3ed. 1744; 4ed. 1754; 5ed. 1769.

165. —— *An attempt towards the natural history of the polype.* 8vo, 218 pp, 1 pl., text figs, Dodsley, London 1743.

166. —— *Employment for the microscope.* 8vo, xiv+442 pp. 17 pls, Dodsley, London 1753; 2ed. 1764; 1785.

—— *Of microscopes and the discoveries made thereby*, 8vo, 2 vols, 31 pls, Dodsley, London 1754, 53. This is only an issue of Nos 164 & 166, with an additional title page.

167. BAKER, John Gilbert *The flowering plants and ferns of Great Britain; an attempt to classify them according to their geognostic relations*, 8vo, 30 pp, Cash, London 1855.

168. —— *North Yorkshire: studies of its botany*. 8vo, xii+353 pp, 4 col. maps, Longman, London 1863; 2ed. [1888–] 1906. Reprinted from *Trans. Yorks. Nat. Union, Bot. Ser.*

169. —— *Review of the British roses*. 8vo, privately printed, Huddersfield 1864. Reprinted from *The Huddersfield Nat.*

170. —— *Elementary lessons in botanical geography*. 8vo, viii+110 pp, Reeve, London 1875.

171. —— *A flora of the English Lake District*. 8vo, viii+262 pp, Bell, London 1885.

172. —— *Handbook of the fern-allies*. 8vo, 159 pp, Bell, London 1887.

—— *see* Baines, H.

—— *see* Hooker, *Sir* William J.

173. BAKER, *Sir* Samuel White *Wild animals and their ways, reminiscences of Europe, Asia, Africa and America*. 8vo, 2 vols, text figs, Macmillan, London 1890.

174. BAKER, Thomas Barwick Lloyd *An ornithological index, arranged according to the Synopsis avium of Mr. Vigors*. 8vo, vii+187 pp, Wood, London 1835. Said to have been prepared by Mr Prince.

175. BAKER, W. B. *Fields and flowers or the Stockton naturalist*. 8vo, 229 pp, 1 portr., Stockton on Tees 1867.

176. BALFOUR, Alexander *Weeds and wild flowers*. 8vo, xciv+280 pp, Lizars, Edinburgh 1830.

177. —— *Illustrations of botany*. 4 pls, Edinburgh [?1850]. All publ.

178. BALFOUR, John Hutton *A manual of botany*. 8vo, xvi+641 pp, text figs, Griffin, Edinburgh 1849; 2ed. London 1851; 3ed. London 1855; 5ed. 1859. 2ed. & 3ed. are *Encyclopaedia Metropolitana*, 2ed., 2 div., Vol. 13. Later editions.

179. —— *Phyto-theology: or, botanical sketches intended to illustrate the works of God*. 12mo, xvi+241 pp, Johnstone & Hunter, Edinburgh 1851; 3ed. 1859 title *Botany and religion*; 4ed. 1882.

180. —— *Class book of botany*. 8vo, 2pts, xi+1114 pp, text figs, Black, Edinburgh 1852, 54: 3ed. 1870.

181. —— *Outlines of botany*. 8vo, xvi+616 pp, text figs, Black, Edinburgh 1854; 2ed. 1862.
182. —— *The botanist's vade-mecum*. 12mo, pp 56, 670–698, Griffin, London 1855. Extracts from *A manual of botany*, 3ed.
—— *Botany and religion*. 1859. *see* No.179.
183. —— *The botanist's companion*. 8vo, 74 pp, text figs, Black, Edinburgh 1860; 2ed. 1875.
184. —— *Account of botanical excursions made in the Island of Arran during . . . 1869*. 16mo, 32 pp, text figs, Edinburgh 1869.
185. —— *The elements of botany*. 8vo, xiii+321 pp, Black, Edinburgh 1869; 3ed.1876; 4ed.1884.
186. —— *Lessons in elementary botany*. 12mo, Longman, London 1869.
187. —— *First book of botany*. 8vo, vii+118 pp, text figs, Collins, London 1872. Elementary Science Series.
188. —— & BABINGTON, Charles Cardale *Vegetation of the Outer Hebrides*. 8vo, 24 pp, Edinburgh [1841].
189. ——, —— & CAMPBELL, William Hunter *A catalogue of British plants*. Fol., Botanical Society of Edinburgh, Edinburgh 1836; 2ed. 8vo 1841; 3ed.1851; 4ed.1865.
190. —— & SADLER, John *Flora of Edinburgh*. Sm. 8vo, vii+174 pp, 1 map, Black, Edinburgh 1863; 2ed. 1871.
191. BALL, Robert *Rough sketches intended to aid in developing the natural history of the seals (Phocidae) of the British Islands.* Fol., 12 pls with litho text, Dublin [1837].
192. BALL, William Platt *Are the effects of use and disuse inherited? An examination of the view held by Spencer and Darwin*. Sm. 8vo, xii+156 pp, Macmillan, London 1890. Nature Series.
192a BALLARD, Edward George *Microscopic amusements; or, a complete companion to the miscroscope . . . with a calendar of the most interesting British insects*. 12mo, vi+55 pp, 2 pls, West, London 1829.
193. BALLARD, Stephen *A treatise on the nature of trees*. 12mo, 67 pp, Ward, Ledbury 1833.
194. BANKS, George *An introduction to the study of English botany*. 8vo, 75 pp, 37 pls, 1 portr., Longman, London 1823; 2ed. Devonport 1832.
195. —— *The Plymouth and Devonport flora.* 8vo, [206] pp, [Devonport 1830–32]. 8 numbers, all publ.
Barbauld, *Mrs* Anna Letitia *see* Aikin, John.

196. BARBUT, James *The genera insectorum of Linnaeus exemplified by various specimens of English insects*. 4to, xvi+471 pp, 20 col.+2 plain pls, Dixwell for the author, London 1781; 2ed. 1783. Text in English and French.
197. —— *The genera vermium . . . the orders of the Intestina et Mollusca Linnaei*. 4to, 2 pts, xx+101+xxvii+76 pp, 24 col.+1 plain pls, Dixwell for the author, London 1783, 88; issue pt. 1 1794. Text in English and French.
198. BARKER, Ernest E. *Rothesay Royal Aquarium and its contents*. 8vo, Rothesay 1877.
199. BARKER, John Theodore *The beauties of flowers in field and wood, containing the natural orders or families of British wild plants*. 8vo, xii+228 pp, 10 col. pls, Binns & Goodwin, Bath [1852]; 2ed. [1857].
200. BARLOW, Thomas Worthington *A chart of British ornithology, designed for popular use*. 4to, 1 folding chart, Robinson, London [1847].
201. BARRETT, Charles Golding *Lepidoptera of Norfolk*. 4to, 18 pp, [London 1884]. Reprinted from Mason's *History of Norfolk*.
202. —— *The Lepidoptera of the British Islands*. Roy. or demy 8vo, 11 vols, 504 col. pls, Reeve, London [1892–] 1893–97. Demy copies have no plates.

Barrett-Hamilton, Gerald Edwin Hamilton *see* Hamilton, G. E. H. B.

203. BARRINGTON, *Hon.* Daines *Miscellanies*. 4to, viii+557 pp, 2 portrs, 1 map, 5 tabs, White, London 1781.
204. BARRINGTON, Richard Manliffe *The migration of birds as observed at Irish lighthouses and lightships*. 8vo, xxv+667 pp, 1 map, text figs, Porter, London [1900]. Earlier reports are in *Rep. Brit. Assoc. Adv. Sci*.

Barron, Charles *see* Adams, A.

205. BARROW, John Henry *Characteristic sketches of animals drawn from the life and engraved by Thomas Landseer*. Imp. or Roy. 4to, x pp, 32 pls with letterpress, text figs, Moon Boys & Graves, London [1829–] 1832.
206. BARRY, George *History of the Orkney Islands*. 4to, viii+509 pp, 11 pls, 1 map, for the author, Edinburgh 1805; 2ed. London 1808; ed. 8vo, Kirkwall 1867.
207. BARRY, Martin *Birds that have bred in the Isle of Arran since 1835*. 12mo, 4 pp, [n.p., n.d.] Mid 19c. Said to be a forgery.

208. BARTHOLOMAEUS ANGLICUS *De proprietatibus rerum*. Fol., 478 11, Wynken de Worde, [Westminister ?1495]; ed. London 1535; ed. London 1582 title *Batman uppon Bartholome, his booke De proprietatibus rerum*; extracts London 1893 title *Medieval lore*. Translated by John Trevisa. Editio princeps is [Cologne ?1472].

209. BARTLETT, Abraham Dee *Wild animals in captivity*. 8vo, 373 pp, 3 pls, text figs, Chapman & Hall, London 1898.

210. —— *Life among wild beasts in the zoo; being a continuation of Wild animals in captivity*. 8vo, xv+373 pp, text figs, Chapman & Hall, London 1900 [=1899].

211. BARTON, Benjamin Smith *Elements of botany: or, outlines of the natural history of vegetables*. Roy. 8vo, xii+344+35 pp, 30 col. pls, Johnson, London 1804. First ed. is Philadelphia 1803.

212. BARTON, John *A lecture on the geography of plants*. 12mo, 94 pp, 4 maps, Harvey & Darton, London 1827.

213. BARY, Heinrich Anton de *Comparative anatomy of the vegetative organs of the phanerogams and ferns*. Roy. 8vo, xvi+659 pp, text figs, Clarendon, Oxord 1884. Translation of *Vergleichende Anatomie der Vegetations-Organe der Phanerogamen und Farne*, Leipzig 1877, by Frederic Orpen Bower & Dukinfield Henry Scott.

214. —— *Comparative morphology and biology of the Fungi, Mycetozoa, and Bacteria*. 8vo, xviii+525 pp, text figs, Clarendon, Oxford 1887. Translation of *Vergleichende Morphologie und Biologie der Pilze*, Leipzig 1884, by Henry Edward Fowler Garnsey & Isaac Bayley Balfour.

215. —— *Lectures on Bacteria*. Sm. 8vo, xii+193 pp, text figs, Clarendon, Oxford 2ed. 1887. Translation of *Vorlesungen über Bacterien*, Leipzig 1855; 2ed. 1887, by Henry Edward Fowler Garnsey & Isaac Bayley Balfour.

216. BASKERVILLE, Thomas *Affinities of plants, with some observations on progressive development*. 8vo, x+114 pp, 1 pl., Taylor & Walton. London 1839.

217. BASTIAN, Henry Charlton *Flora of Falmouth*. 8vo, [?Falmouth ?1850].

218. —— *The modes of origin of lowest organisms*. 8vo, xii+109 pp, Macmillan, London 1871.

219. —— *The beginnings of life: being some account of . . . lower organisms*. Sm. 8vo, 2 vols, text figs, Macmillan, London 1872.

220. —— *Evolution and the origin of life*. 8vo, xv+186 pp, Macmillan, London 1874.
221. BATE, Charles Spence & WESTWOOD, John Obadiah *A history of the British sessile-eyed Crustacea*. Roy. or demy 8vo, 2 vols, text figs, Van Voorst, London [1861–] 1863–68. Natural History of the British Isles.
222. BATEMAN, Gregory Climenson *Fresh-water aquaria*. Sm. 8vo, vii+316 pp, text figs, Upcott Gill, London 1890; 2ed. 1892. Later editions.
223. —— *The vivarium*. 8vo, 424 pp, text figs, Upcott Gill, London [1897].
—— & BENNETT, Reginald Arthur Renaud *The aquarium book*. Sm. 8vo, 2 pts, text figs, Upcott Gill, London 1889. A combination volume of Nos 222 & 278.
224. BATESON, William *Materials for the study of variation treated with especial regard to discontinuity in the origin of species.* 8vo, xvi+598 pp, text figs, Macmillan, London 1894.

Bath Literary and Scientific Institution. *see* Woods, H.

225. BATH, William Harcourt *The young collector's handbook of ants, bees, dragonflies, earwigs, crickets, and flies.* Sm. 8vo, 108 pp, 1 pl., text figs, Sonnenschein, London 1885; 2ed. 1890. Later editions. Young Collector Series.
226. —— *The young collector's handbook of British birds, and their nests and eggs*. Sm. 8vo, 112 pp, text figs, Sonnenschein, London 1888. Young Collector Series.
227. [——] *The naturalist's calendar and weather guide*. 24mo, 48 pp, Naturalists' Publishing Company, Birmingham 1890. By the editor of *The Naturalists' Gazette.*
228. [——] *An illustrated handbook of British dragonflies.* 12mo, 98 pp, 8 pls, text figs, Naturalists' Publishing Company, Birmingham 1890. By the editor of *The Naturalists' Gazette*.
229. BATTERS, Edward Arthur Lionel *A list of the marine Algae of Berwick-on-the-Tweed*. 8vo, 171 pp, 5 pls, Alnwick 1889. Reprinted from *Trans. Berwick Nat. Club.*
—— *see* Holmes, E. M.
230. BATTYE, Aubyn Bernard Rochfort Trevor *Pictures in prose of nature, wild sport, and humble life*. 8vo, x+243 pp, Longman, London 1894.
231. BAXTER, Thomas *A key to the natural orders of British wild flowering plants*. 8vo, 11 pp, Deighton, Worcester 1871.
232. BAXTER, William *British phaenogamous botany.* 8vo, 6

vols, 509 col. pls, Parker for the author, Oxford [1832–] 1834–43; 2ed. Vol. 1 1834; 3ed. 1856. 2ed. Vol. 1 is a title page edition.

233. BAXTER, William Edwin *The feathered vocalist. or history of British singing birds . . . and a short account of stuffing and mounting birds.* 12mo, x+86 pp, text figs, Sussex Press, Lewes 1830; 2ed. 1832.

234. BEALE, Lionel Smith *How to work with the miscroscope*. 8vo, 2 pts, 124 pp, 1 pl., text figs, Churchill, London 1857; 2ed. 1861; 3ed. [1864–] 1865; 4ed. [1867–] 1868; 5ed. [1879–] 1880.

235. —— *Illustrations to How to work with the microscope*. 8vo, 28 pls, Churchill, London 1859. Companion to No. 234.

236. —— *Protoplasm; or, life, form, and matter*. 8vo, 92 pp, 8 pls, Churchill, London 1870; 2ed. 1870; 3ed. 1874 title *Protoplasm: or, matter and life*; 4ed. 1892 title *Protoplasm: physical life and law.*

237. —— *Bioplasm*. 8vo, xvi+345 pp, text figs, Churchill, London 1872.

238. *Beauties and wonders of vegetable life*. 8vo, vi+280 pp, text figs, RTS, London [1866].

*The beauties of creation*. 1790. *see* Riley, G.

239. *The beauties of natural history*. 12mo, 72 pp, text figs, Mozley, Derby [c.1830].

240. *Beauties of natural history; or elements of zoography*. 12mo, xxxi+340 pp, 24 pls, text figs, Richardson & Urquhart, London 1777.

*Beauties of natural history. Selected from Buffon's History of quadrupeds. see* Buffon, G. L. Leclerc, *Conte de* [c.1790].

241. *The beauties of nature and art displayed, in a tour through the world*. 18mo, pls, portr., maps, Payne, London 1763–64; 2ed. 1774–75.

*Beauty in common things.* [1874]. *see* Chambers, *Miss* A. C.

242. BECHSTEIN, Johann Mattaeus *Cage birds: their natural history, and management*. 8vo, vi+394 pp, Orr, London 1848. Translation of *Naturgeschichte der Staubenvögel*, Halle 1840, by William Edward Shuckard. Another translation title *Cage and chamber birds*. 8vo, xv+500 pp, 38 pls, Bohn, London 1853. Translated by Henry Gardiner Adams. Many later printings.

243. BECK, Mary E. *Glimpses into the secrets of nature*. Sm. 8vo, viii+160 pp, pls, text figs, RTS, London 1889.

244. BECK, Richard *A treatise on the construction, proper use, and capabilities of Smith, Beck, and Beck's achromatic microscopes.* Roy. 8vo, viii+144 pp, 26 plain & 2 col. pls, text figs, Van Voorst for Smith, Beck & Beck, London 1865.

245. BECKWITH, William Edmund *Birds of Shropshire*. 8vo, 30+3+146 pp, Adnitt & Naunton, Shrewsbury [1892]. 25 copies. Three reprints with new title leaf from *Trans. Shropshire Archaeol. nat. Hist. Soc.* There is an issue of 1879 of the first paper alone, 30 pp.

—— *see* Phillips, W.

246. BEDDARD, Frank Evers *Animal colouration*. 8vo, vii+288 pp, 4 col. pls, text figs, Sonnenschein, London 1892; 2ed. 1895.

247. —— *A monograph of the order Oligochaeta*. 4to, xii+769 pp, text figs, Clarendon, Oxford 1895.

248. —— *Text book of zoogeography*. 8vo, viii+246 pp, 5 maps, text figs, Cambridge University Press, Cambridge 1895. Natural Science Manuals, Biological Series.

249. —— *Elementary practical zoology*. 8vo, vi+210 pp, text figs, Longman, London 1898. Practical Elementary Series.

250. —— *The structure and classification of birds*. 8vo, xx+548 pp, text figs, Longman, London 1898.

251. —— *The book of whales*. 8vo, xv+320 pp, 21 pls, Murray, London 1900. Progressive Science Series.

—— *see* Hudson, W. H.

252. BEETON, Samuel Orchart *Dictionary of natural history*. 8vo, iv+290 pp, 148 pls, Ward Lock, London 1871.

253. BEHRENS, Wilhelm Julius *Text-book of general botany*. 8vo, viii+374 pp, Pentland, Edinburgh 1885. Translation of *Methodisches der allgemeinen Botanik*, Braunschweig 1882, by *Miss* E. Harris Smith & Patrick Geddes.

Beilby, Ralph *A general history of quadrupeds*. 1790. *see* Bewick, T.

—— *A history of British birds. Vol. 1 Land birds*. 1797. *see* Bewick, T.

254. BELCH, William *History of birds*. 12mo, 8 11, col. text figs, Belch London [c.1820]; ed. 96 pp, Darton Harvey [ ]. First ed. engraved throughout and on one side only.

255. —— *History of fishes and insects*. 12mo, 8 11, col. text figs, Belch, London [c.1820]. Engraved throughout and printed on one side.

256. BELFAST NATURALISTS' FIELD CLUB *Systematic lists illustrative of the flora, fauna, palaeontology, and archaeology*

*of the North of Ireland*, 8vo, 2 vols, Belfast 1870–1906. An appendix to *Proc. Belfast Nat. Fld Club*.

257. —— *Guide to Belfast and the adjacent Counties*. 8vo, 327 pp, 47 pls, 1 map, Belfast 1874. British Association Guide.

258. BELL, Nancy, *Mrs* Arthur *Lowly water animals*. Sm. 8vo, ix+126 pp, text figs, Philip, London 1896. Later editions. Science Ladders No. 1.

259. —— *Links in a long chain: from worms to birds*. Sm. 8vo, x, 7–176 pp, text figs, Philip, London 1896. Later editions. Science Ladders No. 2.

260. —— *Mammals of the land and sea.* Sm. 8vo, xii, 7–191 pp, text figs, Philip, London 1896. Later editions. Science Ladders No. 3.

261. —— *Vegetable life and its lowest forms.* Sm. 8vo, 164 pp, text figs, Philip, London [1896]. Later editions. Science Ladders No. 7.

262. —— *Flowering plants*. Sm. 8vo, 7–204 pp, text figs, Philip, London [1897]. Later editions. Science Ladders No. 8.

263. [BELL, *Sir* Charles] *Animal mechanism, or proof of design in the animal frame*. 8vo, 63 pp, Baldwin, London 1822; 3ed. 1838. Library of Useful Knowledge, Natural Philosophy 4.

264. —— *The hand, its mechanism and vital endowments as evincing design.* 8vo, xvi+314 pp, text figs, Pickering, London 1833; 2ed. 1833; 3ed. 1834; 4ed. 1837; 5ed. 1852; 6ed. 1860; 7ed. 1865; 9ed. 1874; 1885. Bridgewater Treatise No. IV.

265. BELL, Francis Jeffrey *Comparative anatomy and physiology*. 8vo, ix+555 pp, text figs, Cassell, London 1885. Manuals for Students of Medicine.

266. —— *Catalogue of the British echinoderms in the British Museum*. 8vo, xvii+202 pp, 14 plain + 2 col. pls, text figs, British Museum (Natural History), London 1892.

267. BELL, Thomas *A history of British quadrupeds including the Cetacea.* 4to, Imp., Roy. or demy 8vo, xviii+526 pp, text figs, Van Voorst, London 1837; 2ed. 1874. Natural History of the British Isles. *see also* Newman, E. 1845. 4to very few copies. Imp. 8vo 50 copies.

268. —— *A history of British reptiles*. 4to, Imp., Roy., or demy 8vo, xiv+142 pp, text figs, Van Voorst, London 1839; 2ed. 1849. Natural History of the British Isles. *see also* Newman, E. 1845. 4to very few copies. Imp. 8vo 50 copies.

269. —— *A history of British stalk-eyed Crustacea*. Roy., or demy

8vo, lxv+386 pp, text figs, Van Voorst 1853; [ ]. Natural History of the British Isles.

270. —— & BAIRD, William *Tabular view of the Reptilia & Amphibia.* 4to, no text, 4 pls, SPCK or Stanford, London 1862. Diagrams of Natural History No. XI.

271. BELLAMY, John Kremer *The natural history of South Devon*. 8vo, xxvii+viii+455 pp, 9 pls, 5 maps, text figs, Thomas, Plymouth 1839.

272. BELLARS, Henry John *Illustrated catalogue of British land and freshwater shells*. 8vo, 30 pp, 4 col. pls, Minshull & Hughes, Chester 1858.

273. BENEDEN, Pierre Joseph Van *Animal parasites and messmates*. Sm. 8vo, xxvii+274 pp, text figs, King, London 1876; 2ed. 1876; 3ed. 1883. Later editions. International Scientific Vol. XX.

274. BENNETT, Alfred *An introduction to the study of flowerless plants*. 8vo, 86 pp, London 1891. Reprinted from Henfrey, A. *Elementary course of botany*.

Bennett, Alfred William *see* Saunders, W. W.

275. —— & MURRAY, George Robert Milne *A handbook of cryptogamic botany*. 8vo, viii+473 pp, text figs, Longman, London 1889.

276. BENNETT, Edward Turner *The Tower Menagerie.* 8vo, xviii+241 pp, text figs, Jennings, London 1829. Some copies on India paper.

277. —— *The Gardens and Menagerie of the Zoological Society delineated. Quadrupeds. (Birds)*. 8vo, 2 vols, text figs, Tegg for the proprietors, London [1829–] 1830; 1831; 1835.

278. BENNETT, Reginald Arthur Renaud *Marine aquaria: their construction, arrangement and management*. Sm. 8vo, vi+135 pp, text figs, Upcott Gill, London 1889. Also occurs as Part 2 of *The aquarium book* by Bateman, G. C. & B., with title *The marine aquarium.*

279. BENOIT, T. T. W. *The botanist's pocket companion*. 16mo, 108 pp, Churchill, London 1838.

280. BENSON, Charles William *Our Irish song birds*. Sm. 8vo, xi+190 pp, 2 col. pls, Hodges Figgis, Dublin 1886; 2ed. 1901.

281. BENTHAM, George *Handbook of the British flora*. Sm. 8vo, xvi+655 pp, Reeve, London 1858; ed. demy 8vo, 2 vols, text figs, 1865; 3ed. 1873; 4ed. 1878; 5ed. 1887; 6ed. 1892; 7ed. 1900; abridged ed. 12mo 1866. Several later reprints. From

5ed. onwards revised by Hooker, *Sir* Joseph D., and known as Bentham & Hooker. Crown Series. *see also* Fitch, W. H. & Smith, W. G.

282. —— *Outlines of elementary botany, with special reference to local floras*. 8vo, 45 pp, Reeve, London 1861; 4ed. [1879].

283. —— & HOOKER, *Sir* Joseph Dalton *Genera plantarum*. 8vo, 3 vols, 3692 pp, Reeve, London 1862–83; reprint 1965.

—— & —— *Handbook of the British flora. see* No. 281.

284. BENTLEY, Robert *A manual of botany*. 8vo, xxxii+811 pp, text figs, Churchill, London 1861; 2ed. 1870; 3ed. 1873; 4ed. 1882; 5ed. 1887.

285. —— *Botany*. 8vo, vi+128 pp, SPCK, London 1875. Later editions. Manuals of Elementary Science.

286. —— *The students' guide to structural, morphological and physiological botany*. 8vo, xiii+482 pp, Churchill, London 1883.

287. —— *The students' guide to systematic botany*. 8vo, xii+178 pp, Churchill, London 1884.

288. BERGE, Friedrich *Berge's Complete natural history of the animal, mineral, and vegetable kingdoms*. 8vo, 266 pp, 16 col. pls, text figs, Dean, London 1890. Translation of *Illustrierte Naturgeschichte für die Jügend*, Stuttgart, by Richard Frederick Crawford *et al*.

—— *see* Kirby, W. F.

Berger, Alexander Mal *see* Stillingfleet, B.

289. Berkeley, Miles Joseph *Gleanings of British Algae . . . being an appendix to the Supplement to English botany*. Roy. 8vo, 50 pp, 20 col. pls, Sowerby, London 1833. *see* Smith, *Sir* James E. *English botany*.

290. —— *British Fungi*. 8vo, 32+386+xv pp, Longman, London 1836. This is Vol. 5, pt. 2 of Smith, *Sir* James E. *The English flora*, and Vol. 2 of Hooker, *Sir* William J. *The British flora*, 4ed. 1838. *see also* No. 291.

291. —— *British Fungi: consisting of dried specimens of the species described in Vol. V., Part II of the English flora; together with such as may hereafter be discovered as indigenous to Britain*. 4to, 4 vols, 350 specimens with letterpress, Longmans, London 1836. Very few copies.

292. —— *Introduction to cryptogamic botany*. 8vo, viii+604 pp, text figs, Baillière, London 1857. Library of Illustrated Standard Works.

293. —— *Outlines of British fungology*. 8vo, xvii+442 pp, 1 plain

+ 23 col. pls, Reeve, London 1860; supplementum 1891 by Smith, W. G. q.v.

294. —— *Handbook to British mosses.* 8vo, xxxvi+324 pp, 1 plain + 23 col. pls, Reeve, London 1863; 2ed. 1895.

295. BERKENHOUT, John *Clavis Anglica linguae botanici; or, a botanical lexicon.* 12mo, xii+[214] pp, for the author, London 1764; 2ed. 1789 title . . . *to which is added Calendarium botanicum.*

296. —— *Outlines of the natural history of Great Britain and Ireland.* 12mo, 3 vols, Elmsley, London 1769–72; ed. 2 vols 1789 title *Synopsis* . . .; 3ed. 2 vols 1795.

—— *Synopsis of the natural history of Great Britain and Ireland. see* No. 296.

297. BERNARD, Henry Meyners *The Apodidae: a morphological study.* Sm. 8vo, xvii+316 pp, 1 pl., text figs, Macmillan, London 1892. Nature Series.

Bernardin de Saint Pierre, Jacques Henri *see* Saint Pierre, J. H. B. de

*The berries and heaths of Rannoch.* 1881. *see* Cross, J. E.

298. BERT, Paul *First year of scientific knowledge.* Sm. 8vo, 344 pp, text figs, Relfe, London 1885. Many later editions. Translation of *La première année d'enseignement scientifiques*, Paris 1882, by Josephina Clayton.

299. BERTRAM, James Glass *The harvest of the sea.* 8vo, xv+519 pp, 1 pl., text figs, Murray, London 1865; 2ed. 1869; 3ed. 1873; 4ed. 1885.

300. BEST, *Mrs* C. *Six thousand years ago: or, the works of creation illustrated.* 12mo, vi+168 pp, 6 pls, text figs, Binns & Goodwin, Bath [1843]; 2ed. [ ].

301. BETTANY, George Thomas *Simple lessons in botany.* 16mo, 45 pp, Murby, London [1878]. Extra Subject Manuals.

302. —— *First lessons in practical botany.* 8vo, viii+103 pp, Macmillan, London 1881.

303. —— *Botany.* 8vo, 127 pp, Ward Lock, London [1882]. Science Primers for the People.

—— *see* Parker, W. K.

304. [BEVERLEY, Robert Mackenzie] *The Darwinian theory of the transmutation of species examined.* 8vo, viii+336 pp, Nisbet, London 1867; 2ed. 1868. By a graduate of the University of Cambridge.

305. [BEWICK, Thomas] *A general history of quadrupeds.* Roy. or demy 8vo, viii+456 pp, text figs, Beilby & Bewick, New-

castle upon Tyne 1790; 2ed. 1791; 3ed. 1792; 4ed. 1800; 5ed. 1807; 6ed. 1811; 7ed. 1820; 8ed. 1824; ed. 1887. Text by Beilby, R., wood engravings by Bewick. 1887 ed. is Memorial edition Vol. 3.

306. [——] *History of British birds . . . Vol. I . . . Land birds.* (*Vol. II . . . Water birds*). Imp., Roy., or demy 8vo, 2 vols, text figs, Beilby & Bewick, Newcastle upon Tyne, 1797, 1804.
a. *Land birds*. ed. 1797 [=1798]; ed. 1804 or L. P. 1805; ed. 1804 [=?1814]; ed. 1816; ed. 1821 with supplement 46 pp; 6ed. 1826; 7ed. 1832; 8ed. 1847; ed. 1887; supplements 1821 43 pp; 1821 [=1822] 50 pp.
b. *Water birds*. ed. 1805; ed. 1809; ed. 1816; ed. 1821 with supplement 43 pp; ed. 1826; 7ed. 1832; 8ed. 1847; ed. 1887; supplements 1821 43 pp; 1821 [=1822] 49 pp.
c. *Addenda to both vols.* 7 pp, [1822].
Text to Vol. I by Beilby, R. & Bewick; of Vol. II by Bewick & Cotes, H; wood engravings by Bewick. 1887 eds are Memorial edition Vols I & II. *see also* Eyton, T. C.

307. —— *Figures of British land birds . . . to which are added a few foreign birds, with their vulgar and scientific names. Vol. I.* Roy. 8vo, 133 11, figs, Beilby & Bewick, Newcastle upon Tyne 1800. Printed and numbered on recto only. No text. All publ.

308. —— *Bewick's British birds*. 4to, 239 11, figs, Walker printed, Newcastle upon Tyne 1817; another issue 1817 title *Figures to British land and water birds*; supplements to Land (Water) birds 1821; ed. 2 vols 1825 title *British land (water) birds*. These are editions of the wood engravings alone without text or captions. 1817 25 copies; 1825 100 copies.

309. —— *Figures of quadrupeds drawn from life*. 4to, 225 11, figs, Bewick & Son, Newcastle upon Tyne 1818; ed. 1824 title *The figures of Bewick's quadrupeds*. These are editions of the wood engravings only, without text or captions. 1818 25 copies; 1824 100 copies.

310. —— *A memoir of Thomas Bewick written by himself, embellished by numerous wood engravings . . . for a work on British fishes*. 4to, xi+344 pp, text figs, Jane Bewick, Newcastle upon Tyne 1862; ed. Roy. 8vo 1887; ed. 8vo 1924; ed. 1961; another ed. 1961. 1887 ed. is Memorial edition Vol. V.

—— *The works of Thomas Bewick*. Imp., Roy. or demy 8vo, 5 vols, text figs, Charnley, Newcastle upon Tyne 1822; ed. Roy. 8vo 1887. 1822 ed. is merely added title pages to any editions. Birds, 2 vols, is usually 1821; Quadrupeds usually 7ed. 1820. Imperial paper copies may have 3 portraits. 1887 ed. is Memorial edition.

311. —— [*Thomas Bewick's wood engravings.*] Roy. 4to, 7 vols, figs, Ward, Newcastle upon Tyne 1883. This is an edition of the figures only, with title pages, but no text or captions. Very few copies.

[——] *A natural history of British quadrupeds, foreign quadrupeds, British birds, water birds, fishes, reptiles, serpents, and insects*. 1809. *see A natural history of British quadrupeds* etc.

312. BICHENO, James Ebenezer *An address delivered at the anniversay meeting of the Zoological Club of the Linnean Society*. 8vo, 30 pp, Taylor printed, London 1826.

313. BICKNELL, W. I. *The natural history of the sacred scriptures and a guide to general zoology*. 16mo, 2 vols, 316 col. pls, London Printing and Publishing Company, London [c.1845]; ed. of Vol. 1 in 2 vols [1850, 51] title *Tallis's Scriptural natural history*.

314. BIGGE, Edward *Observations on the natural history of two species of wasps*. 8vo, 28 pp, Ashmolean Society, Oxford 1835. Reprinted from *Trans. Ashmolean Soc*., Vol. 1.

315. [BIGGIE, George R.] *A catalogue of British birds*. 8vo, 32 pp, Andrews, Durham 1846.

316. BIGLAND, John *Letters on natural history exhibiting a view of the power, wisdom and goodness of the Deity*. Sm. 8vo, xix+448 pp, 50 pls, portr., Longman, London 1806.

317. BINET, Alfred & FÉRÉ, Charles *Animal magnetism*. 8vo, 378 pp, Kegan Paul, London 1887; 3ed. 1891. Translation of *Le magnétisme animal*. Paris 1887. International Scientific Series.

318. BINGLEY, Thomas *Tales about birds, illustrative of their nature, habits and instinct*. 16mo, vii+198 pp, 8 pls, Tilt, London 1839; 2ed. 1840; ed. 1864.

319. —— *Stories illustrative of the instincts of animals, their characters and habits*. 16mo, vii+197 pp, 8 pls, Tilt, London; 2ed. 1840; 3ed. 1840; ed. 1864.

320. BINGLEY, William *A tour round North Wales . . . and the habitats of above 400 of the more rare native plants*. 8vo, 2 vols, pls, Williams, London 1800.

321. —— *Animal biography*. 8vo, 3 vols, Phillips, London [1802–] 1803; 2ed. 1804; 3ed. 1805; 4ed. 1813; 5ed. 1820; 6ed. 4 vols 1824; 7ed. 1829.

322. —— *North Wales . . . and some sketches of its natural history*. 8vo, 2 vols, 2 pls, 1 map, Longman, London 1804; 2ed. 1814; 3ed. 1839 title *Excursions in North Wales*.

323. —— *Memoirs of British quadrupeds . . . a synopsis of British quadrupeds*. Roy. 8vo, 2 pts, xii+459+70 pp, 71 pls, Darton & Harvey, London 1809. No plates 7, 8 & 22. L. P., coloured copies are recorded, but doubtfully so issued.

324. —— *Animated nature; or, elements of the natural history of animals*. 12mo, xxviii+310 pp, 40 pls, Darton Harvey, London 1814; 2ed. 1826; ed. 1842; ed. [ ] title *Natural history for the use of schools*.

325. —— *Useful knowledge; or a familiar account of the various productions of nature . . . which are chiefly employed for the use of man*. 12mo, 3 vols, 16 pls, Baldwin, London 1816; 2ed. 1818; 3ed. 1821; 4ed. 1825; 5ed. 1831; 6ed. 1842; 7ed. 2 vols 1852.

326. —— *A practical introduction to botany, illustrated by reference to plants of easy access.* 12mo, vi+89 pp, 8 pls, Gale & Fenner, London 1817; 2ed. 1827 title . . . *with some account of the history of the science*, by James Frost; 3ed. 1831. A supplement to Priscilla Wakefield *Introduction to botany*, 1796.

—— *Excursions in North Wales*. 1839. *see* No. 322.

—— *see* Hack, Maria

327. BIRCHALL, Edwin *Catalogue of the Lepidoptera of Ireland.* 8vo, 29 pp. Gill, Dublin 1868.

328. *Bird-catching or, the northern adventurers. Being an account of several methods of taking birds in the Feroe Islands, and in some other places. Intended for children.* 12mo, 82 pp, 3 pls, Harvey & Darton, London 1823.

*The bird picture-book*. 1873. *see* Wood, J. G.

329. *Bird songs and bird pictures*. 16mo, 127 pp, RTS, London [1873].

330. *Birds*. 16mo, 64+32+32+32 pp, col. text figs, RTS, London [1842].

*Birds and bird-life.* [1863]. *see* Buckland, F. T.

331. *Birds, embellished with upwards of thirty neat engravings on wood*. 32mo, [32] pp. text figs, Harrison, Devizes 1824.

332. *Birds of prey*. 12 col. cards with letterpress, Nelson, London 1870.

333. *Birds of the sea shore*. 18mo, 12 col. pls, text figs, SPCK, London [1856]. Later editions.

*Birds' nests and eggs*. *see* Johns, C. A.

334. *Birds' nests and eggs. Also directions for bird-stuffing*. Sm. 8vo, 1 col. pl., text figs, Bickers & Bush, London [1862]. Originally issued as pp. 289–352 of *The book of home pets*. 1862. The separate issue retains this pagination.

335. BISHOP, David *Causal botany*. 8vo, viii+184 pp, Wright, London 1829.

336. [BISHOP, James] *A visit to the Zoological Gardens in the Regent's Park*. 8vo, 64 pp, text figs, Dean & Munday, London 1830; 3ed. [c.1835] title *Henry and Emma's visit to the Zoological Gardens.*

337. [BLACKBURN, Jane, *Mrs* Hugh] *Scenes of animal life and character*. 4to, 20 col. pls with letterpress, Griffith & Farren, London [1858]. Signed J.B.

338. —— *British birds drawn from nature*. Fol., 2 pts, 45 pls, Pt. 1 Edmonston & Douglas, Pt. 2 Maclehose, Edinburgh 1862, 68; 2ed. Glasgow 1868.

339. —— *Birds from Moidart and elsewhere drawn from nature*. Sq. 8vo, viii+191 pp, 88 pls, text figs, Douglas, Edinburgh 1895. Plate are numbered with text.

340. BLACKBURN, Thomas *Outline descriptions of British Coleoptera*. 8vo, 71 pp, Perthshire Society of Natural Science, Perth 1875. Reprinted from *The Scottish Naturalist.*

341. BLACKSTONE, John *Fasciculus plantarum circa Harefield sponte nascentium*. 12mo, viii+118 pp, Woodfall printed, London 1737.

342. —— *Specimen botanicum quo plantarum plurium rariorum Angliae indigenarum loci natales illustrantur*. 8vo, 106 pp, 2 pls, Faden printed, London 1746.

343. BLACKWELL, John *Tables of the various species of periodical birds*. 8vo, 30 pp, for the author, Manchester 1823. Reprinted from *Mem. Manchester lit. phil. Soc.*

344. —— *Observations on the notes of birds, including an inquiry whether or not they are instinctive*. 8vo, 31 pp, for the author, Manchester 1824. Reprinted from *Mem. Manchester lit. phil. Soc.*

345. —— *Observations conducive towards a more complete history of the cuckoo*. 8vo, 36 pp, 1 col. pl., for the author, Manchester 1824. Reprinted from *Mem. Manchester lit. phil. Soc.*

346. —— *Researches in zoology*. 8vo, 434 pp, 3 col. pls, Simpkin & Marshall, London 1834; 2ed. 1873.

347 .—— *A history of the spiders of Great Britain and Ireland*. Roy.

4to, 2 pts, vi+384 pp, 29 col. pls, Ray Society, London 1860, 64; another issue Pt. 1 1861. Ray Society Publ. 31 & 34 for 1859 & 1862.

348. BLADEN, William Wells *List of birds in whose nest the egg of the cuckoo has been found*. 8vo, Stoke on Trent 1896.

349. —— *The cuckoo and its foster parents*. 8vo, 19 pp, Stoke on Trent 1896. Reprinted from *Trans. N. Staffs. Nat. Fld Club*, 1896.

350. BLADH, Andreas Johannes *Fundamenta entomologiae: or, an introduction to the knowledge of insects*. 8vo, viii+90 pp, 2 pls, for the author [i.e. the translator William Curtis], London 1772. Translation of *Fundamenta entomologiae*, Uppsala 1767, a Linnaean dissertation.

351. BLAIR, Patrick *Botanick essays*. 8vo, 414 pp, 5 pls, Innys, London 1720.

352. BLAKE, Charles Carter *Zoology for students*. 8vo, xv+382 pp, text figs, Daldy Isbister, London 1875.

353. BLANCHARD, Émile *The transformations, or metamorphoses of insects . . . and a compilation from the works of Newport, Charles Darwin . . . and others*. 8vo, x+480 pp, pls, Cassell, London [1870]. Later editions. Translation of *Metamorphoses, moeurs et instinctes des insectes*. Paris 1868, with compilation by Duncan, P. M.

354. BLAND, James *Grandpapa's tales about animals*. Sm. 8vo, vi+60 pp, text figs, Milner & Sowerby, Halifax 1851; ed. [ ].

355. —— *Grandpapa's tales about birds and fishes*. Sm. 8vo, 160 pp, text figs, Milner & Sowerby, Halifax [c.1852].

356. BLAND, Samuel King *Biology: being glimpses of universal life*. 8vo, 31 pp, Book Society, London 1860.

357. BLASIUS, Johann Heinrich *A list of the birds of Europe*. 4to, 24 pp, Matchett & Stevenson, Norwich 1862, German original ? not published, translated by Alfred Newton.

358. BLIGHT, John Thomas *A week at the Land's End*. Sm. 8vo, xi+233 pp, 11 pls, 1 map, text figs, Longman, London 1861; ed. Truro 1876.

Blomefield, Leonard *see* Jenyns, L. *later* Blomefield.

359. BLOOMFIELD, Edwin Newson *The Lepidoptera of Suffolk*. 4to, 60 pp, St Leonard's 1860; ed. 8vo, London 1890.

360. —— *Natural history of Hastings and St Leonard's and the vicinity*. 8vo, 68 pp, Hastings 1878; suppl., 53 pp, 1883; 2nd suppl., 24 pp, 1888; 3rd suppl., 32 pp, 1898.

361. —— *The Micro-Lepidoptera of Suffolk: tortrices and tineae*. 8vo, 7 pp, St Leonard's 1888.

362. BLOUNT, *Sir* Thomas Pope *Bart A natural history: containing many not common observations extracted out of the best writers*. 12mo, 469 pp, Bently, London 1693.

363. [BLOXHAM, Andrew] *A description of Bradgate Park . . . with remarks on the natural history of Charnwood Forest*. 8vo, 82 pp, Come, Leicester [?1829]; another issue [?1829] title *Guide to Bradgate Park . . .*, signed.

364. BLUMENBACH, Johann Friedrich *A short system of comparative anatomy*. 8vo, xxxiii+484 pp, Longman, London 1807; 2ed. 1827 title *A manual of comparative anatomy*. Translation of *Handbuch der vergleichenden Anatomie*, Göttingen 1805, by William Lawrence.

365. —— *A manual of the elements of natural history*. 8vo, x+415+xiv pp, 2 pls, Simpkin & Marshall, London 1825. Translation of *Handbuch der Naturgeschichte*, Göttingen, 10ed. 1821, by R. T. Gore.

—— *A manual of comparative anatomy* 1827. *see* No. 364.

366. BOAS, Johan Erik Vesti *Text book of zoology*. 8vo, xviii+558 pp, text figs, Sampson Low, London 1896. Translation of *Lehrbuch der Zoologie*, Jena 1890, itself a translation of *Laerebog i Zoologien*, Copenhagen 1888, by J. W. Kirkaldy & Emma Catherine Pollard.

367. BOATE, Gerard *Ireland's natural history*. 8vo, 186 pp, Wright, London 1652; issue 1657; ed. 4to Dublin 1726; reissue Dublin 1755; ed. 1860, 1861. Eds of 1726 & 1755 in *A natural history of Ireland . . . by several hands*. Ed. of 1860, 61 in *A collection of tracts and treatises illustrative of the natural history . . . of Ireland*. Contains almost no natural history; this was intended for a second volume, never published.

368. BOCKETT, Benjamin Bradney *The lepidopterist's indicator; an alphabetically arranged guide to the species of British Lepidoptera*. 4to, 36 pp, Newman, London 1860.

369. BOHLER, John *Lichenes Britannici*. 8vo, 4 vols, 128 specimens with letterpress, Sheffield 1835–37.

370. BOLTON, James *Filices Britannicae; an history of the British proper ferns*. Roy. 4to, 2 pts, xxii+81 pp, 46 col. pls on 45 11, Pt. 1 Binns, Leeds, Pt. 2 Brook for the author, Huddersfield 1785, 90.

371. —— *An history of the fungusses growing about Halifax*. 4to, 3 vols, xlii+182 pp, 138 col. pls, White for the author, Halifax 1788–89; appendix 44 col. pls, for the author, Huddersfield 1791.

372. —— *Harmonia ruralis; or, an essay towards a natural history of British song birds*. Fol., 2 vol, viii+82 pp, 81 col. pls, for the author, Stannary near Halifax 1794–96; ed. 1823; ed. 4to 1824 unsigned; ed. 4to [1830] unsigned; ed. 4to, London 1845.

373. BOLTON, Thomas *Portfolio(s) of drawings, and descriptions of living organisms (animal and vegetable) illustrative of freshwater and marine life, which have been sent out with the living specimens*. 8vo, 10 numbers, printed wrappers, 168 loose leaves with text figs, David Bogue, London 1879–84. About 50 other leaves, not in portfolios, were issued from 1876.

374. —— *Hints on the preservation of living objects and their examination under the microscope*. 8vo, 7 pp, Herald Printing Offices, Birmingham [1880]. A companion to the *Portfolios*, No. 373. Reprinted from *English Mechanic and World of Science.*

375. BONAPARTE, Charles Lucien Jules Laurent, *Prince* of Canino *A geographical and comparative list of the birds of Europe and North America.* 8vo, vii+67 pp, Van Voorst, London 1838.

376. —— *Systema vertebratorum*. 4to, 58 pp, London 1840.

377. BONAVIA, Emmanuel *Philosophical notes on botanical subjects.* 8vo, v+368 pp, text figs, Eyre & Spottiswoode, London 1892.

378. —— *Studies in the evolution of animals*. 4to, xxiv+362 pp, text figs, Constable, London 1895.

379. BONER, Charles *Forest creatures*. 8vo, viii+245 pp, text figs, Longman, London 1861.

380. BONHOTE, John Lewis & ROTHSCHILD, *Hon.* Nathaniel Charles *Harrow butterflies and moths*. 8vo, 2 vols, 1 pl., Harrow 1895–97.

381. BONNET, Charles *Contemplation of nature*. 12mo, 2 vols, Longman, London 1766. Translation of *Contemplation de la nature*. Amsterdam 1764. For another translation see Wesley, J.

382. *A book about animals*. 16mo, iv+64 pp, 6 col. pls, text figs, RTS, London [c.1853].

383. *A book about birds*. 16mo, iv+68 pp, 6 col. pls, text figs, RTS, London [1851].

384. *A book for the sea-side*. 12mo, viii+275 pp, 1 pl., text figs, RTS, London [c.1860].

385. *The book of animals*. 4to, 8 col. pls with letterpress, Ward Lock, London [1866].

386. *The book of animals, Class Mammalia*. 16mo, 156 pp, text figs, Parker for SPCK, London 1836; 10ed. 1851.

387. *The book of beasts for young persons*. 32mo, 15 pp, text figs, Rusher, Banbury [?1820]. Reprinted many times.

388. *Book of beasts for young readers*. 16mo, 64 pp, text figs, Nelson, London [1867].

389. *The book of birds. Class Aves*. 16mo, 156 pp, text figs, Parker for SPCK, London 1834. Later editions.

390. *Book of birds for young readers*. 16mo, 64 pp, Nelson, London [1867].

391. *A book of birds, their natural history and description*. 12mo, 127 pp.

*The book of botany*. [?1850]. *see* Tyas, R.

392. *The book of fishes*. 16mo, 160 pp, text figs, Parker for SPCK, London 1834; 5ed. 1845.

393. *The book of reptiles. (Class Reptilia)*. 16mo, vii+134 pp, 2 pls, text figs, Parker for SPCK, London 1835; 2ed. 1847; 5ed. 1849.

394. *The book of shells*. 16mo, 144 pp, 1 pl., text figs, Parker for SPCK, London 1836; 1837; 4ed. 1846.

395. *The book of the month, and circle of the seasons*. Sm. 8vo, vii+214 pp, 16 pls, Bogue, London 1844.

*The book of the salmon*. 1850. *see* Fitzgibbon, E.

396. *The book of trees*. 16mo, 216 pp, text figs, Parker for SPCK, London 1837; 2ed. 1839; 4ed. 1852.

397. *The book of wild beasts*. 32mo, 8 pp, text figs, Davison, Alnwick [c.1820].

398. BOOTH, Edward Thomas *Catalogue of the cases of birds in the Dyke Road Museum, Brighton*. 8vo, iv+219 pp, Beal, Brighton 1876; index 7 pp, 1878; 2ed. 1896; 3ed. 1901. Many later editions. Numerous slips were printed and pasted on as cancels in the stock, as exhibits changed. *See also* No. 400.

399. —— *Rough notes on the birds observed . . . in the British Islands*. Fol., 3 vols, 114 col. pls, 2 col. maps, text figs, Porter, London 1881–87.

400. —— *The Booth collection of British birds, Dyke Road, Brighton, index catalogue*. Sm. 8vo, 8 pp, King Thorne & Stace, Brighton 1892. A new edition of the index to No. 398.

401. BOOTH, Henry T. *Collecting and preserving birds' nests and eggs.* 8vo, 8 pp, Naturalists' Publishing Company, Birmingham 1891. Later editions.

402. —— *British birds their nests and eggs*. 8vo, 16 pp, London 1893.

403. [BOREMAN, Thomas] *A description of three hundred animals.* 12mo, 212 pp, 98 pls, Ware & Boreman, London 1730. 43 of the plates are printed on both sides, so that there are 55 plate leaves, and all except one are paginated with the text. For later editions see below. The attribution is uncertain, but traditional.

404. [——] *A description of a great variety of animals and vegetables . . . being a supplement to A description of three hundred animals.* 12mo, 137 pp, 90 pls, Boreman, London 1736; issue Gyles, London 1736; 2ed. 1744; 3ed. 1761; new ed. [c.1775]. 37 of the plates are printed on both sides so that there are 53 plate leaves. 3ed. contains No. 405 as well.

405. [——] *A description of some curious and uncommon creatures omitted in the Description of three hundred animals, and likewise in the supplement to that book.* 12mo, 88 pp, 16 pls, Ware & Boreman, London 1739; ed. 1761. 1761 ed. in No. 404.

Later editions of No. 403, some incorporating parts of Nos 404 & 405:

Title as led and with plates. 12mo, 2ed. 1732; 3ed. 1736; 4ed. 1738; 5ed. 1743; 7ed. 1753; 8ed. 1759; ed. Dublin 1762; 9ed. 1763; 10ed. 1769; 10ed. 1773; 11ed. 1774; ed. Dublin 1776; ed. 1786; ed. 1791.

Title *A description of above three hundred animals*, and with plates. 12mo, Edinburgh 1780; Edinburgh (Leith) 1787; ed. Edinburgh (Leith) 1790; ed. Glasgow (Lumsden) 1794; ed. Glasgow (Robertson) 1794; ed. London 1795; ed. Glasgow (Lumsden, Paisley printed) 1796; ed. Edinburgh 1797; ed. Perth 1807; ed. Edinburgh 1808.

Title *A description of more than* [or *above*] *three hundred animals.* No plates, text figs by S. Williams, 8vo, London 1812; 2ed. 12mo, Alnwick [c.1820]; ed. London 1826; ed. London 1833. 1812 ed. edited by A.D.M.[=Ange Denis Macquin].

Title *A complete natural history containing delineations of upwards of three hundred animals.* 8vo, London 1813.
Title *A compendium of zoology, being a description of more than three hundred animals.* 12mo, London 1818.
Title *Buffon's natural history of quadrupeds, birds etc.* 16mo, Chelmsford 1822.
Title *A natural history of the most remarkable quadrupeds.* 12mo, 2 vols, London 1825; ed. 1826; ed. 1830. By *Mrs* Mary Trimmer. ?Not by *Mrs* Mary Trimmer.
Title *The entertaining naturalist.* 8vo, London 1843; ed. 1850. By *Mrs* Jane Loudon.
Title *Mrs Loudon's entertaining naturalist.* 8vo, London 1867. Edited by William Sweetland Dallas.
Title *Loudon's Natural history.* 8vo, London 1889. An edition of the previous entry.

406. BORLASE, William *The natural history of Cornwall.* Fol., xix+326 pp, 28 pls, 1 map, Jackson for the author, Oxford 1758; additions 66 pp, 1 pl., Truro 1864–66. Additions reprinted from *J. R. Inst. Cornwall*; facsimile 1970.

407. BORRADAILLE, Lancelot Alexander *On a Cancer pagurus with a supernumerary chela.* 8vo, Jersey [1897]. Reprinted from *J. mar. Zool. Micr.*

408. BORRER, William *The birds of Sussex.* Sm. 8vo, xviii+385 pp, 6 col. pls, 1 col. map, Porter, London 1891.
—— *see* Turner, D.

409. [BOSANQUET, Edwin] *A plain and easy account of British ferns.* Sm. 8vo, 64 pp, Hardwicke, London 1854; ed. 1855; 2ed. 1855 signed. For later editions see Lankester, Phebe *Mrs* Edwin. led. is signed E.B.

Boswell, Henry *see* Hobkirk, C. C. P.

*Botanical dialogues.* 1797. *see* Jackson, *Mrs* M. E.

410. *Botanical illustrations of the twenty-four classes in the Linnaean system of vegetables, by select specimens of British plants.* 16mo, 96 pp, 48 pls, Darton Harvey, London 1813. For a companion volume see No. 413.

411. *The botanical keepsake, an arrangement of British plants.* Sm. 8vo, x+233 pp, 28 col. pls, Allman, London 1843; 1846.

412. *A botanical ladder for the young.* 12mo, 176 pp, 1 pl., text figs, RTS, London ?1844].

*Botanical lectures . . . altered from Botanical dialogues.* 1804. *see* Jackson, *Mrs* M. E.

*Botanical pocket book . . . with rarer plants found in the neighbourhood of Tunbridge Wells. see* Forster, T. F.

*Botanical rambles. see* Atkins, *Miss* S.

Botanical Record Club. *see* Hobkirk, C. C. P. & Boswell, H.

413. *Botanical sketches of the twenty four classes of the Linnaean system. With fifty specimens of British plants.* 8vo, xii+126 pp, 50 col. pls, Hatchard, London 1825. Companion volume to No. 410.

414. *Botanical specimens, copied from nature, and designed as simple illustrations of the twenty-four classes into which, according to the Linnaean system of arrangement, all plants are divided.* 4to, 24 col. pls, Smith, Liverpool 1828. Plates in a portfolio.

*A botanical tour through various parts of the Counties of Devon and Cornwall.* 1820. *see* Jones, J. P.

415. BOTANISTA, Theophilus *pseudonym Rural beauties; or, the natural history of . . . Cornwall, Devonshire, Dorsetshire, and Somersetshire.* 12mo, xxxiv+314 pp, 1 pl., Fenner, London 1757.

416. *The botanist's calendar and pocket flora.* Sm. 8vo, 2 vols, White, London 1797.

*The botanist's manual.* [1836]. *see* Deakin, R.

*Botanizing excursions through high and low lands.* 1877. *see* Rogers, C.

*Botano-theology, an arranged compendium.* 1825. *see* Duncan, J. S.

*Botany. In four parts.* [1838]. *see* Lindley, J.

417. *Botany simplified on the natural system.* 8vo, 40 pp, London 1874. By Adoxa, *pseudonym.*

418. BOULENGER, George Albert *The tailless batrachians of Europe.* 8vo, 2 pts, iii+376 pp, 16 col.+8 plain pls, 1 map, text figs, Ray Society, London 1897, 98. Ray Society Publ. Nos 74 & 75 for 1896, 97.

419. BOULGER, George Edward Simonds *Familiar trees.* 8vo, 2 ser., col. pls, Cassell, London [1885–88]; ed. [1891–94]; ed. 3 ser. [1906–07].

—— *see* Owen, *Mrs* Jean A.

420. BOULTER, Daniel *Museum Boulterianum. A catalogue . . . of natural and artificial curiosities in the extensive museum of Daniel Boulter. Yarmouth.* 8vo, viii+165 pp, for the author, Yarmouth [?1810].

421. [BOWDICH, Sarah, *Mrs* Thomas Edward, later *Mrs* Robert Lee] *Taxidermy.* 12mo, 168 pp, 5 pls, Longman, London

1820; 2ed. 1821; 3ed. 1823; 4ed. 1829 signed; 5ed. 1835; 6ed. 1843.

422. —— *The freshwater fishes of Great Britain.* Imp. 4to, 47 col. pls on 45 leaves, Ackermann for the authoress, London 1838. Very few copies.

423. —— *Elements of natural history.* 8vo, viii+485 pp, text figs, Longman, London 1844.

424. —— *British animals.* Roy. 16mo, 2 ser., 14 col. pls with letterpress, Grant & Griffith, London 1852, 53.

425. —— *British birds.* Roy. 16mo, 7 col. pls with letterpress, Grant & Griffith, London 1852.

—— *British animals and birds.* Roy. 16mo, 21 col. pls with letterpress, Griffith & Farren, London [1865]. Reissue of Nos 424 & 425 as a combination volume.

426. —— *Anecdotes of the habits and instincts of animals.* 8vo, viii+312 pp, text figs, Grant & Griffith, London 1852; ed. [1891].

427. —— *Anecdotes of the habits and instincts of birds, reptiles, and fishes.* 8vo, iv+410 pp, 4 pls, text figs, Grant & Griffith, London 1853; 2ed. 1861; ed. [1891].

428. —— *Familiar natural history.* Roy. 16mo, 42 col. pls with letterpress, Grant & Griffith, London 1853 [=1852].

429. —— *Trees, plants, and flowers; their beauties, uses and influences.* 8vo, viii+464 pp, 8 col. pls, Grant & Griffith, London 1854.

430. BOWDICH, Thomas Edward *An analysis of the natural classifications of Mammalia for the use of students and travellers.* 8vo, 115 pp, 16 pls, Smith, Paris 1821. A partial translation of Cuvier, G. L. C. F. D., *Baron* de *La regne animal*, Paris 1817, and of Illiger, J. C. W., *Prodromus systematis mammalium et avium*, Berlin 1811.

431. —— *An introduction to the ornithology of Cuvier, for the use of students and travellers.* 8vo, 86 pp, 21 pls, Smith printed, Paris 1821. A partial translation of Cuvier, G. L. C. F. D., *Baron* de *Le regne animal*, Paris 1817. Contains an index to No. 430.

432. BOWER, Frederick Orpen *Practical botany for beginners.* 8vo, xi+275 pp, Macmillan, London 1894; 2ed. 1902.

433. —— & VINES, Sydney Howard *A course of practical instruction in botany.* Sm. 8vo, 2 pts, Macmillan, London 1885, 87; reissue one vol. 1888.

434. BOWERBANK, James Scott *A monograph of the British*

*Spongiadae.* 8vo, 4 vols, 146 pls, Ray Society, London 1864–82. Ray Society Publ. Nos 36, 38, 51, 57 for 1864, 65, 74 & 79. Vol. 4 by Norman, A. M.

435. BOYLE, *Hon.* Robert *General heads for the natural history of a country, great or small, drawn out for the use of travellers and navigators.* 12mo, 134 pp, Taylor & Holford, London 1692; Latin ed. 4to, Geneva 1696, title *Generalis capita pro historia naturali.*

436. BOYS, William *Collections for an history of Sandwich in Kent.* 4to, viii+877 pp, 31 pls, 4 maps, 6 plans, 4 tab., Simmons for the author, Canterbury 1892 [=1792].

437. —— & WALKER, George *Testacea minuta rariora, nuperrime detecta in arena littoris Sandvicensis: a collection of minute and rare shells lately discovered in the sand of the sea shore near Sandwich.* 4to, v+25 pp, 3 pls, London [1784].

438. *Boys' and girls' book of science.* 8vo, xii+372 pp, text figs, Strahan, London 1881 [=1880]. With Charles Kingsley's address to boys and girls from *Madam How and Lady Why* as preface.

439. BRADBURY, Henry Riley *A few leaves representing nature printing.* 8vo, 21 col. pls, Bradbury & Evans, London 1854.

440. —— *Nature printing: its origin and objects.* 8vo, 28 pp, Bradbury & Evans, London 1856.

441. BRADLEY, Richard *A philosophical account of the works of nature.* 4to, xx+194 pp, 28 col. pls, Mears, London 1721; 2ed. 8vo, 1739.

442. —— *New experiments and observations relating to the generation of plants.* 8vo, 32 pp, Corbett, London 1724.

443. BRADLEY, Samuel Messenger *Manual of comparative anatomy and physiology.* Sm. 8vo, iv+101 pp, text figs, Simpkin Marshall, London 1870 [=1869]; 2ed. 1874; 3ed. 1875.

444. BRADY, George Stewardson *A monograph of the free and semiparasitic Copepoda of the British Islands.* 8vo, 3 vols, 7 col. & 89 plain pls, Ray Society, London 1878–80. Ray Society Publ. Nos 53, 55 & 56 for 1878, 79 & 80. Pls 1–93, 10A, 24A, 24B.

—— *see* Norman, A. M.

Brady, Henry Bowman *see* Siddall, J. D.

445. BRAGG, Jane *Birds and insects.* 16mo, iv+165 pp, text figs, Harvey & Darton, London [1845].

446. BRAITHWAITE, George Foster *Fishes and their food;*

*being a paper on the salmon family.* 8vo, 32 pp, Atkinson, Kendal 1863.

447. BRAITHWAITE, Robert *The British moss-flora.* Roy. 8vo, 3 vols, 128 pls, Reeve for the author, London [1880–] 1887–1905.

448. —— *The Sphagnaceae, or peat mosses of Europe and North America.* Imp. 8vo, 91 pp, 29 col. pls, Bogue London 1880.

Branch, John *see* Holloway, William.

Brand, Fitzjohn *see* Linnaeus, C.

449. BRAY, *Mrs* Anna Elizabeth *A description of the part of Devonshire bordering on the Tamar and Tavey; its natural history.* 12mo, 3 vols, Murray, London 1836; ed. 1838; ed. 2 vols 1879.

450. BREE, Charles Robert *Species not transmutable, nor the result of secondary causes. Being a critical examination of Mr. Darwin's work entitled 'Origin and variation of species'.* 8vo, x+256 pp, Groombridge, London [1860].

451. —— *Popular illustrations of the lower forms of life; comprising I. The protophyton. II. The protozoon. III. The Coelenterata.* 8vo, viii+134 pp, 1 pl., text figs, Cox, London 1868. Field Library No. 5.

452. —— *An exposition of the fallacies in the hypothesis of Mr. Darwin.* 8vo, xviii+418 pp, 4 pls, 1 tab., Longman, London 1872.

Breen, Hugh *see* Ansted, D. T.

453. BREHM, Alfred Edmund *Cassell's book of birds.* 4to, 4 vols, 40 col. pls, text figs, Cassell, London [1869–] 1872; ed. [ ]. Translation of Abt. 2 of *Illustrirtes Thierleben*, Hildenburghhausen 1864–69, by Thomas Rymer Jones.

454. —— *Bird-life; being a brief history of the bird, its structure and habits.* Roy. 8vo, xvi+898 pp, 11 col. pls, Van Voorst, London 1874. Translation of *Das Leben der Vogel.* 2ed. Glogau 1861, by H. M. Labouchere & William Jesse.

455. —— *From North Pole to Equator. Studies of wildlife and scenes.* 8vo, 592 pp, portr, text figs, Blakie, London 1896. Translation of *Vom Nordpol zum Aequator*, Stuttgart 1890, by Margaret R. Thomson.

456. BRENT, Francis & HOLMES, Edward Morell *Mosses of Devon and Cornwall.* 8vo, privately printed, Plymouth [1871].

—— *see* Holmes, E. M.

457. BREWER, James Alexander *A new flora of the neighbour-*

*hood of Reigate, Surrey . . . to which is added . . . list of the fauna.* 8vo, xx+194 pp, 1 map, Pamplin, London 1856.

—— *Flora of Surrey.* 1863. *see* Salmon, J. D.

Bridgewater Treatises (Pickering). *see*

Bell, *Sir* Charles *The hand.* 1833.

Chalmers, T. *Adaptations of external nature.* 1833.

Kirby, W. *The creation of animals.* 1835.

Roget, P. M. *Animal and vegetable physiology.* 1834.

*Brief instructions for making observations in all parts of the world.* 1696. *see* Woodward, J.

458. BRIGGS, Thomas Richard Archer *Flora of Plymouth.* Sm. 8vo, xxxv+432 pp, 1 map, Van Voorst, London 1880.

459. BRIGHTWELL, Thomas *Sketch of a fauna infusoria for East Norfolk.* Sm. 4to, viii+40 pp, 20 col. pls, Fletcher for the author, Norwich 1848. Pls I–XIX, VIIa. 100 copies.

460. BRIGHTWEN, *Mrs* Eliza *Wild nature won by kindness.* Sm. 8vo, 230 pp, text figs, Fisher Unwin, London 1890; 2ed. 1890; 3ed. 1890; 4ed. 1891; 5ed. 1893. Later editions.

461. —— *More about wild nature.* 8vo, xvi+261 pp, 1 pl., text figs, Fisher Unwin, London 1892; 2ed. 1893.

462. —— *Glimpses into plant life; easy guide to the study of botany.* 8vo, 351 pp, 1 pl., text figs, Fisher Unwin, London 1897; 2ed. [1900].

463. —— *Rambles with nature students.* 8vo, 223 pp, 1 pl., text figs, RTS, London 1899.

464. BRISSEAU DE MIRBEL, Charles François *General observations on vegetation . . . intended to inspire young persons with a taste for botany.* 8vo, viii+104 pp, Relfe, London 1833. Translation of Sec. 10, part 1 of *Eleméns de physiologie végètale et botanique*, Paris 1815.

*British angling flies.* 1862. *see* Theakston, M.

465. *British animals.* 16mo, 12 col. pls, SPCK, London [c.1880].

466. BRITISH ASSOCIATION FOR THE ADVANCEMENT OF SCIENCE *Tables for registration of periodic phenomena. List of plants . . . List for the animal kingdom.* 4to, [40] pp, British Association, London [?1845].

467. —— *Bristol and its environs, historical, descriptive and scientific.* 8vo, viii+475 pp, 10 pls, 1 map, text figs, British Association, London 1875.

468. —— *Middlesborough and the district; being notes historical, industrial, scientific.* 8vo, vi+93 pp, 1 col. & 2 plain pls, 1 col. map, British Association, Middlesborough 1881.

469. —— *Handbook of Birmingham.* 8vo, xxi+378 pp, 1 map, British Association, Birmingham 1886.

470. —— *Handbook of Manchester.* 8vo, 109 pp, British Association, Manchester 1887.

—— *see*
Alston, C. R. *West of Scotland.* 1876.
Belfast Naturalists' Field Club. *Guide to Belfast.* 1874.
Bruce, J. C. *Newcastle-on-Tyne.* 1889.
Carr, J. W. *Nottinghamshire.* 1893.
Cordeaux, J. *Birds of the Humber.* 1899.
Cuvierian Society of Cork. *County of Cork.* 1845.
Glasgow Society of Field Naturalists. *Clydeside and the West of Scotland.* 1876.
Herdman, W. A. *Liverpool and the Isle of Man.* 1889.
James, I. *Cardiff.* 1891.
Lebour, G. A. *Newcastle-on-Tyne.* 1889.
Macalister, A. & M'Nab, W. R. *Dublin.* 1878.
McAndrew, R. *List of British marine invertebrate fauna.* 1860.
Miall, L. C. *Leeds and Airedale.* 1890.
Morris, J. W. *Bath.* 1888.
Rogers, B. M. H. *Bristol.* 1898.
Strickland, H. E. *Nomenclature of zoology.* 1842.
Tyndall, J. *Address* [on evolution]. 1874.

471. *British birds.* 16mo, vii+216 pp, 24 pls, RTS, London 1840.

472. *British birds. The land birds.* Sm. 8vo, 282 pp, text figs, RTS, London [?1857].

473. *British birds. The water birds.* Sm. 8vo, 268 pp, 1 col. pl., text figs, RTS, London [1857].

474. *British birds: their haunts and habits.* 8vo, 282 pp, for private circulation, London 1868. LIVINGSTONE, G. [per Alexander Library copy]

*The British botanist.* 1820. *see* Clarke, S. R.

475. *British botany familiarly explained and described in a series of dialogues.* Sm. 8vo, x+233 pp, 28 col. pls, Wakeman, Dublin 1835.

476. *British butterflies.* Sm. 8vo, SPCK, London [?1858].

477. *British butterflies, their distinctions, generic and specific; with lithographic illustrations of each genus.* 24mo, viii+42 pp, 18 pls, Knott, Birmingham 1828.

*The British cyclopaedia of natural history.* 1835–37. *see* Partington, C. F.

478. *British ferns and mosses.* Sm. 8vo, iv+136 pp, text figs, Ward

& Lock, London 1861; ed. [1870]; ed. [1880]. Indispensible Handy Books.

*British fish and fisheries.* [1850]. *see* Martin, W. C. L.

479. *British forest trees.* Sm. 8vo, iv+60 pp, Cradock, London 1843.

*The British Museum.* 1850. *see* Masson, D.

British Museum (Natural History), works published by *see*

Austen, E. E. *How to collect mosquitoes.* [1898].

—— *How to collect Diptera.* [1900].

Bell, F. J. *Catalogue of British echinoderms.* 1892.

Brown, R. *of the British Museum. Directions for collecting and preserving plants.* [ ].

Children, J. G. *Directions for collecting zoological specimens.* [ ].

Crombie, J. M. *A monograph of the lichens found in Britain.* 1894–1911.

Denny, H. *List of British Anoplura.* 1852.

Desvignes, T. *Catalogue of British Ichneumonidae.* 1856.

Gray, G. R. *List of British birds.* 1850.

—— *List of eggs of British birds.* 1852.

—— *Catalogue of British birds.* 1863.

Gray, J. E. *List of British radiated animals.* 1848.

—— *List of British sponges.* 1848.

—— *List of British Mollusca acephala and Brachiopoda.* 1851.

Günther, A. *Instructions for collecting reptiles, batrachians and fishes.* [1891].

Johnston, G. *Catalogue of British non-parasitical worms.* 1865.

Leach, W. E. *Catalogue of the indigenous Mammalia and birds.* 1816.

Lister, A. *A monograph of the Mycetozoa.* 1894.

—— *Guide to the British Mycetozoa.* 1895.

Smith, F. *List of British Hymenoptera Aculeata.* 1851.

—— *Nomenclature of British Hymenoptera.* 1853.

—— *Catalogue of British Hymenoptera. Part 1, Apidae.* 1855.

—— *Catalogue of British fossorial Hymenoptera.* 1858.

Smith, W. *List of British Diatomaceae.* 1859.

Smith, W. G. *Guide to Sowerby's models of British Fungi.* 1893.

Stainton, H. T. *List of British Lepidoptera*, (*completed*). 1854.

Stephens, J. F. *List of British Lepidoptera.* 1850–52.
Thomas, M. R. O. *Directions for preparing small mammal skins.* [1898].
Walton, J. *List of British Curculionidae.* 1856.
White, A. *List of British Crustacea.* 1850.
—— *List of British fish.* 1851.
—— *Nomenclature of British Neuroptera.* 1853.
—— *Nomenclature of British Diptera.* 1, 1853.
—— *Nomenclature of British Anoplura, Euplexoptera and Orthoptera.* 1855.

480. BRITISH NATURAL HISTORY SOCIETY, The *Catalogue of British marine Recent shells.* 8vo, 23 pp, Sotheran for the Society, York 1852. Based on Forbes, E. & Hanley, S. *A history of British Mollusca.* 1853.

*The British naturalist.* 1830. *see* Mudie, R.

481. BRITISH ORNITHOLOGISTS' UNION *A list of British birds.* 8vo, xxxi+229 pp, Van Voorst, London 1883; 2ed. 1915; ed. 1952.

482. *British quadrupeds.* 16mo, xvi+250 pp, 24 pls, RTS, London 1839; 2ed. [ ].

483. *The British sylva.* Sm. 8vo, vi+203 pp, text figs, SPCK, London 1851.

British Zoological Museum. 1795. *see* Ferryman, R.

*The British zoology.* [1761–] 1766. *see* Pennant, T.

484. BRITTAIN, Thomas *Micro-fungi; when and where to find them.* 8vo, 92 pp, 2 pls, Heywood, Manchester 1882.

485. BRITTEN, James *Flora of Buckinghamshire.* 12mo, 8 pp, Wycombe 1867.

486. —— *A list of Lincolnshire plants.* 8vo, 3 pp, [Sheffield] 1872. Reprinted from White, W. *History and directory of Lincolnshire*, 3ed.

487. —— *A list of Suffolk plants.* 8vo, 7 pp, Sheffield 1874. Reprinted from White, W. *History . . . and directory of Suffolk.*

488. —— *Botany of Hampshire.* 8vo, 10 pp, Sheffield 1877. Reprinted from White, W. *History . . . and directory of Hampshire.*

489. —— *Popular British Fungi.* 8vo, 138 pp, The Bazaar, London [1877]; issue 1879.

490. —— *Botany of Devonshire.* 8vo, 7 pp, Sheffield 1878. Reprinted from White, W. *History . . . and directory of Devonshire.*

491. —— *European ferns.* 4to, xliv+196 pp, 30 col. pls, Cassell, London [1879–81]; issue 1883–85; [ ].

492. —— *The young collector's handbook of flowering plants.* Sm. 8vo, 31 pp, text figs, Sonnenschein, London [1883]. Young Collectors' Handbook Series.

493. —— & HOLLAND, Robert *Dictionary of English plant names.* 8vo, xxviii+x+618+28+23 pp, English Dialect Society, London [1878–] 1886.

494. BROADHURST, Henry Tootal *Do young vipers take refuge down the throat of the mother in case of sudden surprise or danger?* 8vo, 29 pp, Maxwell, Dumfries [1895].

495. BROCAS, Frederick York *British mosses. No. 1.* Fol., s.sh., Cottle, Basingstoke [1852]. All publ.

496. —— *A companion to the fernery.* Fol., London 1854.

497. BRODERIP, William John *Hints for collecting animals and their products.* 8vo, 12 pp, Admiralty Hydrographic Department, London 1832.

498. —— *Zoological recreations.* 12mo, viii+380 pp, Colburn, London 1847; 2ed. 1849; 3ed. 1858; ed. [1874].

499. —— *Leaves from the note book of a naturalist.* 12mo, xvi+413 pp, Parker, London 1852.

500. BROMFIELD, William Arnold *A list of plants likely to be found growing wild in the Isle of Wight.* 12mo, 24 pp, Ryde 1840.

501. —— *Flora Vectensis . . . or flowering plants and ferns indigenous to the Isle of Wight.* 8vo, xxxv+678 pp, 1 portr., 1 map, Pamplin, London 1856. Map often absent.

—— *see* More, A. G.

502. BROMHEAD, *Sir* Edward ffrench *Remarks on the arrangement of the natural botanical families.* 8vo, 24 col. pls, London 1836.

503. [BROOKE, Richard] *New cyclopaedia of botany and complete book of herbs.* 8vo, 2 vols, 100 col. pls, Clark, London [1854]; ed. 1 vol. [?1860].

504. BROOKES, Joshua *An address delivered at the anniversary meeting of the Zoological Club of the Linnean Society.* 8vo, 30 pp, Taylor printed, London 1829.

505. BROOKES, Richard *A new and accurate system of natural history.* 12mo, 6 vols, 144 pls, Newbery, London 1763; 2ed. 1766–72; 3ed. 1790–91.

506. BROOKES, Samuel *Introduction to the study of conchology.* 4to, vi+164 pp, 9 col.+2 plain pls, Arch, London 1815.

507. BROUGH, John Cargill *The fairy tales of science.* 8vo, xi+338 pp, 20 pls, Griffith & Farren, London 1859.

508. BROUGHAM, Henry Peter *Baron* Brougham and Vaux *A discourse of natural theology, showing the nature of the evidence and the advantages of the study.* 8vo, 275 pp, Hauman, Brussels 1835; ed. vii+296 pp, Knight, London 1835; 2ed. 1835; 3ed. 1835; ed. 1845. The London editions form Vol. I of the 5 vol. editions of Paley, W. *Natural theology.* Also in *Collected works*, Vol. VI 1856.

509. —— *Dialogues on instinct; with analytical review of the researches on fossil osteology.* 12mo, iv+272 pp, Knight, London 1844. Knight's Weekly Volume. Contains extracts from No. 508.

510. [BROUGHTON, Arthur] *Enchiridion botanicum, complectens characteres genericos et specificos plantarum per Insulas Britannicas sponte nascentium.* 8vo, ii+226 pp, Robinson, London 1782; another issue 1782, signed.

511. BROWN, Alfred *The Mollusca of the Firth of Clyde.* 8vo, xiii+130 pp, Hopkins, Glasgow 1878.

Brown, Edwin *see* Mosley, *Sir* Oswald.

512. BROWN, John Alexander Harvie- *The past and present distribution of some of the rarer animals of Scotland.* 8vo, 50 pp, West Newman, London 1882. Reprinted from *The Zoologist*, 1881–82.

513. —— *The capercaillie in Scotland.* 8vo, xv+155 pp, 2 pls, 1 col. map, text figs, Douglas, Edinburgh 1879; ed. 1888.

514. —— & BUCKLEY, Thomas Edward *A vertebrate fauna of Sutherland, Caithness, and West Cromarty.* Sq. 8vo, x+344 pp, 5 plain+2 col. pls, 1 map, text figs, Douglas, Edinburgh 1887. *The Vertebrate Fauna of Scotland* Vol. 1.

515. —— & —— *A vertebrate fauna of the Outer Hebrides.* Sq. 8vo, xiv+xciii+279 pp, 5 pls, 6 maps, text figs, Douglas, Edinburgh 1888. *The Vertebrate Fauna of Scotland* Vol. 2.

516. —— & —— *A vertebrate fauna of Argyll and the Inner Hebrides.* Sq. 8vo, lxxxiv+262 pp, 6 pls, 7 maps, text figs, Douglas, Edinburgh 1892. *The Vertebrate Fauna of Scotland* Vol. 5.

517. —— & —— *The fauna of the Moray basin.* Sq. 8vo, 2 vols, 23 pls, 1 map, text figs, Douglas, Edinburgh 1895. The Vertebrate Fauna of Scotland Vols 6 & 7.

—— *see* Buckley, T. E.

518. BROWN, Peter *New illustrations in zoology.* 4to, 136 pp, 50

col. pls, White, London 1776. Text in French and English.

519. BROWN, Robert *of the British Museum Brief account of microscopical observations . . . on the particles contained in the pollen of plants.* 8vo, 16 pp, Taylor printed, London [1828]; additional remarks, 7 pp, [1829].

520. —— *The miscellaneous botanical works of Robert Brown.* Roy. 4to & 8vo, 3 vols, 38 pls, Ray Society, London 1866–68. Vol. 3 is 4to of plates. Ray Society Publ. Nos 39, 43 & 44 for 1866, 67 & 1868.

521. [——] *Directions for collecting and preserving plants.* 8vo, 6 pp, British Museum, London [ ].

522. BROWN, Robert *of Campster Manual of botany.* 8vo, xviii+614 pp, text figs, Blackwood, Edinburgh 1874.

523. —— *Science for all.* 4to, 5 vols, col. pls, text figs, Cassell, London [1877–82]. Later printings.

524. BROWN, Thomas *Captain. Elements of conchology.* 8vo, 168 pp, 9 col. pls, Lackington, London 1816.

525. —— *Illustrations of the conchology of Great Britain and Ireland.* Roy. 4to, v+120 pp, 53 col. pls, Lizars, Edinburgh 1827; 2 ed. London [1837–] 1844; issue 1844 [=c.1880] title *Illustrations of the recent conchology* . . . Plates 1–52, 10*.

526. —— *The book of butterflies, sphinges and moths.* Sm. 8vo, 2 vols, 93 col. pls, Whittaker, London 1832; ed. 3 vols 1832–34; ed. 1843. Vols 1 & 2 of 1832–34 ed. are a reissue; this ed. is Constable's Miscellany, Vols 75, 76 & 80.

527. —— *The zoologist's text-book.* 12mo, 2 vols, 578+15 pp, 106 pls, Fullarton, Glasgow 1833.

528. —— *The conchologist's text-book.* 8vo, 180 pp, 19 pls, text figs, Fullarton, Glasgow 1833; 3ed. 1835; 5 ed. 1839; 6ed. 1845. Later editions, 6ed. onwards, bear W. Macgillivray's name and omit Brown's.

529. —— *The taxidermist's manual.* 12mo, xii+150 pp, 6 pls, Fullarton, Glasgow 1833; 22ed. 1872.

530. —— *Anecdotes of the animal kingdom . . . forming an appropriate supplement to Goldsmith's Animated nature.* 12mo, xi+923 pp, Fullarton, Glasgow 1833; 1834; ed. 1840. A companion to Brown's ed. of Goldsmith, O. *A history of earth and animated nature.*

531. —— *Illustrations of the genera of British birds.* Imp. 4to, 4 pts, 16 col. pls, Smith Elder, London 1845. All publ.

532. —— *Illustrations of the land and fresh water conchology of*

*Great Britain and Ireland.* Roy 8vo, xi+142 pp, 27 col. pls, Smith Elder, London 1845; 2ed. 1845.

533. —— *Habits and characteristics of animals and birds.* 8vo, viii+307 pp, 8 pls, Blackwood, London [1869].

534. —— *Illustrative anecdotes of birds, fishes and insects.* 8vo, viii+321 pp, 8 pls, text figs, Blackwood, London [1869].

535. —— *Popular natural history and characteristics of animals.* 8vo, vi+309 pp, col. pls, Blackwood, London [1869].

—— *Illustrations of the recent conchology of Great Britain and Ireland.* 1844 [=c.1888]. *see* No. 525.

536. BROWN, Thomas M.D. *Observations on the Zoonomia of Erasmus Darwin.* 8vo, xxiv+560 pp, Mundell, Edinburgh 1798.

537. BROWN, Thomas *pseudonym Nature's cabinet unlocked.* 12mo, 331 pp, Farnham, London 1657. Not by *Sir* Thomas Browne.

538. BROWNE, Alexander Montagu *List of the British Macro-lepidoptera, scientific and English names.* 8vo, 28 11, Buncher, Birmingham 1871; 2ed. 1877. Printed on recto only.

539. —— *Practical taxidermy.* 8vo, 155 pp, text figs, The Bazaar, London [1878]; 2ed. 1884; 3ed. 1922.

540. —— *Collecting butterflies and moths.* 8vo, 50 pp, text figs, The Bazaar, London [1878].

541. —— *The vertebrate animals of Leicestershire and Rutland.* Sm. 4to, xii+223 pp, 3 plain+1 col. pls, 1 map, Midland Educational Company, Birmingham 1889. 300 copies.

542. —— *Artistic and scientific taxidermy and modelling.* 8vo, xii+463 pp, 22 pls, text figs, Black, London 1896.

Browne, Montagu *see* Browne, A. M.

543. BROWNE, *Sir* Thomas *Pseudodoxia epidemica: or, enquiries into very many received tenets.* Fol., 386 pp, Dod, London 1646; 2ed. 1650; 3ed. 1658; 4ed. 4to 1658; ed. 1659; 5ed. 1669; 6ed. 1672. Also in *Collected works* and often reprinted.

544. —— *Certain miscellany tracts.* 8vo, 215 pp, portr., Meare, London 1683; 1684. Also in *Collected works.*

545. BRUCE, George *The land birds in and around St. Andrews*, 8vo, 563 pp, Leng, Dundee 1895.

546. BRUCE, John Collingwood *Hand-book to Newcastle-on-Tyne and district.* 8vo, x+178 pp, 1 pl., text figs, British Association, Newcastle-on-Tyne 1889. Revision of an earlier

guide book, xii+286 pp, Longman, London 1863; ed. 1864.

547. [BRUCKNER, John] *A philosophical survey of animal creation.* Sm. 8vo, xxiv+166 pp, London 1768; ed. 12mo 1791. Translation of *Theorie de la système animal*, Leiden 1767, by Thomas Cogan.

548. BRUYSSEL, Ernest Jean Van *The population of an old pear-tree; or, stories of insect life.* Sm. 8vo, viii+221 pp, 1 pl., text figs, Macmillan, London 1870. Later editions. Translation of *Les clients d'un vieux poirier*, Paris, by Charlotte M. Yonge.

549. BRYCE, James *Geological notices of the environs of Glasgow, the shores of the Clyde, and the Island of Arran.* 8vo, 24 pp, text figs, Glasgow 1855; ed. 1859 title *Geology of Clydesdale and Arran embracing also the marine zoology . . . flora . . . rarer insects*; 3ed. 1865 title *The geology of Arran and Clydesdale. With an account of the flora and marine fauna*; 4ed. 1872 title *The geology of Arran and the other Clyde Islands, with an account of the botany, natural history, and antiquities.*

550. BUCHAN, Alexander *A description of St. Kilda.* Sm. 8vo, 48 pp, Edinburgh 1727; ed. 1741; ed. 1752; ed. 1773; ed. 1774; ed. 1818. 1774 ed. is in Monro, *Sir* Donald *Description of the Western Isles.* 1818 ed. is in *Miscellanea Scotica*, Vol. 2.

551. BUCKLAND, Anna Jane *The elements of botany on the natural system.* 12mo, 36 pp, 1 pl., Hall, London 1859.

552. BUCKLAND, Francis Trevelyan *known as* Frank *Curiosities of natural history.* [*First series*]. Sm. 8vo, xvi+319 pp, 3 pls, Bentley, London 1857; 2ed. 1858; 4ed. 1859. Very many later editions in 3 or 4 vols with the second and third series.

553. —— *Curiosities of natural history. Second series.* Sm. 8vo, xxiv+336 pp, 1 pl., Bentley, London 1860. Very many later editions in 3 or 4 vols. with first and third series.

554. —— *Fish hatching.* Sm. 8vo, xii+268 pp, 1 pl., text figs, Tinsley, London 1863.

555. —— *Curiosities of natural history. A new series.* Sm. 8vo, 2 vols, 7 pls, Bentley, London 1866 [=1865]. Many later editions in 3 or 4 vols with first and second series.

556. —— *Familiar history of British fishes.* Sm. 8vo, xi+396 pp, text figs, SPCK, London 1873; ed. 1881 title *The natural history of British fishes.* Later editions.

557. —— *Log-book of a fisherman and a zoologist.* 8vo, xiv+407

pp, text figs, Chapman & Hall, London 1875; ed. 1876; eds [ ].

—— *The natural history of British fishes.* 1881. *see* No. 556.

558. —— *Notes and jottings from animal life.* Sm. 8vo, viii+414 pp, 1 portr., text figs, Smith Elder, London 1882; 2ed. 1886; ed. 1894.

559. —— *et al. Birds and bird life.* 16mo, 285 pp, 1 pl., text figs, Leisure Hour Office RTS, London [1863].

560. BUCKLER, William *The larvae of British butterflies and moths.* 8vo, 9 vols, 164 col. pls, Ray Society, London 1886–1901. Ray Society Publ. 63, 64, 66, 68, 69, 71, 72, 73, 77 for 1885, 86, 88, 90, 91, 93, 94, 95 & 1899.

561. BUCKLEY, Arabella Burton *A short history of natural science.* Sm. 8vo, xxv+467 pp, 1 col. pl., text figs, Murray, London 1876; 2ed. 1879; 3ed. 1883, 4ed. 1888; 5ed. 1894.

562. —— *Botanical tables for the use of students.* Obl. fol., 2 11, London 1877; 2ed. 1883. Later editions.

563. —— *The fairy-land of science.* 8vo, viii+244 pp, text figs, Stanford, London 1879. Later editions.

564. —— *Life and her children.* 8vo, xii+312 pp, 2 pls, text figs, Stanford, London 1880. Later editions.

565. —— *Winners in life's race; or, the great backboned family.* 8vo, xv+367 pp, text figs, Stanford, London 1882. Later editions.

566. —— *Through magic glasses, and other lectures.* 8vo, xiv+234 pp, Stanford, London 1890.

Buckley, Thomas Edward *see* Brown, J. A. H.

—— *see* Evans, A. H.

567. —— & BROWN, John Alexander Harvie- *A vertebrate fauna of the Orkney Islands.* Sq. 8vo, xxiv+314 pp, pls, 1 map, text figs, Douglas, Edinburgh 1891. The Vertebrate Fauna of Scotland, Vol. 4.

—— & —— editors *The vertebrate fauna of Scotland.* Sq. 8vo, 11 vols, Douglas, Edinburgh 1887–1911. *see*

Brown, J. A. H. & Buckley, T. E. *Sutherland, Caithness.* 1887.

—— & —— *Outer Hebrides.* 1888.

—— & —— *Argyll and the Inner Hebrides.* 1892.

—— & —— *Moray basin.* 1895.

Buckley, T. E. & Brown, J. A. H–. *Orkney.* 1891.

Evans, A. H. & Buckley, T. E. *Shetland.* 1899.

Graham, H. D. *Birds of Iona and Mull.* 1890.

568. [BUCKMAN, James] *Geology, archaeology, and botany of the most picturesque and interesting spots of Cheltenham.* 8vo, 72 pp, Cheltenham 1842.
569. —— *A botanical guide to the environs of Cheltenham.* 8vo, viii+59 pp, Davies, Cheltenham 1844.
570. —— *On the natural history of meadow and pasture grasses.* 8vo, iv+173 pp, 2 pls, Hamilton Adams, London 1858.
571. —— *On the nature of fairy rings.* 8vo, 15 pp, London [1870]. Reprinted from *The Veterinarian*, 1870.
572. BUCKNILL, *Sir* John Alexander Strachey *The birds of Surrey.* Sm. 8vo, lvi+374 pp, 6 pls, 1 col. map, text figs, Porter, London 1900.
573. BUCKTON, George Bowdler *Monograph of the British Aphides.* 8vo, 4 vols, 9 plain+141 col. pls, Ray Society, London 1876–83. Ray Society Publ. Nos 52, 54, 58 & 60 for 1875, 77, 80 & 1882.
574. —— *Monograph of the British Cicadae or Tettigidae.* 8vo, 2 vols, 7 plain+75 col. pls, Macmillan, London 1890–91.
575. —— *The natural history of Eristalis tenax or the drone-fly.* 8vo, iv+91 pp, 7 plain+2 col. pls, Macmillan, London 1895.
576. [BUDGEN, *Miss* M. L.] *Episodes of insect life.* 8vo, 3 vols, 3 col. pls, col. text figs, Reeve, London 1849–51; ed. 1 vol. 1867; 1879. Vol. 1 not so-called. By *Acheta domestica.*
577. [——] *March winds and April showers; being notes and notions on a few created things.* 8vo, xiii+210 pp, text figs, Reeve, London 1854. By *Acheta.*
578. [——] *May flowers; being notes and notions on a few created things.* 8vo, ix+214 pp, text figs, Reeve, London 1855. By *Acheta.*
579. BÜCHNER, Friedrich Carl Christian Ludwig *Mind in animals.* Sm. 8vo, xii+359 pp, 1 portr., Freethought Publishing Company, London 1880. International Library of Science and Freethought, Vol. 1. Translation of *Aus dem Geistleben der Thiere*, Berlin 1879, by Annie Besant.
580. BUFFON, George Louis Leclerc, *Conte* de [*Natural history.*] There is no complete translation of the *Histoire naturéllès*, 1749–1804, by Buffon et al. Many of the later and smaller works use Buffon's name as synonymous with a natural history of animals.
    1 *The natural history of animals, vegetables, and minerals.* 8vo, 6 vols, pls, Bell, London 1775–76. Edited by William Kenrick.

2 *Natural history general and particular.* 8vo, 8 vols, 301 pls, Creech, Edinburgh 1780 [–1781]; 9 vols, Cadell, London 1781; 2ed. 1785; 3ed. 1791. *The natural history of birds from the French of the Count de Buffon*, 8vo, 9 vols, 262 pls, Strahan, London 1792–93. The above are edited by William Smellie. *The natural history of oviparous quadrupeds and serpents.* 8vo, 4 vols, 48 pls, Smellie, Edinburgh 1802; edited by Robert Kerr. An edition in 20 vols, enlarged by William Wood, 681 pls, 1812. The reptile volumes are from the Conte de Lacépède's 2 volumes 1788–89.

3 *Natural history general and particular. Buffon's Natural history.* Demy or Cr. 8vo, 10 vols, 81 col. pls, Barr, London 1792. *Natural history of birds, fish, insects and reptiles.* Demy or Cr. 8vo, 6 vols, 50 col. pls, Barr, London 1793. The second set is additional to the first and Buffon's name does not appear on the title pages. Later editions of both sets.

4 *The beauties of natural history; selected from Buffon's History of quadrupeds.* 12mo, 92 pp, text figs, Catnach, Alnwick [c.1790]; ed. York & Alnwick [c.1795].

5 *The system of natural history abridged.* 8vo, 2 vols, 40 pls, Morison, Perth 1791; ed. 66 pls, Edinburgh 1800. For a companion volume see *The natural history of insects*, 1792, which is contained in Vol. 2 of the 1800 ed. *See also* No. 9 below.

6 *Buffon's natural history abridged.* 8vo, xii+530+vii pp, 40 pls, Kearsley, London 1791; ed. 2 vols 100 col. pls, 1792; ed. 1821; ed. 1828.

7 *Buffon's natural history of man, earth.* 8vo, text figs, Annereau, Southwark 1801.

8 *The natural history of quadrupeds. To which is prefixed, the history of the earth . . . and man.* 8vo, 84+156 pp, text figs, Kettle, London 1804. Buffon's name only appears on p. 1 of second pagination.

9 *The system of natural history . . . carefully abridged; and the natural history of insects.* 12mo, 4 vols, text figs, Davison, Alnwick 1814. Based on No. 5 (above), with the insects in Vols 3 & 4.

10 *Buffon's natural history of quadrupeds, birds, fishes, serpents, reptiles, and insects.* 12mo, viii+306 pp, 1 pl., text figs, Marsden, Chelmsford 1822.

11 *The history of the whale . . . extracted from Buffon's natural history.* 12mo, Falkirk 1826.

12 *A natural history general and particular . . . to which is added a history of birds, fishes, reptiles, and insects . . . together with an account of the most curious foreign plants by Henry Augustus Chambers.* 4to, 2 vols, 52 col. pls, 2 portrs, London 1828. Later editions.

13 *Buffon's natural history: containing a full and accurate description of the animated beings in nature. In three parts. Quadrupeds – Birds – Fishes.* 32mo, iv+382 pp, 1 pl., text figs, Milner, Halifax [c.1830]. Cottage Library.

14 *Buffon's natural history . . . corrected and enlarged by J. Wright. To which are added elements of botany.* 12mo, 4 vols, text figs, Tegg, London 1831; ed. 1833.

15 *Natural history carefully abridged in four parts. Beasts, birds, fishes, reptiles.* 12mo, 534+xv pp, text figs, Smith, London 1833.

16 *The book of birds: edited and abridged* [by Achille Comte] *from the text of Buffon.* Roy. 8vo, xxxiii=292 pp, 38 col. pls, Tyas, London 1841.

17 *The juvenile Buffon: easy readings in natural history.* 18mo, text figs, London 1846.

18 *Buffon's natural history . . . of all the animated beings in nature.* 24mo, 4 vols in one, 508 pp, 1 pl., text figs, Allman, London 1846; ed. Cornish, London [ ]; ed. Milner, Halifax [ ].

19 *Natural history of beasts, birds, fishes and insects. By Buffon and Trimmer.* 16mo, xvi+496 pp, text figs, Tegg, London 1864.

20 *Buffon's natural history modernized.* 16mo, 252 pp, 32 col. pls, Warne, London 1869.

—— *The wonders and beauties of creation.* [1875]. *see* Wonders

—— *History of singing birds.* Edinburgh 1791, Leith 1825. *see* Albin, E.

581. BULL, Henry Graves *editor Notes on the birds of Herefordshire, contributed by members of the Woolhope Club.* 8vo, xxx+274 pp, 1 portr., Hamilton Adams, London 1888.

582. BULLMORE, William King *Cornish fauna, a short account of all the animals found in the County. Part 1. Vertebrata.* 8vo, 64 pp, Heard printed, Truro 1867. All published. Reprinted from *Rep. R. Cornwall Polytechnic Soc.*, 1866.

583. BULLOCK, William *A companion to Mr. Bullock's Museum, containing a description of upwards of three hundred curiosities.* 8vo, 52 pp, Sheffield 1799; ed. Liverpool [1801] title *A companion to the Liverpool Museum*; 6ed. Hull 1808; 7ed. Bath 1809; 8ed. London 1810; 9ed. 1810; 11ed. 1811; 12ed. 1812; 14ed 1813; 15ed. 1813; 16ed. 1814; 17ed. 1816. Title varies. That of London editions is *A companion to Mr. Bullock's Museum and Pantherion.*

584. —— *A concise and easy method of preserving subjects of natural history.* Sm. 8vo, ix+36 pp, 2 pls, for the proprietor, London 1817; 2ed. 1818. Later editions.

585. BUNBURY, *Sir* Charles James Fox *Botanical notes at Barton and Mildenhall, Suffolk.* 8vo, privately printed, Mildenhall, 1889.

586. BUND, John William Willis *A list of the birds of Worcestershire and the adjoining counties.* 8vo, vi+53 pp, Worcester 1891.

587. —— *The life of a Severn salmon, a paper read at the Victoria Institute.* 8vo, 23 pp, 1 pl., W. Leicester, Worcester 1899.

588. BURBIDGE, Frederick William *The art of botanical drawing.* 8vo, 63 pls, Winsor & Newton, London 1872.

—— *see* Sparkes, J. C. L.

589. BURGESS, Henry W. *Eidodendron, or views of the general character and appearance of trees.* Fol., iii+26 11, 54 pls, Dickinson, London 1827.

590. BURGESS, Joseph Tom *Old English wild flowers to be found by the wayside.* Sm. 8vo, x+291 pp, 8 col. pls, text figs, Warne, London 1868; 3ed. 1878. Later eds with title *English wild flowers.*

591. BURMEISTER, Carl Hermann Conrad *A manual of entomology.* 8vo, xii+654 pp, 25 plain+8 col. pls, Churton, London 1836. Translation of parts of *Handbuch der Entomologie*, 5 vols, Berlin 1832–55, by William Edward Shuckard.

592. BURNETT, Charles Mountford *The power, wisdom, and goodness of God, as displayed in the animal creation.* 8vo, xv+549 pp, Burns, London 1838.

593. BURNETT, Gilbert Thomas *Outlines of botany.* 8vo, 2 vols, viii+vii+1190 pp, Churchill, London 1835.

594. BURR, Malcolm *British Orthoptera.* 8vo, iv+68 pp, 7 col. pls, text figs, Economic & Educational Museum, Huddersfield

1897; errata 1898. Plates I–VI, IIa. Reprinted from *The Naturalists' Journal*, Vol. 6, 1897.

595. BURROWS, *Mrs* E. *Odd stories about animals.* Sm. 4to, Griffith & Farran, 2ed. London 1871.

596. BURROW, Edward John *Elements of conchology, according to the Linnaean system.* Sm. 8vo, xv+248 pp, 28 col. pls, for the author, London 1815; 2ed. 1818. Later editions.

597. BURTON, John Barlow *Lectures on entomology.* 8vo, 48 pp, 4 col. pls, Simpkin & Marshall, London 1837.

598. BUSHNAN, John Stevenson *Introduction to the study of nature.* 8vo, vii+310 pp, Longman, London 1834.

599. —— *The natural history of fishes.* Sm. 8vo, 219 pp, 31 col. pls, 1 portr., Lizars, Edinburgh 1840; issues 1845; 1848. Jardine's Naturalist's Library Vol. XXVII; Vol. XXV in reissues.

600. [——] *Physiology of animal and vegetable life.* 8vo, Orr, London 1854; ed. 1855.

Bute, *Earl of. see* Stuart, John.

601. BUTLER, Arthur Gardiner *British birds' eggs.* Sm. 8vo, viii+219 pp, 38 col. pls, Janson, London [1886].

602. —— *editor British birds with their nests and eggs.* Roy. 4to, 6 vols, 318 plain+24 col. pls, Brumby & Clarke, Hull [1896–98]; issue Vols 1 & 2 only 1907–08 title *Birds of Great Britain and Ireland. Order Passeres.* Plates by Frohawk, F. W. and the work is often attributed to him.

603. BUTLER, Edward Albert *Pond life: insects.* Sm. 8vo, 127 pp, 1 pl., text figs, Sonnenschein, London 1886. Young Collector Series. Later editions.

604. —— *Our household insects.* 8vo, vi+344 pp, 7 pls, text figs, Longman, London 1893; ed. 1896.

605. BUTLER, Samuel *Life and habit.* 8vo, 307 pp, Trübner, London 1878 [=1877]: ed. 1910.

606. —— *Evolution old and new. Op. 4.* Sm. 8vo, xii+384 pp, Hardwicke & Bogue, London 1879; 2ed. 1882; 1892; ed. 1911.

607. —— *Unconscious memory . . . Op. 5.* 8vo, viii+288 pp, Bogue, London 1880; ed. 1910; 3ed. 1920.

608. —— *Luck, or cunning, as the main means of organic modification. Op. 8.* 8vo, xi+328 pp, Trübner, London 1887 [=1886]; [?1910]; 2ed. 1920.

609. BUTT, J. M. *The botanical primer, being an introduction to English botany.* 8vo, 186 pp, London 1825; ed. 1827.

*Butterflies in their floral homes.* [1858]. *see* Humphreys, H. N.

610. BUXTON, Edward North *Epping forest.* 12mo, xii+142 pp, 6 maps, text figs, London 1884; 2ed. 1885; 3ed. 1890; 4ed. 1897; 5ed. 1898; 6ed. 1901; 7ed. 1905; 8ed. 1911.

611. BUXTON, Richard *A botanical guide to the . . . plants . . . indigenous within sixteen miles of Manchester.* 12mo, xxi+168 pp, Longman, London 1849; 2ed. 1859 title . . . *eighteen miles . . .*

612. BYERLEY, Isaac *The fauna of Liverpool.* 8vo, 125 pp, Hall Virtue, London 1856. Reprinted from *Proc. Liverpool lit. phil. Soc.* Appendix, 1854.

## C

C, E. M. *Popular geography of plants.* 1855. *see* Catlow, *Miss* Maria E.

C, Sy. *Ferns of the British Isles.* 1877. *see* Courtauld, S.

*A cabinet of natural history.* 1809. *see The natural history of British quadrupeds.*

613. CADDY, Florence *Through the field with Linnaeus.* 8vo, 2 vols, 2 pls, 2 maps, Longman, London 1887.

614. CAIRNCROSS, David *The origin of the silver eel.* 8vo, viii+96 pp, 1 pl., Shield, London 1862.

615. CAIUS, John . . . *De Canibus Britannicis, liber unus: de rariorum animalium et stirpium historia, liber unus: de libris propriis, liber unus.* 8vo, 3 pts, 30+62+72 pp, Seres, London 1570; ed. 1729; trans. of pt.1 1576 title *Of English dogges*, translated by Abraham Fleming; ed. 1880; ed. 1912 in *Works of John Caius*, Cambridge.

*The calendar of nature.* 1784. *see* Aikin, J.

*The Cambridge natural history.* 1895–1906. *see* Harmer, S. F. & Shipley, A. E. *editors.*

Cambridge Philosophical Society. *see* Jenyns, L.

Cambridge University Anatomical Museum, 1862. *see* Clark, W. *of Cambridge.*

617. CAMBRIDGE, Octavius Pickard- *The spiders of Dorset,*

*with an appendix . . . of those British species not yet found in Dorsetshire*. 8vo, 2 pts, xliii+625+xi pp, 6 pls, Ruegg, Sherborne 1879, 81.

618. —— *List of British & Irish spiders*. 8vo, 86 pp, Sime for the author, Dorchester 1900.

619. CAMERON, John *Gaelic names of plants, Scottish and Irish and Manx.* 8vo, ix+130 pp, Blackwood, Edinburgh 1883; 2ed. Glasgow 1900. Reprinted from *The Scottish Naturalist.*

620. CAMERON, Peter *A catalogue of the British Tenthredinidae*. 8vo, 22 pp, Glasgow 1878.

621. —— *Monograph of the British phytophagous Hymenoptera*. 8vo, 4 vols, 29 plain + 51 col. pls, Ray Society, London 1882–93. Ray Society Publ. 59, 62, 67 & 70 for 1881, 84, 89 & 1892.

622. CAMPBELL, *Mrs* C. C. *Natural history for young folks*. Sm. 9vo, 228 pp, text figs, Nelson, London 1884.

623. CAMPBELL, Douglas Houghton *The structure and development of the mosses and ferns – Archegoniatae*. 8vo, viii+544 pp, text figs, Macmillan, London 1895.

Campbell, W. H. *see* Balfour, J. H.

Campbell-Hyslop, C. W. *see* Hyslop, C. W. Campbell-.

624. CANDOLLE, Alphonse Louis Pierre Pyramus de *Laws of botanical nomenclature adopted by the International Botanical Congress . . . 1867.* 8vo, 72 pp, Reeve, London 1868. Translation of *Lois de la nomenclature botanique*, Paris 1867, by Hugh Algernon Weddell.

625. CANDOLLE, Augustin Pyramus de *Vegetable organography*. 8vo, 2 vols, 38 pls, Houlston & Stoneman, London 1839–40; 2ed. 1841. Translation of *Organographie vegetale*, Paris 1827, by Broughton Kingdon.

626. —— & SPRENGEL, Curt Polycarp Joachim *Elements of the philosophy of plants*. Roy. 8vo, xxxiii+486 pp, 8 pls, Blackwood, Edinburgh 1821. Translation of *Grundzug der wissenschaftlicken Pflanzenkunde*, 1820, and adaptation of Candolle's *Theorie élèmentaire de la botanique*, Paris 1813; 2ed. 1819.

627. CAPPER, James *Meteorological and miscellaneous tracts, applicable to navigation, gardening and farming, with calendars of flora for Greece, France, England and Sweden*. 8vo, xix+211 pp, Bird, Cardiff [?1810].

628. CARDIFF NATURALISTS SOCIETY *The birds of*

*Glamorgan*. 4to, 163 pp, 5 pls, Cardiff 1900. Printed on one side only.

629. CARLISLE, *Sir* Anthony *The Hunterian oration* [on the anatomy of the oyster]. 4to, ii+47 pp, Booth, London 1826.

Canino, *Prince of. see* Bonaparte, C. L. J. L.

630. CARPENTER, George Herbert *Insects their structure and life. A primer of entomology*. 8vo, xi+404 pp, text figs, Dent, London 1899; 2ed. 1924.

630a [CARPENTER, Philip] *Elements of zoology; being a concise account of the animal kingdom, according to the system of Linnaeus. Intended for the use of young persons, and as a companion to the new copper-plate sliders . . . and a description of an improved phantasmagoria lantern.* 12mo, 168 pp, Bywater for the maker, Liverpool 1823; *A short account of the copper-plate sliders, and . . . phantasmogoria lantern*, 12 pp, 2 pls, Rushton & Melling printed, Liverpool [1823]; issue, signed, Rowatt & Hunter, London 1823. A text to accompany 50 sliders.

630b —— *A companion to the miscroscopes, being a complete account of the insects . . . seen in the microscopes of that interesting exhibition*. 12mo, 23 pp, P. Carpenter, London 1827.

631. CARPENTER, William Benjamin *Principles of general and comparative physiology, intended as a guide . . . to the philosophical pursuit of natural history*. 8vo, xii+478 pp, 6 pls, London 1839; 2ed. 1841; 3ed. 1851; 4ed. 1854.

632. —— *Vegetable physiology and systematic botany*. 8vo, viii+576 pp, text figs, Orr, London 1844; ed. 1848; 1850; ed. 1858; 3ed. 1865; 1873; 1883; 1890.

633. —— *Zoology*. 8vo, 2 vols, text figs, Rees, London 1844–45. Later editions. Popular Cyclopaedia of Natural Science.

634. —— *The microscope and its revelations*. Sm. 8vo, xx+778 pp, text figs, Churchill, London 1856; 2ed. 1857; 3ed. 1862; 4ed. 1868; 5ed. 1875; 6ed. 1881; 7ed. 1891; 8ed. 1891; 1901.

635. —— *Introduction to the study of Foraminifera*. Roy. 4to, xxii+319+42 pp, 22 pls, text figs, Ray Society, London 1862. Ray Society Publ. No. 32 or 1860.

636. CARR, Daniel C. *The life of Linnaeus . . . to which is added a short account of the botanical systems of Linnaeus and Jussieu*. 12mo, xi+111 pp, Shalders, Holt 1837.

637. —— *Linnaeus and Jussieu, or, the rise and progress of systematic botany*. 12mo, 112 pp, Parker, London 1844.

638. [CARR, Edmund Donald] *Flowering plants, ferns, and fern allies*. 8vo, Liverpool [?1865].

639. —— *List of British flowering plants . . . in the south-west of Westmorland*. 8vo, Liverpool [1868].

640. CARR, John Wesley *editor A contribution to the geology and natural history of Nottinghamshire*. Sm. 8vo, 90 pp, Nottingham 1893. British Association Handbook.

641. CARRINGTON, Benjamin *British Hepaticae*. Imp. 8vo, 4 pts, 88 pp, 16 col. pls, Hardwicke, London [1874–75]. All publ.

—— *see* Miall, L. C.

642. CARRINGTON, Edith *Workers without wages*. 8vo, 2 vols, Griffith & Farren, London 1893.

643. —— *The extermination of birds*. 8vo, 44 pp, Reeve, London 1894. Humanitarian League Publ. No. 10.

644. —— *From many lands*. Sm. 8vo, 210 pp, 7 maps, text figs, Bell for the Humanitarian League, London 1895. Animal Life Readers.

645. —— *Animals in the wrong places*. 16mo, 83 pp, Bell, London 1896.

646. —— *Nature's wonders*. Sm. 8vo, 220 pp, text figs, Bell for the Humanitarian League, London 1896. Animal Life Readers.

647. —— *Animal ways and claims*. Sm. 8vo, xx+172 pp, text figs, Bell, London 1896.

648. —— *Wonderful tools*. Sm. 8vo, 171 pp, 1 pl., text figs, Bell for the Humanitarian League, London 1897.

649. CARRINGTON, John T. *The lepidopterist's register*. 4to, 340 pp, Van Voorst, London 1868.

650. CARROLL, W. *The anglers' vademecum . . . containing a descriptive account of the water flies*. 8vo, vii+128 pp, 12 col. pls, Constable, Edinburgh 1818.

651. CARTMEL, Richard *The sensation of the human mind when in a state of fever. Local and early associations. Personal observations on the habits of insects*. 8vo, 31 pp, [n.p.] 1833.

652. CARUS, Carl Gustav *An introduction to the comparative anatomy of animals compiled with constant reference to physiology*. 8vo+4to atlas, 3 vols, 20 pls, Longman, London 1827. Translation of *Lehrbuch der vergleichenden Zootomie*. Leipzig 1818, by R. T. Gore.

653. CASH, James *Where there's a will there's a way; or science in the cottage*. 16mo, vi+224 pp, Hardwicke, London 1873.

*Cassell's book of birds*. 1872. *see* Brehm, A. E.

*Cassell's natural history*. [1876–82]. *see* Duncan, P. M. *editor*

654. CASSELL, *publisher Cassell's natural history. The feathered tribes*. 8vo, viii+632 pp, Cassell, London 1854. All publ.

655. —— *Cassell's popular natural history*. 8vo, 4 vols, 1 plain + 47 col. pls, text figs, Cassell, London [1859–63]; ed. [1863–70]; ed. [1871–75]. See also Duncan, P. M. *editor*.

656. CASTEELS, Peter *Icones avium*. Fol., 12 col. pls, Bowles, London [1726].

657. CASTLE, Thomas *An introduction to systematical and physiological botany*. 12mo, xviii+285 pp, 9 col. pls, Cox, London 1829.

658. —— *The Linnaean artificial system of botany illustrated and explained*. 4to, 3 pts, 30 pls, Cox, London 1836.

*A catalogue of British birds*. 1846. *see* Biggie, G. R.

*Catalogue of British flowering plants and ferns*. [1835]. see Francis, G.

*Catalogue of British insects*. [c.1825]. *see* Hobson, E.

659. *Catalogue of British mosses*. 4to, 3 pp, [n.p.] 1866. Lithograph.

660. *Catalogue of British plants*. Fol., [?Edinburgh ?1834].

*A catalogue of British quadrupeds and birds*. 1795. *see* Ferryman, R.

*A catalogue of British vertebrated animals*. 1845. *see* Newman, E.

*A catalogue of plants copyed from nature in paper mosaick*. [1778]. *see* Delaney, Mrs M.

661. *Catalogue of British land and fresh water Mollusca*. 24mo, 10 pp, [n.p.] 1847.

662. *Catalogue of British phaenogamous plants and ferns*. 8vo, 48 pp, Booth, Hyde 1856.

*Catalogue of the cases of birds in the Dyke Road Museum, Brighton*, 2ed. 1896. *see* Booth, E. T.

*A catalogue of the flowering plants and forms growing wild in the Isle of Wight*. [1868]. *see* More, A. G.

*A catalogue of the plants growing wild in the environs of London*. 1774. *see* Curtis, W.

*Catalogue of the plants of Newbury*. 1839. *see* Russell, Anna.

*A catalogue of the scarce plants of London*. 1813. *see* Cockfield, J.

*Catalogue of the vasculares or phaenogamous plants of Great Britain*. [1842] *see* Coxhead, H.

*Catalogus plantarum circa Cantabrigiam nascentium*. 1660. *see* Ray, J.

*Catalogus plantarum circa Scarborough nascentium*. [?1850]. *see* Travis, W.

663. *A catechism of botany, being a pleasing description of the vegetable kingdom*. 12mo, 70 pp, Pinnock & Maunder, London [?1820]. By a friend to youth. = No. 2945.

664. *A catechism of natural history*. 12mo, 70 pp, Pinnock & Maunder, London [?1820]; ed. Bristol [?1825]. By a friend to youth. = No. 2944.

*A catechism of nature*. 1790. *see* Martinet, J. F.

665. CATHRALL, William *The history of Oswestry, with notices of botany*. Sm. 8vo, xvi+295 pp, 1 pl., Lewis, Oswestry [1855].

666. CATLOW, *Miss* Agnes *Popular conchology; or, the shell cabinet arranged*. Sm. 8vo, x+300 pp, 2 tab., text figs, Longman, London 1843; 2ed. 1854.

667. —— *Popular field botany*. Roy. 16mo, xxxii+380 pp, 20 col. pls, Reeve, London 1848; 2ed. 1849; 3ed. 1852; 4ed. 1860. Popular Natural History Series.

668. —— *Drops of water*. Sq. 12mo, xviii+194 pp, 4 col. pls, Reeve, London 1851.

669. —— & REEVE, Lovell Augustus *The conchologist's nomenclature*, 8vo, viii+326 pp, Reeve, London 1845.

670. CATLOW, *Miss* Maria E. *Popular British entomology*. Roy. 16mo, xx+269 pp, 16 col. pls, Reeve, London 1848; 2ed. 1852; ed. 1860. Popular Natural History Series.

671. [——] *Popular geography of plants*. Roy. 16mo, xl+370 pp, 20 col. pls, Reeve, London 1855; ed. 1865 title *Plants of the world and where they grow*. By E. M. C. Popular Natural History Series.

Catton, Charles *see Animals drawn from nature*. 1788.

672. Cely, *Madlle de The science of botany according to the system of Linnaeus*. 10 × 6.5 cm, t.p. & 52 col. cards with letterpress, London [c.1810].

673. CHADWICK, Herbert Clifton *L.M.B.C. Memoirs on typical British marine plants and animals . . . III. Echinus*. 8vo, viii+28 pp, 5 pls, Dobb for the Liverpool Marine Biology Committee, Liverpool 1900.

674. CHALMERS, Thomas *On the power, wisdom and goodness of God as manifested in the adaptation of external nature to the moral and intellectual constitution of man*. 8vo, 2 vols, Pickering, London 1833; ed. 1834; ed. 1 vol. 1853. Bridgewater Treatise No. 1.

675. [CHAMBERS, *Miss* A. C.] *Life in the walls, the hearth and the eaves*. 16mo, 123 pp, 4 pls, SPCK, London [1871].

676. [——] *Life underground, in the church tower, the woods and the old keep*. 16mo, 204 pp, SPCK, London [1873].
677. [——] *Beauty in common things*. Roy. 4to, 55 pp, 12 col. pls, SPCK, London [1874].
678. CHAMBERS, Richard *An introduction to the study of botany*. 8vo, 27 pp, text figs, Chambers, London 1847.
679. [CHAMBERS, Robert] *Vestiges of the natural history of creation*. 12mo, vi+390 pp, Churchill, London 1844; 2ed. 1844; 3ed. 1845; 4ed. 1845; 5ed. 1846; 6ed. 1847; reprinted 6ed. 1847; 7ed. 1847; 8ed. 1847; 9ed. 1851; 10ed. 1853; 11ed. 1860; 12ed. 1884 signed; ed. 1887. *See also* No. 680.
680. [——] *Explanations; a sequel to Vestiges of the natural history of creation*. 8vo, vii+198 pp, Churchill, London 1845; 2ed. 1845; 1846. The matter is incorporated in editions of *Vestiges* from 1853.
681. [——] *An expository outline of the Vestiges of the natural history of creation*. 16mo, London 1846. Reprinted from *The Atlas*.
682. CHANTER, *Mrs* Charlotte *Ferny coombes; or rambles after ferns in . . . Devonshire*. Sm. 8vo, 118 pp, 8 col. pls, Reeve, London 1856; 2ed. 1856; 3ed. 1857.
683. CHAPMAN, Abel *Bird life of the borders*. 8vo, xii+286 pp, 15 pls, text figs, Gurney & Jackson, London 1889; 2ed. 1907.
684. CHAPMAN, Thomas Algernon *The genus Acronyota and its allies; an examination of the early stages of the British species*. 8vo, iv+116 pp, 6 col. + 3 plain pls, London 1893 [–1896]. Reprinted from *The Entomologists' Record* and *Trans. Woolhope Fld Club*.

*Chapters on flowers. see* Phelan, *Mrs* C. E.

*Characteres genera plantarum*. 1793. *see* Hope, J.

685. CHARLESWORTH, Edward *Catalogue of the British marine recent shells*, 8vo, York 1856.
686. CHARLETON, Walter *Onomasticon zoicon; cui accedunt Mantissa anatomica.* 4to, 309+50 pp, 12 pls, text figs, Allestry, London 1668; issue 1671; ed. 1674; 2ed. fol., Oxford 1677 title *Excercitationes de differentiis et nominibus animalium. Mantissa*, pp. 197–213 is by *Sir* George Ent.

Charlotte Elizabeth *see* Phelan, Mrs C. E.

687. CHESELDEN, William *Osteographia, or the anatomy of bones*. Fol., [50] pp, 56+56 pls, text figs, London 1733. The plates are in duplicate, one set with descriptions on verso, one without. There are later editions of the human bones only.

688. CHILDREN, John George *An address delivered at the anniversary meeting of the Zoological Club of the Linnean Society*. 8vo, 21 pp, Taylor printed, London 1827.

689. —— *Ochsenheimer & Treitsche: a short abstract of the characters of their genera of the Lepidoptera of Europe; with a list of the species.* 8vo, 150 pp, London 1829 [–1830]. Reprinted from *Phil. Mag.* Vols IV–VIII, 1828–30.

690. [——] *Directions for collecting zoological specimens*. 8vo, 12 pp, British Museum, London [ ]; ed. [ ].

691. *The children's cabinet: or, a key to natural history*. Obl. 12mo, 47 pp, 25 pls, Laurie & Whittle, London 1798.

692. CHILDREY, Joshua *Britannia Baconica: or the natural rarities of England, Scotland and Wales*. 8vo, xxvi+183 pp, text figs, H. E. for the author, London 1660; issue 1661; issue 1662.

693. CHILDS, Archibald Prentice *British botanist's field book*. 12mo, xvi+160 pp, Longman, London 1857.

694. CHILDS, George *Woodland sketches, a series of characteristic portraits of trees*. Fol., 32 pp, 16 pls, London 1839.

695. [CHILDS, Robert Walker] *Animal intellect, and its relation to matter and spirit.* 8vo, vii+127 pp, Wertheim & Macintosh, Edinburgh 1851.

696. *The child's botany*. 16mo, 124 pp, 8 pls, Limbird, London [?1835].

697. *A child's first book about birds*. Sm. 8vo, 123 pp, 5 col. pls. Hogg, London [1861]. By a country clergyman.

698. *The child's first book of natural history*. Sm. 8vo, 93 pp, pls, Ward Lock, London [1879].

*Child's guide to knowledge. see* Umphelby, Frances.

699. CHIMMO, William *On Euplectella aspergillum. Also on a new form of Globigerina; phosphorescent animalcules; see-saw-dust; new forms of Foraminifera and Polycistina*. 4to, 22 pp, 6 pls, Taylor & Francis, London 1878.

700. CHRISTY, Robert Miller *The birds of Essex*. 8vo, viii+302 pp, 1 pl., 2 plans, text figs, Essex Field Club, Buckhurst Hill 1890. Essex Field Club Special Memoir II.

—— *Birdnesting and bird-skinning*. 2ed. 1888. *see* Newman, E.

701. —— *A catalogue of local lists of British birds*. 8vo, 42 pp, Porter, London 1891.

702. CHURCH, John *A cabinet of quadrupeds*. Fol. or Roy. 4to, 2 vols, 84 pls, Darton & Harvey, London [1795–] 1805; ed. 4to 1848–49.

*Circle of the sciences*. 1854–56. *see* Orr, W. S. *editor*.
*The circle of the sciences*. [1862–67]. *see* Wylde, J. *editor*.

703. CLARK, Bracy *An essay on the bots of horses and of other animals*. 8vo, 72 pp, 2 col. pls, for the author, London 1815; ed. 4to 1815.

704. —— *An appendix or supplement to a treatise on the Oestri and Cuterebrae of various animals*. 4to, 4 pp, for the author, London 1841, 48. Reprinted from *Proc. & Trans. Linn. Soc. Lond.*

Clark, Hamlet *see* Dawson, J. F.

705. CLARK, James Henry *The flora of Monmouthshire*. Sm. 8vo, xliv pp, County Observer Office, Usk [c.1868].

706. [CLARK, Samuel] *The little conchologist*. 12mo, viii–70 pp, 1 col. & 3 plain pls, Darton & Clark, London [1837]. By T. Wilson *pseudonym*. Also in a combination volume with two others by the same author, and *The little entomologist* and *The little marine botanist*, with title *The little book of nature*.

707. CLARK, William *of Bath. Mollusca testacea marium Britannicarum: a history of the British marine testaceous Mollusca*. 8vo, xii+536 pp, Van Voorst, London 1855.

708. CLARK, William *of Cambridge Catalogue of the osteological portion of specimens in the Anatomical Museum of the University of Cambridge*. 8vo, viii+132 pp, University Press, Cambridge 1862.

709. CLARKE, Benjamin *A new arrangement of phanerogamous plants, with especial reference to relative position*. Obl. fol., iv+56 pp, 1 pl., tabs, by the author, Hampstead 1866; issue London 1866; 3ed. 1888. 2ed. in No. 710.

710. —— *On systematic botany and zoology, including a new arrangement of phanerogamous plants*. Fol., 56 pp, 1 pl., 8 tabs, Bale for the author, London 1870. Includes the 2ed. of No. 709.

711. —— *A new arrangement of the classes of zoology . . . including a new mode of arranging the Mammalia*. 4to, 20 pp, Williams & Norgate, London 1879. Derived from the zoological part of No. 710.

712. CLARKE, Charles Baron *A list of the flowering plants, ferns, and mosses, collected in the immediate neighbourhood of Andover*. 8vo, 114 pp, 1 col. map, Calcutta 1866.

713. CLARKE, J. *The naturalist. A treatise on the growth of the horns of the red deer*. 8vo, 12 pp, 1 pl., Barnstable 1866.

714. CLARKE, Louisa Lane *The microscope: being a popular

*description of the most interesting and beautiful objects for exhibition*. 12mo, xi+231 pp. 8 col. pls, Routledge, London 1858; 2ed. 1863 title *Objects for the microscope*; 3ed. 1870; 4ed. 1871; 8ed. 1889.

715. —— *The common seaweeds of the British coasts and Channel Islands*. 8vo, 140 pp, 10 col. pls, Warne, London [1865].

716. [CLARKE, Stephen Reynolds]. *The British botanist, or a familiar introduction to the science of botany*. Sm. 8vo, 260 pp, 10 col. pls, Rivington, London 1820.

717. CLARKE, William Ambrose *First records of British flowering plants.* 8vo, 103 pp, West Newman, London 1897; 2ed. 1900. Reprinted from *J. Bot*. 1892–96.

718. CLARKE, William Barnard *Notice of a species of seal apparently new to the British shores*. 4to, [2]pp, 1 pl., [Ipswich 1847].

719. CLARKE, William Eagle & ROEBUCK, William Denison *A handbook of the vertebrate fauna of Yorkshire*. 8vo, xlviii+149 pp, Reeve, London 1881.

*The classes and orders of the Linnaean system of botany*. 1816. *see* Duppa, R.

*A classified list of the names and latin synonyms of British birds*. 1853. *see* Sealy, A. F.

720. CLAUS, Carl Friedrich Wilhelm *Elementary text book of zoology*. 8vo, 2 vols, text figs, Sonnenschein, London 1884–85; 2ed. 1890; 3ed. 1891; 1893. Translation of *Grundzüge der Zoologie*, Marburg 4ed. 1880–82, by Adam Sedgwick & Frederick Granville Heathcote.

Clennell, Luke *see Recreations in natural history*. 1815.

Clerck, Carl Alexander *see* Martyn, Thomas *draughtsman*.

721. CLERMONT, Thomas Fortescue *Baron A guide to the quadrupeds and reptiles of Europe*. Sm. 8vo, viii+277 pp, Van Voorst, London 1859.

722. CLIFFORD, J. R. S. *Insect lights and insect sounds*. 8vo, 94 pp. Wesleyan Conference Office, London [1879].

723. CLODD, Edward *The story of creation. A plain account of evolution*. 8vo, xv+242 pp, text figs, Longman, London 1888. Later editions.

724. —— *A primer of evolution*. 8vo, xii+186 pp, Longman, London 1895; ed. 1908. Abridged from No. 723.

725. —— *Pioneers of evolution from Thales to Huxley*. 8vo, x+250 pp. Grant Richards, London 1897. Later editions.

726. CLOUSTON, Charles *Guide to the Orkney Islands, includ-*

*ing their scenery . . . natural history*. 16mo, iv+76 pp, 1 map, Black, Edinburgh 1862. Reprinted from Anderson's *Guide to the Highlands*. 1862.

727. COBBOLD, Thomas Spencer *On the scope, tendency, and educational value of the natural sciences*. 8vo, 17 pp, Davies, London 1861.

728. —— *Catalogue of the specimens of Entozoa in the collection of the Royal College of Surgeons*. 8vo, iv+24 pp, London 1866.

729. —— *Entozoa; an introduction to the study of helminthology*. Roy. 8vo, xvi+480 pp, 21 col. pls, text figs, Groombridge, London 1864; suppl. viii+124 pp, 1869.

730. —— *Parasites; a treatise on the Entozoa of man and animals, including some account of the Ectozoa*. 8vo, xi+508 pp, text figs, Churchill, London 1879.

731. COCKERELL, Theodore Dru Alison *The British Naturalist catalogue of land and Fresh water Mollusca of the British Islands*. 8vo, London [1890].

732. [COCKFIELD, Joseph] *Catalogue of the scarce plants of London*. 8vo, London 1813.

733. COCKLE, Mary *The fishes grand gala, a companion to the 'Peacock at home'*. [With natural history notes.] 16mo, 2 parts, 16+16 pp, 11 pls, Chapple, London 1808.

734. COCKS, John *The sea-weed collector's guide*. Sm. 8vo, viii+120 pp, 1 col. pl., Van Voorst, London 1853.

735. —— *Algarum fasciculi: or, a collection of British sea-weeds*. 4to, 18 fasc., 179 specimens, Dublin & Plymouth 1855–60.

736. COCKS, W. P. *Specimens of natural history procured in Falmouth and neighbourhood*. 4to, 15 pp, [n.p.] 1849. Lithographed.

737. —— *Contributions to the flora of Falmouth*. 8vo, 1857.

738. COE, Charles Clement *Nature versus natural selection: an essay on organic evolution*. 8vo, xiii+591+xx pp, Sonnenshein, London 1895.

739. COETLOGON, Denis de *A tour through the animal world: or, an . . . account of near four hundred animals*. 12mo, v+173 pp, 172 pls, London 1746.

740. COKER, John *A survey of Dorsetshire. Containing the antiquities and natural history of that county*. Fol., 128 pp, 6 pls, 1 col. map, Wilcox, London 1732.

741. COLE, Arthur Charles *editor Studies in microscopical science*. 8vo. 4 vols, 187 col. pls, Baillière, London 1883–87.

742. —— *The methods of microscopical research*. 8vo, viii+207

pp, 2 pls, text figs, Baillière, London 2ed. 1895. First edition is in No. 741, Vol. 2.

743. [COLE, John] *The Scarborough natural historians; or, a visit to that celebrated watering-place*. 16mo, 70 pp, text figs, Cole, Scarborough 1829.

Cole, Martin J. *see* Cross, M. I.

Coleman, William Higgins *see* Webb, R. H.

744. COLEMAN, William Stephen *Our woodlands, heaths and hedges*. Sm. 8vo, viii+198 pp, 8 col. pls, text figs, Routledge, London 1859. Many later editions. Country Books.

745. —— *British butterflies*. Sm. 8vo, vii+179 pp, 16 col. pls, text figs, Routledge, London 1860. Many later editions. Country Books.

746. COLES, William *Adam in Eden: or, natures paradise. The history of plants, fruits, herbs and flowers*. Fol., 1–144, 75–78, 145–165, 116–396, 1–66, 551–629 pp, Brooke, London 1657.

Colgan, Nathaniel & Scully, Reginald William *Contributions towards a Cybele Hibernica*. 2ed. 1898. *see* Moore, D. & More, A. G.

*Collecteana for a flora of Moray*. 1839. *see* Gordon, G.

*A collection of curious insects*. [c.1760]. *see* Roesel von Rosenhof, A. J.

*A collection of tracts and treatises illustrative of the natural history of Ireland*. 1860, 61. *see* Boate, G.

Collin, James E. *see* Verrall, G. H.

747. COLLINGE, Walter Edward *A catalogue of the slugs of the British Isles*. 8vo, 4 pp, Ore. Hartlepool 1892. Reprinted from *The British Naturalist*.

748. —— *A review of the Arionidae of the British Isles*. 8vo, 11 pp, Sonnenschein, London [1892]. Reprinted from *The Conchologist*, Vol. 2, 1892.

749. —— *The skull of the dog: with a glossary of osteological terms*. 8vo, 124 pp, text figs, Dulau, London 1896.

750. COLLINGWOOD, Cuthbert *On the scope and tendency of botanical study*. 8vo, 31 pp, Longman, London 1858.

751. —— *A comparison between the ancient and modern views on the migration and habits of birds*. 8vo, Liverpool 1859. Reprinted from *Proc. Liverpool lit. phil. Soc.* 1859.

752. —— *The fauna of Blackheath and its vicinity. Part 1. Vertebrate animals*. 8vo, 46 pp, London 1859. ? All publ.

753. —— *On homomorphism; or, organic representative form*. 8vo, Liverpool 1860.

754. —— *On the opportunities of advancing science enjoyed by the mercantile marine*. 8vo, Liverpool 1862.

755. —— *Twenty-one essays on various subjects, scientific and literary*. 8vo, 21 pts, Liverpool 1865. A collective issue of reprints and offprints with preliminaries.

756. —— *Instinct and reason . . . a paper read before the Victoria Institute*. 8vo, 33 pp, Harrison, London [1889].

757. COLLINS, Charles *Icones avium, cum nominibus anglicis: twelve engravings of British birds.* Obl. fol., no text, 12 col. pls, Collins & Lee, London 1736. Also occurs in 8vo format as Book 21 of *Bowles' Complete drawing book*.

758. COLLINS, Frederick Howard *Epitome of the synthetic philosophy*. 8vo, 2 pts, xviii+636 pp, Williams & Norgate, London 1889; 2ed. 1890; 3ed. 1894; 4ed. 1897; 5ed. 1901. Verbatim potted from Herbert Spencer.

759. COLLINS, Samuel *A systeme of anatomy treating of the body of the man, beasts, birds, fish, insects, and plants*. Fol., 2 vols, lvi+1263 pp, 73 pls, Vol. 1 Newcomb; Vol. 2 Newcomb for the author, London 1685.

760. *Coloured pictures of natural history*. Fol., col. pls with letter-press, RTS, London [1875].

*Coloured specimens to illustrate the natural history of butterflies. see* Lee, J.

761. COLQUHOUN, John *The moor and the loch*. 8vo, ii+ii+128 pp, pls, Blackwood, Edinburgh 1840. Many later editions.

762. —— *Lectures on the natural history of the British Islands*. 8vo, 20 pp, Patch & Ritchie printed, Edinburgh 1856.

763. —— *Lecture on the ferae naturae of the British Islands.* 8vo, 46 pp, Blackwood, Edinburgh 1873.

764. —— *Lecture on instinct and reason*. 8vo, 31 pp, Blackwood, Edinburgh 1874.

765. COMBER, T. *On the world distribution of British plants*. 8vo, [Liverpool 1874].

766. COMENIUS, Johann Amos *Orbis sensualium pictus, hoc est omnium fundamentalium in mundo rerum, et in vita actionum, pictura et nomenclatura.* 8vo, [xxii]+309+[xii] pp, portr., text figs, Kirton, London 1659; ed. 1664; ed. 1672; ed. 1685; ed. 1689; ed. 1700; ed. 1705; 11ed. 1728; 12ed. 1777. Text in Latin and English from Nuremburg ed. 1658.

English ed. unsigned, title *The visible world*, London 1792. See also Greenwood, James. Translated by Charles Hoole.

767. *A companion to the hydro-oxygen microscope; being a description of the living animals ... shown daily at the Adelaide Street Gallery, West Strand*. 12mo, 16 pp, 2 pls, London 1836.

*The companion to the microscope*. 1827. *see* Gould, C.

*A comparative view of the state and faculties of man with those of the animal world*. 1765. *see* Gregory, J.

*Comparisons in structure of animals. The hand and arm*. [1848]. *see* Martin, W. C. L.

*A compendium of natural philosophy*. 1809. *see* Wesley, J.

*A compendium of zoology*. 1812. *see* Boreman, T.

*A complete natural history containing delineations of upwards of three hundred animals*. 1813. *see* Boreman, T.

768. *A comprehensive view of the animal kingdom, from the highest to the lowest grades*. 8vo, Reynolds, London [1873]. [A text to accompany wall charts for schools, one of a set of 7.] *see also The structure, physiology, and classification of plants*.

Comte, Achille *The book of birds*. 1841. *see* Buffon, J. L. L. *Conte de*

*The conchologist's companion*. 1824. *see* Roberts, Mary

*Conchology, or natural history of shells*. [1771]. *see* Mendes da Costa, E.

769. *A concise abridgement of natural history*. 12mo, 5 vols, 5 pls, text figs, Marshall, London 1800. The Juvenile; or, Child's Library.

*The concise knowledge natural history*. 1897. *see* Miles, A. H.

770. CONN, Herbert William *Evolution of today*, 8vo, ix+342 pp, Putnam, London 1886. New York printed.

771. —— *The living world. Whence it came and whither it is drifting*. 8vo, v+195 pp, Putnam, London 1891. New York printed.

772. —— *The story of germ life: bacteria*. Sm. 8vo, 212 pp, text figs, Newnes, London 1897. Library of Useful Stories, Vol. XIII.

773. —— *The method of evolution*. 8vo, ix+408 pp, text figs, Putnam, London 1900. New York printed.

774. CONNOLD, Edward *Wasps and how they live, or British social wasps and the habitations they construct*. 8vo, 17 pp, Burfield & Pennells printed, Hastings [1894].

*The contemplative philosopher*. 1800. *see* Lobb, R.

*A contribution towards a complete list of the fauna and flora of*

*Clydesdale and the west of Scotland*. 1876. *see* Young, H. C.

*Contributions to the fauna and flora of Repton*. 1881. *see* Wyatt, W. & Thornton, C. G.

*Conversations on botany*. 1817. *see* Fitton, *Miss* E. & Fitton, *Miss* M.

775. *Conversations on gardening with incidental remarks on natural history*. 8vo, viii+216 pp, 1 pl., text figs, Parker, London 1834; 2ed. 1838. By the author of *The elements of botany*.

776. *Conversations on instinct. A first book of natural history for the young*. 12mo, 72 pp, text figs, Foster, London [1844].

*Conversations on natural history*. [?1815]. *see* Sandham, *Mrs* E.

*Conversations on vegetable physiology*. 1829. *see* Marcet, *Mrs* Jane.

777. COOK, James *Beautiful seaweeds, illustrated by natural specimens . . . and a complete list of all the British species*. Roy. 8vo, [182]pp, 34 specimens, Paisley 1877; ed. 1883. 50 copies.

778. COOKE, Mordecai Cubitt *Manual of botanical terms*. Sm. 8vo, iv+90 pp, 27 pls, Hardwicke, London [1862]; issue 1865; 2ed. 1873; 3ed. 1875. Later editions.

779. —— *A manual of structural botany*. 8vo, iv+123 pp, text figs, Hardwicke, London 1862; 1871; 2ed. 1876. Many later editions.

780. —— *A plain and easy account of British Fungi*. Sm. 8vo, vi+148 pp, 24 col. pls, text figs, Hardwicke, London 1862; 1866; 2ed. 1871; 3ed. 1876; 4ed. 1878; 5ed. 1884; 6ed. 1898. Later editions.

781. —— *Index fungorum Britannicarum. A complete list of the Fungi found in the British Islands*. Roy. 8vo, 3 pts, 58 pp, Hardwicke, London [1863]. Printed on one or both sides of the paper.

782. —— *Hardwicke's Science Gossip Easy guide to the study of British Hepaticae*. 4to, 27 pp, Hardwicke, London [1865].

783. —— *Our reptiles, A plain and easy account of lizards, snakes, newts, toads, frogs, and tortoises*. Sm. 8vo, vii+199 pp, 11 col. pls, Hardwicke, London 1865; ed. 1893 title *Our reptiles and batrachians*.

784. —— *Rust, smut, mildew, and mould*. 8vo, vii+238 pp, 12 col. pls, Hardwicke, London 1865; 2ed. 1870; 3ed. 1872; 3ed. 1875; 4ed. 1878; 5ed. 1886; 6ed. 1898; 7ed. 1902.

785. —— *A fern book for everybody*. 8vo, 124 pp, 12 col. pls, text figs, Warne, London 1867. Later undated editions.

786. —— *One thousand objects for the microscope.* 8vo, iv+123 pp, 12 pls, Warne, London 1869. Later editions.

787. —— *Handbook of British Fungi.* Sm. 8vo, 2 vols, 981 pp, 7 col. pls, text figs, Macmillan, London 1871; 2ed. 1883 [=1883–91]. 2ed is unfinished.

788. —— *Freaks and marvels of plant life; or, curiosities of vegetation.* Sm. 8vo, viii+463 pp, text figs, SPCK, London 1872. Later editions.

789. —— *Fungi, their nature, influence and uses.* Sm. 8vo, xii+209 pp, text figs, King, London 1875; 2ed. 1875. Several later editions. International Scientific Series, Vol. XIV.

—— *Contributions to Mycologia Britannica.* 1877. *see* Rostafinski, J. T.

790. —— *Mycographia seu icones fungorum: Vol. 1. Discomycetes Pt. 1.* Roy. 8vo, 267 pp, 113 col. pls, Williams & Norgate, London [1875–] 1879 All publ.

791. —— *Natural History Rambles. The woodlands.* Sm. 8vo, 288 pp, SPCK, London 1879. Natural History Rambles Series. Later editions.

792. —— *Natural History Rambles, Ponds and ditches.* Sm. 8vo, 254 pp, text figs, SPCK, London 1880. Natural History Rambles Series. Later editions.

793. —— *Illustrations of British Fungi (Hymenomycetes) . . . to serve as an atlas to the Handbook.* 8vo, 8 vols, 20 pp, 1198 col. pls, Williams & Norgate, London 1881–91. See No. 787 for *Handbook*

794. —— *British freshwater Algae.* 8vo, 2 vols, viii+329 pp, 130 col. pls, Williams & Norgate, London 1882–84.

795. —— *British desmids. A supplement to British freshwater Algae.* 8vo, xiv+205 pp, 66 col. pls, Williams & Norgate, London [1886–] 1887.

796. —— *Toilers in the sea.* 8vo, viii+373 pp, 4 pls, text figs, SPCK, London 1889.

797. —— *Introduction to fresh-water Algae, with an enumeration of all the British species.* Sm. 8vo, vi+339 pp, 13 pls, Kegan Paul, London 1890. Later editions. International Scientific Series, Vol. LXIX.

798. —— *British edible Fungi.* 8vo, 237 pp, 12 col. pls, Kegan Paul, London 1891.

799. —— *Vegetable wasps and plant worms.* Sm. 8vo, v+364 pp, 4 pls, text figs, SPCK, London 1892.

800. —— *Romance of low life among plants. Facts and phenomena*

*of cryptogamic vegetation.* 8vo, vii+320 pp, text figs, SPCK, London 1893.

—— *Our reptiles and batrachians.* 1893. *see* No. 783.

801. —— *Edible and poisonous mushrooms; a popular history of entomogenous Fungi.* Sm. 8vo, 126 pp, 18 pls, SPCK, London 1894. Later editions.

802. —— *Handbook of British Hepaticae.* Sm. 8vo, vii+310 pp, 7 pls, text figs, Allen, London 1894 [=1893]. Later editions.

803. —— *Introduction to the study of Fungi.* 8vo, x+360 pp, text figs, Black, London 1895.

804. [——] *Across the common after wild flowers.* 8vo, 98 pp, Nelson, London 1895. By Uncle Matt.

805. [——] *Around a cornfield, in a ramble after wild flowers.* 8vo, 98 pp, Nelson, London 1895. By Uncle Matt.

806. [——] *Down the lane and back, in search of wild flowers.* 8vo, 114 pp, Nelson, London 1895. By Uncle Matt.

807. [——] *A stroll in a marsh, in search of wild flowers.* 8vo, 94 pp, Nelson, London 1895. By Uncle Matt.

808. [——] *Through the copse. Another ramble after wild flowers.* 8vo, 106 pp, Nelson, London 1895. By Uncle Matt.

809. —— & GUELET, L. *Clavis synoptica hymenomycetum Europaeorum.* Sm. 8vo, 240 pp, Hardwicke & Bogue, London 1878.

810. COOKE, Thomas *A letter to Mark Milbank Esq., M.P. . . . descriptive of the character of the whistling swan.* 4to, 11 pp, 2 pls, Rodwell & Martin, London 1823.

811. COOPER, Daniel *Flora Metropolitana; or, botanical rambles within thirty miles of London.* 12mo, xvi+139 pp, Highley, London 1836; suppl. 36 pp [1837]; issue with suppl. 1837.

812. —— *A list of land and fresh-water shells found in the environs of London.* 12mo, 12 pp, Stewart printed, London 1836. Reprinted from *Flora Metropolitana.*

813. —— *The little book of botany.* 24mo, xii+148 pp, 6 pls, Darton & Clark, London 1839.

814. [COOPER, William White] *Zoological notes and anecdotes.* Sm. 8vo, x+362 pp, Bentley, London 1852; ed. 1855. By Sestertius Holt *pseudonym.*

815. COPE, Edward Drinker *The origin of the fittest, essays on evolution.* 8vo, xix+467 pp, text figs, Macmillan, London 1887.

816. COPE, Richard *Natural history; or, description of earth and*

*of animated nature.* Roy. 8vo, vi+730 pp, text figs, Smith, London 1840; issue Milner, London [?1840].

817. CORDEAUX, John *Birds of the Humber district.* Sm. 8vo, xii+231 pp, 1 pl., text figs, Van Voorst, London 1872.

818. [——] *A full account of the monster Royal whale.* 8vo, 16 pp, Leeds 1892.

819. —— *A list of British birds belonging to the Humber district having a special reference to their migrations.* 8vo, viii+40 pp, Porter, London 1899. Revised edition of a paper in *Reports on migration*, Brit. Assoc. Adv. Sci.

820. CORDINER, Charles *Remarkable ruins, and romantic prospects of North Britain, with ancient monuments and singular subjects of natural history.* 4to, 2 vols, 98 pls, Vol. 1 Mazell, Vol. 2 Taylor, London 1788, 95.

821. CORNISH, Charles John *Wild England today and the wild-life in it.* 8vo, xiii+310 pp, pls, Seeley, London 1895; 2ed. 1895; 3ed. 1900. Later editions.

822. —— *Life at the Zoo. Notes and traditions of the Regent's Park Gardens.* 8vo, vi+340 pp, 16 pls, text figs, Seeley, London 1895; 2ed. 1895. Later editions.

823. —— *Animals at work and play. Their activities and emotions.* 8vo, xi+323 pp, text figs, Seeley, London 1896.

824. —— *Animals of to-day, their life and conversation.* 8vo, xii+319 pp, text figs, Seeley, London 1898.

Cornwall Royal Institution. *see* Couch, J. & Couch, R. Q.

Corry, Thomas Hughes *see* Stewart, S. A.

Costa, Emanuel Mendes da *see* Mendes da Costa, E.

Cotes, Henry *A history of British birds. Vol. 2 Water birds.* 1804. *see* Bewick, T.

825. COTTON, John *The resident song birds of Great Britain. (The song birds of Great Britain)*. Roy. 8vo, 2 pts, 33 col. pls, privately printed, London 1835; issue 1836. Second issue has no title page to Part 2 and cancel title page to Part 1.

826. —— *Beautiful birds: their natural history.* 8vo, 3 vols, 36 col. pls, text figs, Houlston & Stoneman, London [1853–] 1854–56.

827. COUCH, Jonathan *Illustrations of instinct deduced from the habits of British animals.* 12mo, xii+343 pp, Van Voorst, London 1847.

828. —— *A history of the fishes of the British Islands.* Roy. 8vo, 4 vols, 252 col. pls, Groombridge, London [1860–] 1861–65; reissues Vol. 1 1862, 1867, 1868, 1877; Vol. 2 1863, 1867,

1877; Vol. 3 1864, 1866, 1868, 1877; Vol. 4 1865, 1867, 1869, 1877.

828a —— & COUCH, Richard Quiller *A Cornish fauna.* 8vo, 3 pts, 23 pls, Royal Institution of Cornwall, Truro 1838–44; ed. pt. 1 1878.

Couch, Richard Quiller *see* Couch, J.

Coues, Elliott *see* Gurney, J. H.

829. COULTAS, Harland *Animals and their young.* 4to, 174 pp, 24 pls, Partridge, London [1874].

830. —— *Our zoological friends.* 4to, 192 pp, pls, Partridge, London [1876].

831. —— *The home naturalist; with practical instructions for collecting, arranging, and preserving natural objects.* 8vo, xii+234 pp, text figs, RTS, London [1877]. Rough translation of Klasing, Otto *Das Buch der Sammlungen*, 2ed. Leipzig 1875.

832. [COURTAULD, Sydney] *Ferns of the British Isles described and photographed.* 8vo, ii+65 pp, 20 pls, text figs, Van Voorst, London 1877. Signed Sy C.

833. COURTNEY, John Sampson *A guide to Penzance and its neighbourhood, including the Isles of Scilly, with an appendix containing the natural history of western Cornwall.* 8vo, 223+80 pp, Rowe, Penzance 1845.

834. COVOLO, Giovambatista dal, *Count A discourse concerning the irritability of some flowers.* 8vo, xliii pp, 1 pl., Dodsley, London [1797]; ed. 1811. Translation of *Discorso della irritabilita d'alcuni fiori*, by Benjamin Stillingfleet. 1811 ed. in Stillingfleet's *Literary life and select works.*

835. COWARD, Thomas Alfred & OLDHAM, Charles *The birds of Cheshire.* 8vo, 278 pp, 6 pls, 1 map, Sherratt & Hughes, Manchester 1900.

836. COWELL, Matthew Henry *A floral guide for east Kent.* 8vo, xiii+98 pp, 1 col.+1 plain pls, maps, Ratcliffe, Faversham 1839.

837. —— *A series of botanical labels for the herbarium.* 8vo, vii+327 pp, Faversham 1841. Based on Smith, *Sir* James E. *The English flora.*

838. COWLEY, Abraham *A.Couleii Plantarum libri duo.* 8vo, 118 pp, Brooks, London 1662; ed. 1668 title *Poemata latina in quibus continentur, sex libri plantarum.* 8vo, 420 pp, portr., Martyn, London; ed. 1678; English ed. 1795. Also in collected works, in Latin and English.

839. COX, Herbert Edward *Handbook of the Coleoptera or beetles, of Great Britain and Ireland.* 8vo, 2 vols, Janson, London 1874.

840. COX, Rosalinda Alicia, *Mrs* Edward William *Our common insects; first steps in entomology.* 12mo, 120 pp, text figs, Hardwicke, London 1864; 2ed. 1869.

841. COX, Thomas *A topographical, ecclesiastical, and natural history of Great Britain.* 4to, Nutt, London 1700.

842. COXHEAD, Henry *Catalogue of the Vasculares or phaenogamous plants of Great Britain.* 8vo, 15 pp, London [1842].

843. COYTE, William Beeston *Hortus Britannicus Gippovicensis.* 4to, 158 pp, Jermyn, Ipswich 1796.

844. —— *Index plantarum, or an alphabetical arrangement of all the genera and species of plants hitherto described.* 4to, 445 pp, London 1807. All publ.

845. CRADOCK, *Hon. Mrs* Harriet Grove *The calendar of nature, or the seasons of England.* Fol., 4 pts, 24 col. pls, text figs, Dickinson, London [1849–50].

846. CRAWFORD, James Hunter *Wild life of Scotland.* 8vo, 280 pp, 1 pl., text figs, Macqueen, London 1896.

847. —— *Wild flowers of Scotland.* 8vo, 228 pp, 1 pl., text figs, Macqueen, London 1897; ed. 1909.

848. CRAWFORD, Oswald John Frederick *editor A year of sport and natural history.* 4to, xii+331 pp, text figs, Chapman & Hall, London 1895.

849. CRICHTON, Arthur *The festival of flora. A poem with botanical notes.* 8vo, vii+73 pp, 8 pls, London 1818; 2ed. 1818.

850. CRICHTON, Arthur William *A naturalist's ramble to the Orcades.* Sm. 8vo, iv+132 pp, 1 tinted pl., Van Voorst. London 1866.

Croall, Alexander *see* Johnstone, W. G.

851. CROMBIE, James Morrison *Braemar, its topography and natural history.* 8vo, x+152 pp, 1 map, Smith, Aberdeen 1861.

852. —— *Lichenes Britannici, seu lichenum in Anglia, Scotia, et Hibernia vigentium enumeratio.* 8vo, vii+138 pp, Reeve, London 1870; issue 1871.

853. —— *A monograph of the lichens found in Britain.* 8vo, 2 pts, 59 pls, British Museum (Natural History), London 1894, 1911; 2ed. 1918–26. Part 2 of led and all 2ed. by Smith, Annie Lorrain.

854. CROMPTON, Sarah *The scholar's book of beasts.* 16mo, viii+119 pp, text figs, Darton, London 1858.
855. —— *The scholar's book of birds, fishes, insects.* 16mo, viii+159 pp, pls, Darton, London 1858.
856. CROSFIELD, George *A calendar of flora, composed during the year 1809 at Warrington.* 8vo, 34 pp, Haddock, Warrington 1810.
857. [CROSS, J. E.] *The berries and heaths of Rannoch.* Roy. 8vo, 24 pp, 13 col. pls, Bell, London 1881. By a snowdrop.
858. CROSS, M. I. & COLE, Martin J. *Modern microscopy. A handbook for beginners.* 8vo, 109 pp, text figs, Baillière, London 1893; 2ed. 1895; 3ed. 1903; 4ed. 1913; 5ed. 1922.
859. CROTCH, George Robert *A catalogue of British Coleoptera.* 8vo, 45 pp, for the author, Cambridge 1863; 2ed. [1871].
860. —— *Synopsis Coleopterorum Europeae et confinium anno 1868 descriptorum.* 8vo, 68 pp, Williams & Norgate, London [1871].
861. CROUCH, Edmund A. *An illustrated introduction to Lamarck's conchology.* 4to, iv+47 pp, 22 col. pls, Longman, London 1826. Largely a translation of the relevant parts of *Histoire naturelle des animaux sans vertèbres*, Paris 1815–22.

Crown Series of Natural History (Reeve). *see*

Bentham, G. *Handbook of the British flora.* 1858.
Fitch, W. H. & Smith, W. G. *Illustrations of the British flora.* 1880.
Gray, S. O. *British seaweeds.* 1867.
Green, G. C. *Natural history and sport.* 1886.
Hobkirk, C. C. P. *Synopsis of British mosses.* 1873.
Lovell, M. S. *Edible mollusks.* 1867.
Massee, G. E. *British Fungi.* 1891.
Pennington, A. S. *British zoophytes.* 1885.
Plues, *Miss* M. *British ferns.* 1866.
—— *British grasses.* 1867.
Rye, E. C. *British beetles.* 1866.
Shuckard, W. E. *British bees.* 1866.
Stainton, H. T. *British butterflies and moths.* 1867.
Staveley, *Miss* E. F. *British spiders.* 1866.
—— *British insects.* 1871.
Sterland, W. J. *Birds of Sherwood Forest.* 1869.

862. CROWTHER, James *Solomon's little people: a story about the ants.* 8vo, 189 pp, text figs, Sunday School Union, London [1882].

863. —— *The microscope and its lessons. A story of the invisible world.* 8vo, 286 pp, 1 pl., text figs, Cauldwell, London 1891.

Cryptogamic Society of Scotland. *see* Stevenson, J.

864. CULLEN, William Henry *Flora Sidostiensis; or, a catalogue of the plants . . . of Sidmouth.* 16mo, xix+57 pp, Hoyte, Sidmouth 1849.

865. [CULLUM, *Sir* Thomas Gery *Baronet*] *Florae Anglicae specimen . . . anno 1774 inchoatum.* 8vo, 104 pp, [?London ?1775].

866. CULPEPPER, Edward *The description and use of a set of portable microscopes, made and sold by Ed. Culpepper.* 24mo, 16 pp, 1 pl., [London c.1715].

867. CUNDALL, James *The every day book of natural history.* 8vo, xii+509 pp, text figs, Warne, London 1866. Later eds.

868. CUNNINGHAM, Joseph Thomas *A treatise on the common sole* (*Solea vulgaris*). 4to, viii+147 pp, 18 col. pls, Marine Biological Association, Plymouth 1890.

869. —— *The natural history of the marketable marine fishes of the British Islands.* 8vo, xvi+375 pp, 2 col. maps, text figs, Macmillan, London 1896.

870. —— *Sexual dimorphism in the animal kingdom, a theory of the evolution of secondary sexual characters.* 8vo, xi+317 pp, text figs, Black, London 1900.

871. *Curiosities and wonders of nature, science and art; or the intellectual observer.* 8vo, iv+474 pp, 3 col.+9 plain pls, Groombridge, London [c.1865].

872. *Curiosities and wonders of the vegetable kingdom.* 16mo, 184 pp, text figs, SPCK, London [1852].

873. *Curiosities for the ingenious: selected from the most authentic treasures of nature, science and art.* 12mo, 192 pp, 12 pls, Boys, London 1821; 2ed. 1822; 3ed. 1831.

*Curiosities of animal life, with recent discoveries of the microscope.* [1848]. *see* Williams, C. of St John's Wood.

*Curiosities of entomology.* [1871]. *see* Wood, T. W.

*Curiosities of ornithology.* [1871]. *see* Wood, T. W.

874. CURLEY, E. *Floral calendar.* 4to, London 1869.

875. CURTIS, John *British entomology.* Roy. 8vo, 16 vols, 770 col. pls with letterpress, for the author, London 1823–40; ed. 1823 [=1829] –1840; ed. 8 vols 1862. Plates 1–769, 205*. Also available in 8 vols made up by orders. In the 2ed. pts 1–8 rewritten, 9–30 reprinted, and 31–192 the same as in the first edition. See also Nos 877 & 878.

876. —— *A guide to the arrangement of British insects.* 8vo, vi+256 columns, Westley for the author, London 1829; 2ed. 1837.

877. —— *The genera of British Lepidoptera, transferred from the original figures in 193 plates.* 4to, vii pp, 35 col. pls, Reeve, London 1858. Contains the Lepidoptera of *British Entomology* without the plants or dissections.

878. —— *The genera of British Coleoptera, transferred from the original figures in 256 plates.* 4to, viii pp, 29 col. pls, Reeve, London 1858; ed. 1863 title *British beetles: transferred from Curtis's British entomology.* Contains the Coleoptera of *British entomology* without the plants or dissections.

879. —— *Farm insects.* Roy. 8vo, xiv+528 pp, 16 col. pls, text figs, Blackie, Glasgow 1860; ed. London 1870; issue 1883.

880. [CURTIS, William] *Instructions for collecting and preserving insects, particularly moths and butterflies.* 8vo, iv+44 pp., 1 pl., for the author, London 1771. Sold by George Peach or by B. White.

881. [——] *A catalogue of the plants growing wild in the environs of London; with some explanation of the Linnaean system of botany.* 8vo, 40 pp, London 1774; issue 1774 without 2nd part of title, and not including 2nd part. Attribution doubtful.

882. —— *Linnaeus system of botany so far as it relates to his classes and orders of plants.* 4to, 4+15 pp, 2 col.+2 plain pls, for the author, London 1777; ed. [1803]. Plates are duplicates, coloured and plain.

883. —— *Flora Londinensis: or, plates and descriptions of such plants as grow wild in the environs of London.* Fol., 2 vols, 435 col. pls on 432 leaves, White for the author, London [1775–] 1777–98; another issue 1777–98; ed. 5 vols 1817–28; issue 1835; abridgement 8vo 1792. Vols 4 & 5 of 1817–28 edition are by Hooker, *Sir* William J. Bibliography is complex.

884. —— *A catalogue of certain plants . . . in the environs of Settle, in Yorkshire.* Fol., 6 pp, [White, London 1782]. Issued with *Flora Londinensis.*

885. —— *A short history of the brown-tail moth.* 4to, 13 pp, 1 col. pl., White, London 1782; reprint 1969.

886. [——] *General observations on the advantage which may result from the introduction of the seeds of our best grasses.* Fol., 4 pp, 1 pl., Lambeth Marsh 1787; 2ed. Roy. 8vo, London 1790 title *Practical observations on the British*

*grasses . . . likewise an enumeration of the British grasses*; 3ed. 1798; 4ed. 1805; 5ed. 1812; 6ed. 1824; ed. 1834. All except the first edition are signed. There is a *Hortus siccus gramineus*, 2 vols, 1802–06 by William Salisbury to accompany the third edition; 1 vol. 1812.

887. —— *Companion to the botanical magazine; or a familiar introduction to the study of botany.* 8vo, 33 pp, 8 pls, for the author, London 1788.

888. —— *Proposals for a course of herbarizing excursions for Londoners for the summer months.* Fol., s.sh., London 1792.

889. —— *Lectures on botany, as delivered to his pupils.* 8vo, 3 vols, 119 col. pls, 1 portr., S. Curtis, London 1803–04; 1805 title *Lectures . . . as delivered in the Botanical Garden at Lambeth*; 2ed. 1807.

—— *see* Bladh, A.

890. CUVIER, George Leopold Chrétien Frederic Dagobert, *Baron de An introduction to the study of the animal economy.* 8vo, Ross & Blackwood, Edinburgh 1801. Translation of the introduction to *Leçons d'anatomie comparée*, Paris 1800–05.

891. —— *Lectures on comparative anatomy.* 8vo, 2 vols, 5 tabs, Longman, London 1802. All publ. Translation of part of *Leçons d'anatomie comparée*, Paris 1800–05, by William Ross.

892. —— [The animal kingdom].

1 Griffith *The animal kingdom arranged in conformity with its organization.* 4to, Roy. or demy 8vo, 16 vols, 799 col. pls, Whittaker, London [1824–] 1827–35; index and synopsis 1844. 4to copies have plain plates on India paper; Roy. 8vo only are coloured. This translation is much edited, by Edward Griffith et. al. Vol. 14/15 has no plate 139.

2 [McMurtrie] *The animal kingdom arranged according to its organization.* 8vo, 4 vols & 4 vols atlas, 743 pls mostly col., 7 portr., Henderson, London [1833–] 1834–36 [–1837]. First issue is New York, 1831.

3 [McMurtrie abridged] *Cuvier's Animal kingdom . . . abridged for the use of students.* 8vo, xx+508 pp, text figs, Orr, London 1834. First edition is New York 1833.

4 Macgillivray *The animal kingdom of the Baron Cuvier, enlarged and adapted to the present state of zoological science.* Fol., 2 vols, 128 col. pls, Edinburgh [1835–] 1839–40. Forms Vols 1 & 2, all publ., of *Edinb. Jl nat. Hist.* Translated by William Macgillivray

5 *Cuvier's Animal kingdom arranged according to its organization.* Roy. 8vo, vii+670 pp, text figs, Orr, London 1840; ed. 36 col. pls, text figs, Orr, London 1849. Many later editions.
The above are translations or derivatives of *Le regne animale*, Paris 1817; 2ed. 1829.
—— *see* Bowdich, T. E.
—— *see* Lacépéde, B. G. E., *Conte de*

893. CUVIERIAN SOCIETY OF CORK *Contributions towards a fauna and flora of the County of Cork.* 8vo, 3 pts, iv+24+24+v+130 pp, Van Voorst, London 1845. Prepared for British Association meeting by Harvey, J. R., Humphreys, J. D. & Power, T.

# D

D–. *The botanic garden; a poem.* 1791, 89. *see* Darwin, E.
D., P. B. *On the voice of birds.* [?1830]. *see* Duncan, P. B.
D., R. *Elements of the science of botany.* 1809. *see* Duppa, R.
D., R. W. *Principia botanica.* 1787. *see* Darwin, R. W.
Da Costa, Emanuel Mendes *see* Mendes da Costa, E.

894. DALE, Charles William *The history of Glanville's Wootton, in the County of Dorset, including its zoology and botany.* Sm. 8vo, viii+392 pp, 2 pls, Hatchard's, London 1878.
895. —— *The Lepidoptera of Dorsetshire.* 8vo, 8+76 pp, Ling, Dorchester 1886; 2ed. 1891.
896. —— *The history of our British butterflies.* 8vo, xli+232 pp, Kempster, London [1890]. Originally published as a suppl. to *The Young Naturalist*, 1887.

Dale, Samuel *see* Taylor, S.

897. DALLAS, William Sweetland *A natural history of the animal kingdom.* Sm. 8vo, 820 pp, text figs, Houlston & Stoneman, London 1856; ed. 1860. Originally published in Orr, W. S. *editor Circle of the sciences*, Vol. 3, 1855.
898. —— *Elements of entomology.* 12mo, vi+424 pp, text figs, Van Voorst, London 1857. Later editions.

—— *editor Mrs Loudon's Entertaining naturalist.* 1867; ed. 1889, title *Loudon's Natural history. see* Boreman, T.

899. —— *and others Short studies from nature.* 8vo, vi+335 pp, 1 col. pl., text figs, Cassell, London [?1885].

900. DALYELL, *Sir* John Graham *Baronet Observations on some interesting phenomena in animal physiology exhibited by several species of Planariae.* Roy. 8vo, ix+150 pp, 1 col. pl., Constable, Edinburgh 1814.

901. —— *Rare and remarkable animals of Scotland.* 4to, 2 vols, 110 col. pls, 1 portr., Van Voorst, London 1847–48.

902. —— *The power of the creator displayed in the creation.* 4to, 3 vols, 145 col. pls, 1 portr, Van Voorst, London 1851–58.

903. DAMON, Robert *Handbook on the geology of Weymouth . . . with notes on the natural history.* 8vo, xii+199 pp, 1 col.+1 plain pls, 1 col. map, Stanford, London 1860; suppl. 4 pp, 9 pls, 1860; ed. 1880 title *Geology of Weymouth*; suppl. 1880; ed. Weymouth 1884; suppl. 1888.

904. DANIELL, William *Interesting selections from animated nature.* Obl. fol., 2 vols, 120 pls, Cadell for the author, London [1809]. Plates mostly from Wood, W. *Zoography*, 1807.

Darley Dale *pseudonym see* Steele, F. M.

905. DARLINGTON, Ralph *The birds, wild flowers, ferns, mosses and grasses of North Wales.* Sm. 8vo, 5 pts, Darlington, Llangollen [1895]; ed. London 1896. Darlington's Handbooks.

*906. DARWIN, Charles Robert *A monograph of the sub-class Cirripedia.* 8vo, 2 vols, 3 col. +39 plain pls, text figs, Ray Society, London 1851, 54; reprint 1965. Plates Vol. 1, I–X, 3*, Vol. 2, I–XXX, 3*. Ray Society Publ. 21 & 25 for 1851 & 1853.

907. —— *On the origin of species by natural selection, or the preservation of favoured races in the struggle for life.* 8vo, ix+512 pp, 1 plan, Murray, London 1859; 5th thousand 1860; 3ed. 1861; 4ed. 1866; 5ed. 1869; 6ed. 1872 title *The origin of species*; ed. 1875. Very many later editions and continuously in print. Facsimilies 1964, 1969.

908. —— *On the various contrivances by which British and foreign orchids are fertilised by insects.* 8vo, vi+365 pp, 1 pl, text figs,

* The editions, issues and reprints of the works of C. R. Darwin are too numerous to enter here; only editions and issues altered by him, and facsimiles of first editions are entered here.

Murray, London 1862; 2ed. 1877 title *The various contrivances by which orchids are fertilised by insects.* Later editions.

909. —— *On the movements and habits of climbing plants.* 8vo, 118 pp, text figs, Taylor & Francis for the Linnean Society, London 1865; 2ed. 1875; ed. 1882. First edition is reprint from *Proc. Linn. Soc. Lond., Bot.* Vol. 9, and also sold as a book by Longman & Williams & Norgate. Later editions. Facsimile 1969.

910. —— *The variation of animals and plants under domestication.* 8vo, 2 vols, text figs, Murray, London 1868; issue 1868; 2ed. 1875. Later editions. Facsimile 1969.

911. —— *The descent of man and selection in relation to sex.* 8vo, 2 vols, text figs, Murray, London 1871; 3 other issues 1871, 2ed. 1 vol. 1874; 1875; 1877. Many later editions. Facsimile 1969.

912. —— *The expression of the emotions in man and animals.* 8vo, vi+374 pp, 7 pls, text figs, Murray, London 1872; 1873; 2ed. 1890. Later editions. Facsimile 1969.

913. —— *Insectivorous plants.* 8vo, x+462 pp, text figs, Murray, London 1875; 2 other issues 1875; 2ed. 1888. Later editions. Facsimile 1969.

914. —— *The effects of cross and self-fertilization in the vegetable kingdom.* 8vo, viii+482 pp, text figs, Murray, London 1876; 2ed. 1878. Later editions. Facsimile 1969.

915. —— *The different forms of flowers on plants of the same species.* 8vo, viii+352 pp, text figs, Murray, London 1877; 2ed. 1880; issue 1884. Later editions. Facsimile 1969.

916. —— *The power of movement in plants.* 8vo, x+592 pp, text figs, Murray, London 1880. Later editions. Facsimile 1969.

917. —— *The formation of vegetable mould through the action of worms.* 8vo, vii+326 pp, text figs, Murray, London 1881; 3 other issues 1881; issue 1882; issue 1883. Later editions. Facsimile 1969.

—— *An essay on instinct.* 1883 *see* Romanes, G. J. *Mental evolution in animals.*

918. DARWIN, Erasmus *The loves of the plants.* 4to, 184 pp, 7 pls, Jackson, Lichfield 1789. Second and later editions form Pt 2 of *The botanic garden.*

919. —— *The botanic garden; a poem.* 4to, 2 pts, 21 pls, Johnson, London 1791, 90; Facsimile 1973; ed. 1791; 3ed. 1795, 94; ed. Dublin 1793–96; 4ed. 8vo 1799. In 1791, 90 edition Pt. 2

is 2ed; in 1791 edition Pt. 2 is 3ed. in 3ed. Pt. 2 is 4ed. For 1ed. of Pt. 2 see No. 918. Later 8vo editions.

920. —— *Zoonomia; or, laws of organic life.* 4to, 2 vols, 6 col. pls, Johnson, London 1794–96; 2ed. of Vol. 1 1796; ed. 8vo 2 vols Dublin 1794–96; ed. 8vo 2 vols Dublin 1800; 3ed. 8vo 4 vols 1801.

921. DARWIN, Francis *The elements of botany.* 8vo, xv+235 pp, text figs, University Press, Cambridge 1895; ed. 1896; ed. 1899.

922. [DARWIN, Robert Waring] *Principia botanica; or, a concise and easy introduction to the sexual botany of Linnaeus.* 8vo, vii+280 pp, Allin, Newark 1787; 2ed. 1793; 3ed. 1810 signed. Preface signed R.W.D.

923. *The Darwinian theory examined.* 8vo, viii+131 pp, Bickers, London 1878.

*The Darwinian theory of the transmutation of species examined.* 1867. *see* Beverley, R. M.

924. DAUBENY, Charles Giles Bridle *The study of botany.* 8vo, Oxford 1834.

925. —— *Remarks on the final cause of the sexuality of plants.* 8vo, 34 pp, 1 pl., Parker, Oxford 1860.

926. —— *Climate . . . its influence on vegetable life.* 8vo, 4+vi+144 pp, 1 plan, 1 map, text figs, Parker, Oxford 1863.

927. —— *Miscellanies.* 8vo, 2 vols, Parker, Oxford 1867.

Davies, Arthur Morley *see* Wells, H. G.

928. DAVIES, George Christopher *Mountain, meadow, and more; a series of outdoor sketches . . . of natural history.* 8vo, viii+226 pp, 4 pls, text figs, King, London 1873; 2ed. 1874.

929. —— *The swan and her crew, or the adventures of three young naturalists and sportsmen on the broads and rivers of Norfolk.* 8vo, xx+267 pp, text figs, Warne, London [1876]; 2ed. [1877]; 3ed. [1877]. 4ed. [1878]. Later editions.

930. —— *Norfolk broads and rivers, or the waterways, lagoons and decoys of East Anglia.* 8vo, viii+290 pp, pls, Blackwood, Edinburgh 1883.

931. DAVIES, Hugh *Welsh botanology. Part the first. A systematic catalogue of the native plants of the Isle of Anglesey . . . Part II. Llysieuiaith Gymreig yr ail ran . . . Catalogue of the Welsh names of vegetables.* 8vo, 2 pts, xvi+152+xv, 153–255 pp, 1 pl., for the author, London 1813.

932. DAVIES, James Boyd *The practical naturalist's guide . . . for collecting, preparing and preserving specimens.* 8vo, iv+82

pp, text figs, Maclachlan & Stewart, Edin. 1858; ed. 1872.

933. DAVIES, Thomas *The preparation and mounting of microscopic objects.* Sm. 8vo, iv+156 pp, Hardwicke, London [1863]; 2ed. [1873]. Many later editions.

934. DAVIS, Edward *A list of the most interesting animal, vegetable, mineral . . . objects for the microscope.* 12mo, 11 pp, Cheltenham 1833.

935. DAVIS, George Edward *Practical microscopy.* 8vo, viii+335 pp, 1 col. pl., text figs, Bogue, London 1881; 2ed. 1882; ed. 1889.

936. DAVIS, James Richard Ainsworth *A text book of biology.* 8vo, xiii+462 pp, text figs, Griffin, London 1888; 2ed. 2 vols 1893; title *An elementary text book of biology.*

937. —— *The flowering plants as illustrating the first principles of botany.* 8vo, x+181 pp, text figs, Griffin, London 1890; 2ed. 1892; 3ed. 1900.

938. DAVIS, James William & LEES, Frederic Arnold *West Yorkshire: an account of its geology . . . and botany.* 8vo, 2 pts, xl+414 pp, 21 col. pls, 2 maps, text figs, Reeve, London 1878; 2ed. 1880.

939. [DAVY, *Sir* Humphry *Baronet*] *Salmonia: or days of fly fishing . . . with some account of the habits of fishes belonging to the genus Salmo.* Sm. 8vo, xiii+335 pp, 6 pls, Murray, London 1828; 2ed. 1829; 3ed. 1832 signed; 4ed. 1851. By an angler.

940. DAVY, John *Researches, physiological and anatomical.* 8vo, 2 vols, pls, Smith Elder, London 1839.

940a. —— *Physiological researches.* 8vo, viii+448 pp, Williams & Norgate, London 1863. Largely corrected reprints of papers in serials, but some original.

941. DAWSON, John Frederic *Geodephaga Britannica. A monograph of the carnivorous ground beetles indigenous to the British Isles.* 8vo, xx+224 pp, 3 col. pls, Van Voorst, London 1854.

942. —— & CLARK, Hamlet *A rearrangement of the nomenclature and synonymy of those species of British Coleoptera which are comprised under the sections Geodephaga, Hydradephaga and part of Philhydrida.* 8vo, 10 pp, Van Voorst, London 1860.

943. DAY, Francis *The fishes of Great Britain and Ireland.* Roy. 8vo, 2 vols, 180 pls, Williams & Norgate, London 1880–84. Plates 1–179, 172*.

944. —— *On the food of fishes.* 8vo, 36 pp, Clowes, London 1883. Reprinted from *Great International Fisheries Exhibition*, Vol. VI.

945. —— *British and Irish Salmonidae.* Roy. 8vo, viii+298 pp, 9 col. & 3 plain pls, text figs, Williams & Norgate, London 1887. Also special edition, 25 copies.

946. DEAKIN, Richard *The botanist's manual: comprising a complete list of the British flowering plants and ferns.* 16mo, iv+107 pp, Groombridge, London [1836].

947. —— *Ferns of Great Britain.* 8vo, pls, Tunbridge Wells 1870.

948. —— *The flowering plants of Tunbridge Wells and neighbourhood.* 8vo, v+371 pp, 1 col. pl., text figs, Stidolph & Bellamy, Tunbridge Wells [1870–] 1871; 2ed. 1876.

949. —— & MARNOCK, Robert *Florigraphia Britannica; or engravings and descriptions of the flowering plants and ferns of Britain.* 8vo, 4 vols, 297 col. pls, Groombridge, London 1837–48; 2ed. 1857. Vols 2–4 are by Deakin alone.

De Bary, Heinrich Anton *see* Bary, Heinrich Anton de

950. DEBY, Julien Marc *Analysis of the diatomaceous genus Campylodisous: being a prelude to a monograph of the same.* Roy. 8vo, 96 pp, 15 pls, for the author, London [1891].

Decaisne, Joseph *see* Le Maout, E.

951. DE CRESPIGNY, Eyre Champion *A new London flora.* Sm. 8vo, xxiv+179 pp, Hardwicke, London 1877.

952. DEDE, James *The English botanist's pocket companion.* 12mo, vii+148 pp, Hatchard, London 1809.

953. DEERING, George Charles *Catalogus stirpium . . . or, a catalogue of plants naturally growing . . . in diverse parts of England, more especially about Nottingham.* 8vo, 231+6+24 pp, for the author 1738; another issue 1738. Second issue has cancel dedication.

De Gray, Thomas *see* Walsingham, *Baron*

954. [DELANY, *Mrs* Mary] *A catalogue of plants copyed from nature in paper mosaick.* 8vo, 47 ll, London [1778].

De Mandeville, Bernard *see* Keogh, J.

955. DEMOOR, Jean, MASSART, Jean & VANDERVELDE, Emile *Evolution by atrophy in biology and sociology.* Sm. 8vo, 322 pp, Kegan Paul, London 1899. International Scientific Series No. LXXXVII. Translation of *L'évolution régressive*, Paris 1897, by Lilian Chalmers Mitchell.

956. DENNE, John *The wisdom of God in the vegetable creation . . . a sermon.* 8vo, 30 pp, Knaplock, London 1730.

957. —— *The wisdom and goodness of God in the vegetable creation . . . a sermon.* 8vo, viii+29 pp, Pemberton, London 1733.

Dennes, George Edgar *see* Watson, H. G.

Denny, Alfred *see* Miall, L. C.

958. DENNY, Henry *Monographia Pselaphidarum et Scydmenidarum Britanniae.* 8vo, vi+74 pp, 14 col. pls, Wilkin, Norwich 1825.

959. —— *Monographia Anoplurorum Britanniae; or, an essay on the British species . . . of . . . Anoplura.* 8vo, xxi+262 pp, 26 col. pls, Bohn. London 1842.

960. [——] *List of the specimens of British animals in the collection of the British Museum: Part XI. Anoplura or parasitic insects.* 12mo, 51 pp. British Museum, London 1852.

961. DERHAM, William *Physico-theology; or, a demonstration of the being and attributes of God, from his works of creation.* 8vo, [xv]+483 pp, 1 pl., Innys, London 1713; 2ed. 1716 [=1714]; 2ed. 1714; 3ed. 1714; 4ed. 1716; 5ed. 1720; 6ed. 1723; 7ed. 1727; 8ed. 1732; 9ed. 1737; ed. 1739; 10ed. 1742; 11ed. 12mo Glasgow 1745; 11ed. 2 vols 12mo 1749; 12ed. 1754; 13ed. 1768; ed. 12mo Edinburgh 1773; ed. 2 vols 1786; ed. 2 vols 1798. Boyle Lectures 1711 & 1712.

—— *see* Albin, E. *A natural history of British insects.* 1724.

—— *see* Albin, E. *A natural history of birds.* 1738.

*A description of a great variety of animals and vegetables.* 1736. *see* Boreman, T.

962. *The description of a new universal microscope . . . also the improved solar microscope . . . to which is added a catalogue of all the principal microscopic objects.* 8vo, 16 pp, 2 pls, Gregory & Wright, London 1786.

*A description of above three hundred animals.* 1787. *see* Boreman, T.

*A description of Bradgate Park.* [?1829]. *see* Bloxham, A.

963. *A description of England and Wales. Containing a particular account of each County, with its antiquities . . . plants.* Sm. 8vo, 10 vols, 218 pls, text figs, Newbery & Carnan, London 1769–70; 2ed. 1775. Title page calls for 240 pls.

*A description of more than three hundred animals.* 1826. *see* Boreman, T.

*A description of some curious and uncommon creatures, omitted.* 1739. *see* Boreman, T.

964. *A description of some curious and wonderful animals.*

*Intended as an introduction to the study of natural history.* 12mo, 72 pls, Pratt, Coventry [c.1830].

*A description of three hundred animals.* 1730. *see* Boreman, T.

965. DESVIGNES, Thomas *Catalogue of the British Ichneumonidae in the British Museum.* 12mo, 120 pp, British Museum, London 1856.

966. DE TABLEY, John Byrne Leicester Warren, *Baron The flora of Cheshire.* 8vo, cxiv+399 pp, 1 portr., 1 map, Longman, London 1899.

967. DE VARIGNY, Henry *Experimental evolution.* Sm. 8vo, ix+271 pp, Macmillan, London 1892.

968. DEWAR, George Albemarle Bertie *Wild life in Hampshire highlands.* Roy. or Sm. 8vo, x+304 pp, 2 col.+5 plain pls, Dent, London 1899. Haddon Hall Library Series. Royal 8vo 150 copies.

969. DEWHURST, Henry William *A dictionary of anatomy and physiology. Part 1.* 8vo, vii+160 pp, Callow & Wilson for the author, London 1827. All publ.

970. —— *A lecture introductory to the study of anatomy and physiology.* 8vo, 20 pp, the author, London 1827; 2ed. 1830 title *A dissertation on the component parts of an animal body.*

971. —— *An essay on the study of natural history. Adapted for the perusal of youth.* 32mo, 22 pp, the author, London 1831.

972. —— *Observations on the comparative anatomy of the Mus musculus or common mouse.* 12mo, 14 pp, Alexandre for the author, London 1831. Reprinted from *The Scientific Gazette.*

Diagrams of natural history. (SPCK., or Stanford). *see*
Baird, W. *Mammalia.* [c.1862].
Bell, T. & Baird, W. *Reptiles and Amphibia.* [1862].
Gosse, P. H. *Fishes.* [1856].
Gray, G. R. *Birds.* [c.1860].
Hanfrey, A. *Vegetable kingdom.* 1855.
White, A. *Insects.* [c.1860].
—— *Myriapods, Arachnida, Crustacea.* [c.1860].
Woodward, S. P. *Mollusca.* [c.1860].

Dialect Society. *see* Friend, H.

—— *see* Swainson, C.

*Dialogues on botany for the use of young persons.* 1819. *see* Edgeworth *Miss* M.

973. *Dialogues on curious subjects in natural history.* 12mo, viii+151 pp, text figs, Darton, London 1815.

974. *Dialogues on entomology, in which the forms and habits of insects are familiarly explained.* 12mo, xii+408 pp, 25 col. pls, text figs, Hunter, London 1819.

975. *Dialogues on the microscope intended for the instruction and entertainment of young persons.* 12mo, 2 vols, 10 pls, Baldwin, Cradock & Joy, London.

976. DICK, Archibald Hastie *Outlines of natural history.* 8vo, 192 pp, Collins, London 1874. Collins' School Series.

Dick, Robert *see* Smiles, S.

977. DICK, Thomas *The telescope and the microscope.* 12mo, RTS, London [1850]. Monthly Volume No. 67. See also No. 978.

978. —— *The atmosphere and its phenomena, and the telescope and the microscope.* 12mo, RTS, London [1851]. Monthly Series, double volume. See also No. 977.

979. DICKIE, George *Flora Abredonensis: comprising a list of the flowering plants and ferns . . . of Aberdeen.* 8vo, v+70 pp, Gray, Aberdeen 1838.

980. —— *The botanist's guide to the Counties of Aberdeen, Banff and Kincardine.* Sm. 8vo, xxxii+344 pp, 2 col. maps, Brown, Aberdeen 1860.

981. —— *A flora of Ulster, and botanist's guide to the North of Ireland.* 16mo, xix+176 pp, Aitchison, Belfast 1864.

—— *see* M'Cosh, J.

982. DICKINSON, Joseph *Flora of Liverpool.* 8vo, 166 pp, Deighton & Laughton, Liverpool 1851; suppl. 24 pp, 1855. Suppl. in *Proc. lit. phil. Soc. Liverpool*, Vol. 9.

983. DICKSON, James *J. Dickson fasciculus primus* [*-quartus*] *plantarum cryptogamicarum Britanniae.* 4to, 4 fasc., 111 pp, 12 pls, Nicol, London 1785–1801; ed. Lisbon 1800.

984. DICQUEMARE, Jacques François *An essay towards elucidating the history of the sea-anemonies.* 4to, 44 pp, 2 pls, Bowyer printed, London 1774; *A second essay . . .*, 4to, 42 pp, 1 pl., 1775; *A third essay*, 1777. All reprinted from *Phil. Trans.*

985. *A dictionary of natural history; or, a complete summary of zoology.* 16mo, xxxii+[496] pp, 48 col. pls, Scatcherd, London 1802: ed. 1815.

986. DIGBY, *Sir* Kenelm *A discourse concerning the vegetation of plants.* 12mo, 100 pp, Dakins, London 1661; ed. 1669.

987. DILLENIUS, Johann Jacob *Historia muscorum.* 4to, xvi+576 pp, 85 pls, Sheldonian, Oxford 1741; ed. Edinburgh

1811; English ed. without main text, London 1763; ed. 1768. 250 copies, 50 on F.P. For index see Giseke, P. D.

988. DILLWYN, Lewis Weston *British Confervae.* 4to, 2 vols, 87 pp, 116 col. pls with letterpress, Phillips, London [1802–] 1809. Plates 1–39, 40–44 unnumbered, 45–107, 109, 108, A–G.

989. —— *A descriptive catalogue of Recent shells, arranged according to the Linnaean method.* 8vo, 2 vols, xii+1092 pp, Arch, London 1817.

990. —— *Memoranda relating to coleopterous insects found in the neighbourhood of Swansea.* Roy. 8vo, 75 pp, Murray & Rees printed, not published, Swansea [1829].

991. —— *Materials for a fauna and flora of Swansea and the neighbourhood.* Roy. 8vo, 44 pp, Rees printed, not published, Swansea 1848.

—— *An index to the Historia conchyliorum of Lister.* 1823. *see* Lister, M. L. *Historia conchyliorum.* 1685–92.

—— *see* Turner, D.

992. *Directions for catching and preserving insects, in England or foreign countries.* Fol., 2 11, text figs, London [c.1815].

*Directions for collecting and preserving animals.* 1835. *see* Royal College of Surgeons of England.

*Directions for collecting and preserving plants.* [ ]. *see* Brown, R. *of the British Museum.*

*Directions for collecting zoological specimens.* [ ]. *see* Children, J. G.

*Directions for preparing small mammal skins.* 1898. *see* Thomas, M. R. O.

*Directions to the Captains of ships. see* Mawe, J.

*A discourse on the emigration of British birds.* 1780. *see* Legg, J.

*District flora, as compiled by the Ashton-under-Lyme Botanical Society.* 1888. *see* Ashton-under-Lyme Botanical Society.

993. DIXON, Charles *Rural bird life; being essays on ornithology, with instructions for preserving objects relating to that science.* Sm. 8vo, xiv+374 pp, 1 col. pl., text figs, Longman, London 1880.

994. —— *A contribution to the philosophy of birds' nests and eggs.* 8vo, 24 pp, text figs, Taylor & Francis printed, London 1884.

995. —— *Evolution without natural selection.* 8vo, 80 pp, Porter, London 1885.

996. —— *Our rarer birds.* Sm. 8vo, xiv+373 pp, 1 pl., text figs, Bentley, London 1888.

997. —— *Annals of bird life.* 8vo, viii+352 pp, 4 pls, Chapman & Hall, London 1890.

998. —— *Stray feathers from many birds.* 8vo, viii+231 pp, 1 pl., Allen, London 1890.

999. —— *The birds of our rambles, a companion for the country.* 8vo, x+429 pp, 1 pl., text figs, Chapman & Hall, London 1891.

1000. —— *Idle hours with nature.* 8vo, xii+278 pp, Chapman & Hall, London 1891.

1001. —— *The migration of birds; an attempt to reduce avian season-flight to law.* 8vo, xvi+300 pp, tab., text figs, Chapman & Hall, London 1892; ed. 1895 title *The migration of British birds*; ed. 1897.

1002. —— *The game birds and wild fowl of the British Islands.* 8vo, xv+468 pp, 13 col. pls, Chapman & Hall, London 1893; 1895; 2ed. 4to, Sheffield 1900.

1003. —— *Jottings about birds.* 8vo, vi+239 pp, 1 col. pl., Chapman & Hall, London 1893.

1004. —— *The nests and eggs of British birds.* 8vo, xii+371 pp, Chapman & Hall, London 1893; ed. 1894.

1005. —— *The nests and eggs of non-indigenous British birds.* 8vo, xii+368 pp, Chapman & Hall, London 1893; ed. 1894.

—— *The migration of British birds.* 1895. *see* No. 1001.

1006. —— *British sea birds.* 8vo, ix+295 pp, 8 pls, Sands & Foster, London 1896.

1007. —— *Curiosities of bird life.* 8vo, xii+322 pp, Redway, London 1897.

1008. —— *Our favourite song birds.* 8vo, x+287 pp, Lawrence & Bullen, London 1897.

1009. —— *Lost and vanishing birds.* 8vo, 2 pts, 295 pp, 10 pls, Macqueen, London 1898.

1010. —— *Bird life in a southern County, being eight years gleanings among the birds of Devonshire.* 8vo, 303 pp, 10 pls, 1 portr., Walter Scott, London 1899.

1011. —— *Among the birds in northern Shires.* 8vo, x+303 pp, 1 col. pl., text figs, Blackie, London 1900.

1012. —— *The story of the birds.* Sm. 8vo, xiv+304 pp, 1 col. pl., Allen, London 1900.

1013. DIXON, George *Hand book to the herbarium, containing botanical and English names of all the British flowering plants and ferns.* 12mo, Simpkin, London 1845; 2ed. 1888. title *The classbook of botany and hand-book.*

1014. —— *Classified list of British mosses.* Fol., single sheet, [Great Ayton 1857].

—— *The classbook of botany, and hand-book.* 1888. *see* No. 1013.

1015. —— & WATSON, John William *A descriptive manual of British land and fresh-water shells.* 8vo, x+88 pp, 8 col. pls, for the authors, Darlington 1858.

1016. DIXON, Hugh Nevill *The student's handbook of British mosses.* 8vo, xlvi+520 pp, 60 pls, Sumfield, Eastbourne 1896; 2ed. 1904; ed. 1924; reprint of 3ed. 1954.

1017. —— *'Handbook' catalogue of British mosses.* 8vo, 15 pp, Sumfield, Eastbourne 1897.

1018. —— *'Handbook' label list of British mosses.* 8vo, 26 pp, Sumfield, Eastbourne [1897].

1019. DIXON, John Henry *Gairloch in north-west Ross-shire; its . . . inhabitants and natural history.* 8vo, xlviii+435 pp, 3 pls, 1 map, text figs, Co-Operative Printing Co., Edinburgh 1886.

1020. DOBIE, William Henry *Birds of West Cheshire, Denbighshire and Flintshire.* 8vo, 1 map, Chester 1898. Reprinted from *Proc. Chester Soc. nat. Sci.* 1894.

1021. DOBSON, George Edward *A monograph of the Insectivora, systematic and anatomical.* Pt. 1 [–Pt. 3]. 4to, iv+72 pp, 28 col. pls, Van Voorst, London 1882 [–1890]. All publ.

1022. DODD, James Solas *An essay towards a natural history of the herring.* 8vo, 178 pp, 1 pl., Vincent, London 1752.

1023. DODOENS, Rembert *A nieuwe herball or historie of plantes.* Fol., 779 pp, 1 portr., text figs, Dewes, London (Antwerp printed) 1578; ed. 4to 1586; ed. 8vo 1595; ed. 1619; abridged ed. 4to 1606 title *Rams little Dodoen.* Translation of *Cruyde boeck*, Antwerp 1554, by Henry Lyte.

—— *see* Gerard, John *surgeon.*

*The domestic habits of birds.* 1833. *see* Rennie, J.

1024. DON, David *Descriptions of several new or rare native plants found in Scotland.* 8vo, [?Edinburgh ?1815].

1025. —— *Outlines of a course of lectures on botany.* 8vo, 16 pp, London 1836.

1026. DON, George *A general system of gardening and botany.* 8vo, 4 vols, Rivington, London 1831–38. Alternative title *General history of dichlamydeous plants.*

1027. DONCASTER, Arthur *Catalogue of British Lepidoptera.* 8vo, 27 pp, Reeve, London 1887. Printed on one or both sides of the paper.

Donisthorpe, Horace St John Kelly *see* Fowler, William Weekes.

1028. DONKIN, Arthur Scott *The natural history of the British Diatomaceae.* Roy. or demy 8vo, 3 pts, 74 pp, 12 pls, Van Voorst, London [1870–73]. All publ. Natural History of the British Isles Series.

1029. [DONOVAN, Edward] *An essay on the minuter parts of plants in general.* 4to, 5 pts, ii+26+6 pp, 14 col pls, London [1789–90]. All publ. Plates signed E. O. Donovan.

1030. —— *The natural history of British insects.* Roy. 8vo, 16 vols, 569 col.+7 plain pls, Rivington, London 1792–1813; ed. Vol. 1 1793, ed. 1 vol. 1802.

1031. —— *Instructions for collecting and preserving various subjects of natural history.* Roy. 8vo, 86 pp, 2 pls, Rivington for the author, London 1794; 2ed. 1805.

1032. —— *The natural history of British birds.* Roy. 8vo, 10 vols, 244 col. pls with letterpress, for the author and Rivington, London 1794 [–1819]; issues Vols 1–5 1799; issue 1815–20.

1033. —— *The natural history of British shells.* Roy. 8vo, 5 vols, 180 col. pls, for the author and Rivington, London [1799–] 1800–04.

1034. —— *The natural history of British fishes.* Roy. 8vo, 5 vols, 120 col. pls, Rivington, London 1802–08; ed. 1845.

1035. —— *The natural history of British quadrupeds.* Roy. 8vo, 3 vols, 72 col. pls, Rivington, London 1820; ed. 1845.

1036. —— *The natural history of the nests and eggs of British birds.* Obl. 8vo, 5 pts, 17 col. pls, for the author, London 1826. All publ.

1037. DOUBLEDAY, Edward & WESTWOOD, John, Obadiah *The genera of diurnal Lepidoptera.* Fol., 2 vols, 85 col.+1 plain pl, Longman, London 1846–52.

1038. DOUBLEDAY, Henry *A nomenclature of British birds; being a systematic catalogue.* Roy. 8vo, iv+28 pp, Westley, London 1836; 2ed. 1839; 3ed. 1841; 4ed. 1845.

1039. —— *A synonymic list of British Lepidoptera.* 8vo, 27 pp, Van Voorst, London [1847–] 1850; 2ed. 1859 title The *Zoologist Synonymic list of British butterflies and moths*; 2ed. 1866; 1st suppl. 2 pp, 1865; 2nd suppl. 2 pp, 1866; 3rd suppl. [4] pp, 1872. Printed on one side. The 1866 ed. is the same as 1859 with the 2nd suppl. added. The 2nd suppl. contains the 1st.

1040. DOUGLAS, James *Lilium Sarniense: or, a description of the Guernsay-lilly.* Fol., 2 pts, vi+35+liv+26 pp, 2 pls, Strahan, London 1725; 2ed.1729; 2ed. 1737.

1041. DOUGLAS, John William *The world of insects: a guide to its wonders.* 12mo, x+244 pp, Van Voorst, London 1856.

1042. —— & SCOTT, John *The British Hemiptera. Vol. 1. Hemiptera-Heteroptera.* 8vo, xii+628 pp, 21 pls, Ray Society, London 1865. All publ. Ray Society Publ. No. 37 for 1865.

1043. —— & —— *A list of British Hemiptera (Heteroptera).* 8vo, 16 pp, Entomological Society of London, London 1895.

1044. —— & —— *A catalogue of British Hemiptera; Heteroptera and Homoptera.* 8vo, viii+99 pp, Napier for the Entomological Society of London, London 1876.

—— *see* Stainton, H. T.

1045. DOWDEN, Richard *Walks after wild flowers; or the botany of the bohereens.* Sm. 8vo, viii+232 pp, 1 col. pl., Van Voorst, London 1852.

1046. DOWKER, George *A tabulated list and description of the birds of East Kent.* 8vo, 42 pp, Gurney & Jackson, London [1890]. Reprinted from *Trans. N. E. Kent nat. Hist. Soc.* 1889.

*Down the lane and back again, in search of wild flowers.* [1895]. *see* Cooke, M. C.

1047. DRAPER, Bourne Hall *The juvenile naturalist; or, walks in the country in spring, summer, autumn, and winter.* 16mo, xii+234+viii+236 pp, text figs, Bohn, London 1840. Later editions.

1048. —— *Studies of the animal world; arranged so as to form a systematic introduction to zoology.* 16mo, xv+462 pp, text figs, Darton & Clark, London [1842]. Child's Library.

1049. —— *Sketches from the volume of creation, as displayed in the seasons of the year.* 8vo, 202 pp, London 1830; 2ed. 1834.

1050. [DRAPER, B. T.] *Graphic microscopy. Illustrations drawn from nature.* 8vo, 2 pts, London 1856. All publ.

1051. DRESSER, Christopher *The rudiments of botany, structural and physiological.* 8vo, xxiii+433 pp, text figs, Virtue, London 1859.

1052. —— *Unity in variety as deduced from the vegetable kingdom.* 8vo, xv+162 pp, text figs, Virtue, London 1859; ed. 1860.

1053. —— *Popular manual of botany.* 8vo, viii+233 pp, 12 col. pls, Edinburgh 1860.

1054. —— *The art of decorative design, with an appendix giving the hours of the day at which flowers open (the floral clock).* 8vo, xi+241 pp, 28 col. pls, Day, London 1862.

1055. DRESSER, Henry Eeles *A history of the birds of Europe*. Roy. 4to, 8 vols, 632 col. + 2 plain pls, by the author, London 1871–82; suppl. 89 col. pls., 1895–96. Suppl. is also called Vol. 9. Pls 1–722, 599 bis.

1056. —— *A list of European birds*. 8vo, 40 pp, by the author, London 1881.

1057. DRUCE, George Claridge *The flora of Oxfordshire*. Sm. 8vo, lii+451 pp, 1 col. map, Parker, Oxford 1886; 2ed. 1927.

1058. —— *The flora of Berkshire*. Sm. 8vo, cxcix+644 pp, 1 map in pocket, Clarendon, Oxford 1897.

1059. DRUERY, Charles T. *Choice British ferns: their varieties and culture*. 8vo, iv+167 pp, 7 pls, Upcott Gill, London [1888].

1060. DRUMMOND, James Lawson *Thoughts on the study of natural history; and on the importance of attaching museums of the productions of nature to national seminaries of education*. 12mo, 122 pp, Finlay, Belfast, 1820.

1061. —— *First steps to botany*. 12mo, viii+289 pp, text figs, Longman, London 1823; 2ed. 1826; 3ed. 1831; 4ed. 1835.

1062. —— *Letters to a young naturalist on the study of nature and natural theology*. Sm. 8vo, 342 pp, text figs, Longman, London 1831; 2ed. 1832.

1063. —— *First steps to anatomy*. 12mo, 12 pls, Van Voorst, London 1845.

1064. —— *Observations on natural systems of botany*. 16mo, 100 pp, Longman, London 1849.

Drummond-Hay, Henry Maurice *see* Hay, H. M. D.

1065. DRURY, H. *The ferns of Derbyshire*. 8vo, 1861; ed. Derby 1877. ?A ghost.

1066. DUBOIS, Charles *An epitome of La Marck's arrangement of Testacea*. 8vo, xxx+317 pp, Longman, London 1824; ed. 1825.

1067. DUCK, John Nehemiah *The natural history of Portshead*. 12mo, 65 pp, 3 pls, 1 col. map, Bristol 1832.

Duckworth, William *see* Macpherson, H. A.

1068. DULCKEN, Henry William *Animal life all over the world*. Obl. 4to, 20 col. pls, Routledge, London [1870].

1069. DUNCAN, Henry *Sacred philosophy of the seasons; illustrating the perfections of God in the phenomena of the year: winter, spring, summer, autumn*. Sm. 8vo, 4 vols, 4 pls, Oliphant, Edinburgh 1836; 2ed. 1837; 3ed. 1838; 4ed. 1840–41. Later editions.

1070. DUNCAN, James *The natural history of beetles*. Sm. 8vo, vii+269 pp, 30 col. pls, 1 portr., Lizars, Edinburgh 1834; ed. 1845–46; 1848, 1860; 1876. Jardine's Naturalist's Library Vol. VIII; 2ed, etc. Vol. XXXIII.

1071. —— *The natural history of British butterflies*. Sm. 8vo, xv+246 pp, 34 col. + 2 plain pls, 1 portr., Lizars, Edinburgh 1835; ed. 1845–46; 1848; 1860; 1876. Jardine's Naturalist's Library Vol. X; 2ed. etc Vol. XXIX.

1072. —— *The natural history of British moths, sphinxes*. Sm. 8vo, xv+268 pp, 30 col. pls, 1 portr., Lizars, Edinburgh 1836; ed. 1845–46; 1848; 1860; 1876. Jardine's Naturalist's Library Vol. XIV; 2ed. etc. Vol. XXX.

1073. —— *Introduction to entomology*. Sm. 8vo, xv, 17–331 pp, 36 col. pls, 1 portr., Lizars, Edinburgh 1840; ed. 1845–46; 1848; 1860. Jardine's Naturalist's Library Vol. XXIX; 2ed. etc Vol. XXVIII.

—— *see* Wilson, J.

1074. DUNCAN, John *Birds of the British Isles*. Sm. 4to or Sm. 8vo, xvi+456 pp, text figs, Scott, London 1898. 350 copies. Reprinted from *The Newcastle Weekly Chronicle*.

1075. [DUNCAN, John Shute] *Botano-theology, an arranged compendium*. 8vo, iv+112 pp, Parker, Oxford 1825; 2ed. 1826 signed, title *Botanical theology*. 2ed. also published with Paxton, James *Illustrations of Paley's Natural theology*, 1826 with title *Natural theology . . . collected from the appearances of nature*.

1076. [——] *Analogies of organized beings*. 8vo, iv+157 pp, for the author, Oxford 1831.

1077. [DUNCAN, Philip Bury] *Introduction to the Catalogue of the Ashmolean Museum . . . Appendix. Specimen of Aristotle's History of animals*. 8vo, 59 pp, 1 pl., 5 tab., [Oxford 1826].

1078. [——] *On the voice of birds*. 8vo, 48 pp, 1 pl., [?London ?1830]. By P.B.D.

1079. [——] *A catalogue of the Ashmolean Museum, descriptive of the zoological specimens, antiquities, coins, and miscellaneous curiosities*. 8vo, viii+188 pp, 2 pls, Oxford 1836.

1080. DUNCAN, Peter Martin *editor Cassell's Natural History*. 4to, 6 vols, col. pls., Cassell, London [1876–82]. Several later issues.

1081. —— *Natural history rambles. The sea shore*. Sm. 8vo, 256 pp, text figs, SPCK, London 1879. Later eds. Natural History Rambles Series.

—— *see* Blanchard, E.

1082. DUNMAN, Thomas *A glossary of biological, anatomical and physiological terms*. 8vo, vi+161 pp, Griffith & Farran, London 1879; ed. 1889 title . . . *anatomical, physiological and biological terms*.

1083. —— *Talkes about science*. 8vo, xv+239 pp, Griffith & Farran, London [1882].

1084. DUNN, Robert *The ornithologist's guide to the Islands of Orkney and Shetland*. Sm. 8vo, x+128 pp, 1 pl., 2 maps, for the author, Hull 1837.

1085. DUNN, Stephen Troyte *Flora of south-west Surrey*. Sm. 8vo, xv+106 pp, West Newman, London 1893; ed. 1909.

1086. DUNSTER, Henry Peter *The young collector's handybook of botany*. 8vo, vi+168 pp, 16 pls, Reeve, London 1871.

1087. [DUPPA, Richard] *Elements of the science of botany, as established by Linnaeus*. 16mo, 2 vols, 124 col. pls, Murray, London 1809; 2ed. 1809; 3ed. 3 vols 1810; 1812. By R. D.

1088. —— *The classes and orders of the Linnaean system of botany*. 8vo, 3 vols, xxvii+549 pp, 237 col. + 3 plain pls, Bensley, London 1816.

1089. D'URBAN William Stewart Mitchell *A sketch of the natural history of the neighbourhood of Exeter*. 12mo, Exeter 1871; ed. 1876 title . . . *of South Devon*. Reprinted from Besley's *Handbook of Exeter*, 2ed. 1871.

1090. —— & MATHEW, Murray Alexander *Birds of Devon*. 8vo, lxxxvii+459 pp, 5 col. + 4 plain pls, 3 maps, Porter, London 1892; suppl. 31 pp, 1895; 2ed. 1895.

1091. DURHAM, William *Evolution, antiquity of man, bacteria*. 8vo, 127 pp, Black, Edinburgh 1890. Science in Plain Language Series.

1092. DURNFORD, W. Arthur *List of the birds found in the neighbourhood of Walney Island, with notes*. 8vo, 20 pp, Barnsley 1883.

Durrant, John Hartley *see* Walsingham, *Baron*

1093. DUTFIELD, James *A new and complete natural history of English moths and butterflies.* 4to, 6 pts, 12 pp, 12 col. pls, Payne, London 1748 [–1749]; reprint 1969. All publ. An almost unknown work.

1094. DUTTENS, Louis *An inquiry into the origins of the discoveries attributed to the moderns*. 8vo, 4 pts, London 1769. Translation of *Recherches sur l'origines des decouvertes attribuées aux modernes*, Paris 1766. For another edition see Wesley, J.

1095. DYCE, & SIM, G. *Catalogue of fish in the vicinity of Aberdeen . . . with English, scientific, and local names*. 8vo, 12 pp, n.p. [1878].

1096. DYER, Thomas Firminger Thiselton *The folk-lore of plants*. 8vo, 328 pp, Chatto, London 1889.

1097. DYER, *Sir* William Turner Thiselton *On the classification and sexual reproduction of the thallophytes*. 8vo, 23 pp, text figs, London 1875. Revision of articles in *Quart. Jl micr. Sci.*, 1875.

Dyke Road Museum, Brighton. *see* Booth, E. T.

1097a. DYSON, David *The land and freshwater shells of the districts around Manchester*. 12mo, xi+95 pp, 1 pl., Harrison, Manchester 1850.

# E

E., D. *Contributions to natural history chiefly in relation to the food of the people*. 1865. *see* Esdaile, D.

E., H. *Some account of Jura red deer*. 1891. *see* Evans, H.

E., J. *Sylva*. 1664. see Evelyn, J.

1098. EARL, Alfred George *The living organism: an introduction to the problems of biology*. Sm. 8vo, xiii+217 pp, text figs, Macmillan, London 1898.

1099. EARLE, John *English plant names from the tenth to the fifteenth century*. 16mo, cxii+122 pp, Clarendon, Oxford 1880.

1100. *An easy introduction to the study of the animal kingdom, according to the natural method of Cuvier*. Sm. 8vo, xviii+290 pp, 4 pls, text figs, Varty Educational Depository, London [1855].

Eaton, Alfred Edwin *see* McLachlan, R.

*Ebb and flow*. 1865. *see* Fraser, Robert William.

1101. ECKER, Alexander *Anatomy of the frog*. 8vo, xvi+449 pp, 2 col. pls, text figs, Clarendon, Oxford 1889; facsimile 1971. Translation of *Die Anatomie des Frosches*, Brunswick 1864–82. Translations of Foreign Biological Memoirs, Vol. 2, by George Haslan.

*The economy of vegetation (or phenomena of plants)*. 1838. *see* Murray, J.

1102. [EDGEWORTH, *Miss* Maria] *Dialogues on botany for the use of young persons*. 12mo, vii+467 pp, Hunter, London 1819. Attribution uncertain.

1103. EDGEWORTH, Michael Packenham *Pollen*. 8vo, 92 pp, 24 pls, Hardwicke, London 1877; 2ed. 1879.

1104. EDMONDS, Henry *Elementary botany, theoretical and practical*. 8vo, 207 pp, text figs, Longman, London 1882; 2ed. 1886. Many later editions. Elementary Science Manuals.

1105. —— *Botany for beginners*. 12mo, 117 pp, Longman, London 1896; ed. 1902.

1106. EDMONDS, Richard *The Land's End district; its antiquities, natural history*. 8vo, 270 pp, 6 pls, 1 map, text figs, Smith, London 1862.

1107. EDMONSTON, Arthur *A view of the ancient and present state of the Zetland Islands; including their civil, political and natural history*. 8vo, 2 vols, 1 map, Ballantyne, Edinburgh 1809.

1108. EDMONSTON, Biot & SAXBY, Jessie Margaret Edmonston *The home of a naturalist*. 8vo, 395 pp, 4 pls, portr., Nisbet, London 1888; 3ed. [ ].

1109. EDMONSTON, Thomas *A flora of Shetland*. Sm. 8vo, xxvii+67 pp, Clark, Aberdeen 1845; 2ed. 12mo, 1903.

Edward, Thomas *see* Smiles, S.

1110. EDWARDS, Arthur Mead *Life beneath the waters: or, the aquarium in America*. 8vo, xii+168 pp, 10 pls, text figs, Baillière, London 1858. First edition is USA 1858.

1111. EDWARDS, George *The natural history of some uncommon birds and of some other rare and undescribed animals*. 4to, 4 pts, 210 col. pls, portr., for the author, London 1743–51; ed. 1802–03. There are later editions dated 1743 but published c.1752 & 1776. It also occurs with the text in French. Parts are titled *A natural history of birds*. See also No. 1112.

1112. —— *Gleanings of natural history*. 4to, 3 pts, xxxv+vii+347 pp, 152 col. pls, for the author, London 1758–64; ed. 1805–06. There is a later edition of Pt. 1 which is dated 1758 but is actually c.1795, and of Pt. 2 dated 1760. The 1802–03 ed. of No. 1111 and the 1805–06 ed. of this also occur in folio 25 copies.

1113. —— *Essays upon natural history and other miscellaneous subjects*. 8vo, viii+231 pp, portr., Robson, London 1770. Based on No. 1112.

1114. —— *Some memoirs of the life and works of George Edwards*. 4to, 38 pp, 4 pls, Robson, London 1776. Published in No. 1112, 1776, or with the next item, but occurs separately. Contains reprints of articles published in *Phil. Trans. Roy. Soc.*

—— *A catalogue of the birds, beasts, fishes, insects, plants etc. contained in Edwards Natural history*. 1776. *see* Linnaeus, C.

[——] *attributed author. A discourse on the emigration of British birds*. 1814. *see* Legg, J.

1115. EDWARDS, Henri Milne- *A manual of zoology*. Sm. 8vo, xxiii+503 pp, text figs, Rensham, London 1856; 2ed. 1863. Translation of *Elémens de zoologie*, 2ed. Paris 1840.

1116. EDWARDS, Henry *Illustrations of the wisdom and benevolence of the Deity, as manifested in nature*. 12mo, Reeve, London 1845.

1117. EDWARDS, James *The Hemiptera-Homoptera (Cicadina and Psyllina) of the British Islands*. Roy. or demy 8vo, xii+271 pp, 28 col. + 2 plain pls, Reeve, London [1894–]1896. Demy copies have only the 2 plain pls.

1118. EDWARDS, John *The British herbal, containing one hundred plates of the most beautiful and scarce flowers . . . which blow in the open air of Great Britain*. Fol., 50 pp, 100 col. pls, Barnes for the author, [1768–]1770; issue 1775 title *Select collection containing* . . . Alternative title to first edition is *Edward's Herbal containing* . . .

1119. EDWARDS, W. & TOWNDROW, Richard Francis *The butterflies and moths of Malvern*. 8vo, viii+42 pp, Malvern 1899.

1120. EDWARDS, William Frederic *On the influence of physical agents on life*. 8vo, xv+488 pp, 1 pl., Highley, London 1832. Translation of *De l'influence des agen physique sur la vie*, Paris 1824, by Thomas Hodgkin & *Dr* Fisher.

1121. EDWARDS, Zachary James *The ferns of the Axe and its tributaries*. 8vo, 126 pp, 4 pls, text figs, Hamilton London 1862; ed. 1866. ? ed. Chard 1861.

1122. EIMER, H. G. Theodor *Organic evolution as a result of the inheritance of acquired characters according to the laws of organic growth*. 8vo, xxviii+435 pp, Macmillan, London 1890. Translation of *Die Entstehung der Arten*, Jena 1888, by Joseph Thomas Cunningham.

1123. *The elements of botany for families and schools*. Sm. 8vo, text figs, Parker for the SPCK, London 1833; 3ed. 1837. Many later editions with title *The elements of the science of botany*. 10ed. & 11ed. are signed by Thomas Moore; earlier eds are not by him.

*Elements of natural history, being an introduction to the Systema naturae of Linnaeus*. 1801–02. *see* Stewart, C.

*The elements of the science of botany. see The elements of botany.*

*Elements of zoology*. 1823. *see* Carpenter, Philip.

Elizabeth, Charlotte *see* Phelan, *Mrs* C. E.

1124. ELLACOMBE, Henry Nicholson *Common English names of plants*. 8vo, Bath 1870.

1125. ELLIOT, Daniel Giraud *The life and habits of wild animals*. Fol., or 4to, 72 pp, 20 col. pls, Macmillan, London 1874. Plates by Josef Wolf.

1126. ELLIOT, George Francis Scott *Flora of Dumfriesshire and Dumfries district*. 8vo, 2 pts, London [1891–]1892.

1127. —— *A flora of Dumfriesshire, including parts of the Stewartry of Kircudbright*. 8vo, xl+219 pp, 1 map, Maxwell, Dumfries 1896.

1128. ELLIOTT, Jannion Steele *The vertebrate fauna of Bedfordshire* [birds]. 8vo, 5 pts, viii+215 pp, for private circulation, Birmingham [1897–1901]. All publ. 150 copies.

1129. ELLIOTT, Mary *The book of birds*. 16mo, 24 col. pls with letterpress, Darton, London 1826.

1130. ELLIS, Alfred *Notes about birds*. Sm. 8vo, 50 pp, privately printed, Leicester 1868.

1131. ELLIS, John *An essay towards the natural history of the corallines*. 4to, vii+103 pp, 40 pls, for the author, London 1755. Pl. 38 absent in many, pl. 39 in most copies. Plate of microscope unnumbered.

1132. —— *On the nature of the Gorgonia; that it is a real marine animal, and not of a mixed nature, between animal and vegetable*. 4to, 19 pp, 2 pls, London 1776. Reprinted from *Phil. Trans.*

1133. —— & SOLANDER, Daniel Carl *The natural history of many curious and uncommon zoophytes*. 4to, xii+208 pp, 63 pls, White, London 1786.

1134. ELLIS, John William *The coleopterous fauna of the Liverpool district*. 8vo, 2 pts, Turner Routledge, Liverpool 1889. Reprinted from *Proc. Liverpool biol. Soc.*, Vols 2 & 3.

1135. —— *The lepidopterous fauna of Lancashire and Cheshire*.

8vo, 136 pp, McCorquodale, Leeds 1890. Reprinted from *The Naturalist*.

1136. EMERSON, Peter Henry *Wild life on a tidal water*. 4to, xiii+145 pp, text figs, Sampson Low, London 1890. 100 copies.

1137. —— *Birds, beasts and fishes of the Norfolk Broadlands*. 8vo, xix+396 pp, 68 pls, Nutt, London 1895; 2ed. 1895.

1138. EMSLIE, John *Diagram exhibiting a view of nature in all climates*. 4to, 4 pp, card triangle + 1 col. pl., London 1852.

1139. [——] *View of nature in ascending regions. Notes explanatory of the View of nature in ascending regions.* 4to, 4 pp, folding col. pl., London 1853.

1140. *The encyclopaedia of natural history*. 4to, 131 pls, text figs, Griffin. London 1848. Reprinted from *Encyclopaedia Metropolitana*, 2ed.

1141. *England illustrated, or a compendium of the natural history, geography, topography . . . of England and Wales*. 4to, 2 vols, 29 pls, 54 maps, Dodsley, London 1764.

1142. *English forests and forest trees*. 8vo, 406 pp, Ingram Cooke, London 1853. Illustrated London Library, No. 7.

*The English moths and butterflies.* 1747–60. *see* Wilks, B.

1143. ENGLISH, James Lake *A manual for the preservation of the larger Fungi (Hymenomycetes) in their natural condition . . . also a new process for the preservation of wild flowers*. 8vo, viii+41 pp, Davis, Epping 1882.

Ent, *Sir* George *see* Charleton, W.

Entomological Societies of Oxford and Cambridge. *see* Oxford and Cambridge Universities.

Entomological Society of London *see* Douglas, J. W. & Scott, J.; McLachlan, R.; Marshall, T. A.; Smith, F.; Westwood, J. O.

*Episodes of insect life*. 1849–51. *see* Budgen, *Miss* M. L.

Ephemera *pseudonym see* Fitzgibbon, E.

1144. [ESDAILE, David] *Contributions to natural history, chiefly in relation to the food of the people*. 8vo, viii+364 pp, Blackwood, Edinburgh 1865; 2ed. 1867 signed. By a rural D.D.

*An essay on comparative anatomy*. 1744. *see* Monro, A. *first of the name*.

*An essay on the minuter parts of plants in general.* [1789–90]. *see* Donovan, E.

*An essay towards a probable solution to this question. Whence come the stork and the turtle, the crane and the swallow*. 1703. *see* Morton, C.

Essex Field Club. *see* Christy, R. M.

1145. EVANS, Arthur Humble & BUCKLEY, Thomas Edward *A vertebrate fauna of the Shetland Islands*. Sq. 8vo, xxix+248 pp, 14 pls, 1 col. map, text figs, Douglas, Edinburgh 1899. Vertebrate Fauna of Scotland Series, Vol. 8.

1146. EVANS, Ernest *Botany for beginners*. 8vo, viii+290 pp, Macmillan, London 1899; ed. 1900.

1147. [EVANS, H.] *Some account of Jura red deer.* 8vo, 38 pp, 1 map, Derby 1891. By H. E.

1148. EVANS, John *A tour through part of North Wales in the year 1798, with a view to botanical researches*. 8vo, viii+416 pp, White, London 1800: 2ed. 1802.

1149. EVANS, William *The mammalian fauna of the Edinburgh district*. 8vo, 123 pp, M'Farlane & Erskine, Edinburgh 1892; suppl. 17 pp., 1906. 200 copies.

1150. EVANS, William Edward *The song of the birds; or, analogies of animal and spiritual life*. 12mo, viii+264 pp, Rivington, London 1845; 2ed. 1851; ed. 1888 title *The songs* . . .

1151. EVANS, William Frederick *British Libellulinae, or dragon flies.* 8vo, 28 pp, 19 col. + 2 plain pls, printed for private circulation, London 1845.

1152. [EVELYN, John] *Sylva: or, a discourse of forest trees . . . Pomona . . . Kalendarium hortense*. Fol., 3 pts, text figs, Martyn & Allestry, London 1664; 2ed. 1670 signed; 3ed. 1679; 4ed. 1706 title *Silva* . . .; 5ed. 1729; ed. 4to, York 1776; ed. 4to, York 1786; 3ed. 4to, York 1801; 4ed. 4to, York 1812; 5ed. 4to, York 1825; ed. 1908 title *Sylva* . . . The editions of the parts are not the same as those of the whole work. *Terra* and *Acetaria* are added to fourth and later editions. Edition of 1908 is from 4th.
—— *see* Mitchell, J.

1153. [EWART, James Cossar] *Syllabus of the Mammalia.* obl. 4to, 2 pts, 120 pp, 84 pls, 1 map, privately printed, [Edinburgh 188–]. Plates signed J.C.E.

1154. EWER, Samuel *Manuale sive compendium botanices*. 8vo, 470 pp, Bensley, London 1808.

1155. EWING, J. W. *Some materials for a flora of Wrotham and its neighbourhood.* 8vo, West Malling 1882.

1156. EWING, Peter *The Glasgow catalogue of native and established plants.* 8vo, Glasgow 1892; 2ed. 1899.

1157. *Exhibition of some of the wonders of God in the creation*. 12mo, 38 pp, 1 pl., text figs, Evans, London [c.1820].

*Explanations*. 1845. *see* Chambers, R.

*An expository outline of the Vestiges of the natural history of creation*. 1846. *see* Chambers, R.

1157a. EYTON, Thomas Campbell *A history of the rarer British birds*. Roy. or demy 8vo, vi+101 pp, text figs, Longman, London 1836.

1158. —— *A catalogue of British birds*. Roy. or demy 8vo, vi+67 pp, Longman, London 1836. Usually occurs at the end of No. 1157a.

1159. —— *Some account of a dredging expedition on the coast of the Isle of Man . . . 1852*. 8vo, [London 1852].

1160. —— *A catalogue of the skeletons of birds in his possession*. 8vo, 15 pp, London 1858.

1161. —— *A history of the oyster and the oyster fisheries*. 8vo, iv+40 pp, 4 tinted & 2 plain pls, Van Voorst, London 1858.

1162. —— *Osteologia avium; or, a sketch of the osteology of birds*. 4to, vi+x+229 pp, atlas vol. iv pp, 117 pls, Hobson, Wellington, Salop [1858–]1867; 1st suppl. 18 pls, 1869; 2nd suppl. 51 pls, London, [1873–75].

1163. [——] *List of plates of skeletons of birds that have been published*. 12mo, 15 pp, Wellington, Salop 1874.

# F

F., R. W. *Contributions towards a catalogue of plants indigenous to the neighbourhood of Tenby*. 1848. *see* Falconer, R. W.

F., S. M. *The four seasons; a short account of the structure of plants*. 1865. *see* Fitton, *Miss* S. M.

1164. FABRICIUS, Johann Christian *Nomenclator entomologicus; enumerans Insecta omnia in J. C. Fabricii Entomologia systematica*. 8vo, 62 11, Manchester 1796.

*The faculties of birds*. 1835. *see* Rennie, J.

1165. [FALCONER, Randle Wilbraham] *Contributions towards a catalogue of the plants indigenous to the neighbourhood of Tenby*. 8vo, 54 pp, Longman, London 1848. By R.W.F.

1166. FALCONER, William *Miscellaneous tracts and collections relating to natural history, selected from the principle writers of antiquity on that subject.* 4to, v+203 pp, Merrill, Cambridge 1793.

1167. *A familiar history of the British fishes*. Sm. 8vo, iv+317 pp, text figs, SPCK, London [1859].

1168. FARMER, John Bretland *Practical introduction to the study of botany. Flowering plants.* 8vo, viii+274 pp, text figs, Longman, London 1899; ed. 1902.

1169. FAUJAS DE SAINT-FOND, Barthélemi *Travels in England, Scotland, and the Hebrides* [*Staffa*]. 8vo, 2 vols, London 1799; ed. 12mo 1809; ed. Glasgow 1907. The edition of 1809 is in Mavor, W. F. *The British tourist's companion*, Vol. 5. Translation of *Voyage en Angleterre*, Paris 1797.

*Fauna and Flora of Haileybury*. Part 1. 1888. *see* Haileybury Natural History Society.

*The fauna of Scotland*. 1876. *see* Glasgow Natural History Society.

Favart, Émile *pseudonym. see* Hammond, T. W.

1170. FAWCETT, James William *The birds of Durham*. 8vo, 20 pp, Dent for the author, Consett 1890.

1171. FENN, *Lady* Elleanor *A short history of insects . . . designed as an introduction to the study of that branch of natural history, and as a pocket companion to those who visit the Leverian Museum*. 12mo, xxiv+107 pp, 7 col. pls, Stevenson & Matchett, Norwich [1797]; ed. [?1800].

1172. FENN, George Manville *Featherland: bird life in the spring at Greenlawn.* 16mo, Griffith & Farran, London 1866; ed. 1896.

1173. FENNELL, James Hamilton *The child's book of zoology*. 16mo, text figs, Thomas, London 1840.

1174. —— *Drawing room botany*. 8vo, 32 pp, 19 col. pls, Thomas, London 1840.

1175. —— *A natural history of British and foreign quadrupeds*. 8vo, xv+556 pp, text figs, Tegg, London 1841; 1843.

Féré, F. *see* Binet, A.

1176. FERGUSON, D. *The natural history of Redcar and its neighbourhood.* 12mo, 128 pp, 1 pl., Simpkin, London 1860.

1177. FERGUSON, John *The microscope; its revelations and applications in science and art*. 8vo, 160 pp, Hamilton, Edinburgh 1858.

1178. *Fern collector's album*. Fol., Hardwicke, London 1859. Text with spaces for insertion of specimens. L.P. issue without text.

*The ferns of Derbyshire*. 1880. *see* Howe, W. E.
*The ferns of Moffat*. 1893. *see* Moffatt, Jeannie C.

1179. FERRYMAN, Robert *A catalogue of the British quadrupeds and birds contained in the museum of Rev. R. Ferryman*. 8vo, Bristol 1789.

1180. [——] *A catalogue of the British quadrupeds and birds contained in the British Zoological Museum, Oxford Street*. 8vo, 24 pp, London 1795.

1181. *A few leaves represented by nature printing, showing the application of the art for the reproduction of botanical and other natural objects*. Fol., no text, 21 col. pls, Bradbury & Evans, London 1854.

*The field, the garden and the woodland*. 1838. *see* Pratt, Miss A.
Field, William *see* Snaith, W. A.

1182. FIELDING, Henry Cecil *Memories of Malling and its valley; with a flora and fauna of Kent*. 8vo, xiv+291 pp, 2 pls, 1 map, text figs, Oliver, West Malling 1893.

1183. FIGUIER, Guillaume Louis *The vegetable world*. 8vo, viii+576 pp, 24 pls, text figs, Chapman & Hall, London 1867; 1868; ed. [1872]. Translation of *Histoire des plantes*, Paris 1865, by W. L. Q.

1184. —— *The insect world*. 8vo, 519 pp, 12 pls, text figs, Chapman & Hall, London 1868; ed. 1872. Translation of *Les insectes*, Paris 1867, by Y.D.

1185. —— *The ocean world*. 8vo, xii+605 pp, 32 pls, text figs, Chapman & Hall, London 1868; ed. [1869]; ed. [1872]. Translation of *La vie et les moeurs des animaux . . . zoophytes et mollusques*, Paris 1866, by W. S. O.

1186. —— *Reptiles and birds*. 8vo, xiii+648 pp, 1 pl., text figs, Chapman & Hall, London 1870; ed. [1873]; 1892. Translation of parts of *Les poissons, les reptiles et les oiseaux*, Paris 1868, by Parker Gillmore.

1187. —— *Mammalia. Their various orders and habits popularly illustrated by typical species*. 8vo, xi+606 pp, 1 pl., text figs, Chapman & Hall, London 1870; issue [1870]; another translation 1875. Translation of *Les mammifères*, Paris 1869, by Edward Perceval Wright.

—— *Figuier's Popular scientific works*. 8vo, 7 vols, text figs, Cassell, London [1891–93]. A collected edition.

1188. *First natural history reader for standard II. Quadrupeds and birds*. 8vo, viii+84pp, Marshall, London 1882.

1189. *First steps in general knowledge*. 16mo, 5 vols, text figs,

SPCK, London 1848–50. Vol. 3 *The animal kingdom*; Vol. 4 *The vegetable Kingdom.*

1190. *First steps to the British flora: arranged according to the natural system*. 16mo, xv+144 pp, 2 col. pls, Wertheim & Macintosh, London [1855]. By the author of *Wild flowers and their teachings.*

1191. FISCHER, Alfred *The structure and functions of Bacteria*. Roy. 8vo, viii+198 pp, text figs, Clarendon, Oxford 1900. Translation of *Untersuchungen über den Bau der Cyanophyceen und Bakterien*, Jena 1897, by A. Coppen Jones.

1192. FISHER, Arthur Thomas *Outdoor life in England*. 8vo, ix+474 pp, Bentley, London 1896.

Fisher, William Richard *see* Gurney, J. H. *the father*

1193. FITCH, Walter Hood & Smith, Worthington George *Illustrations of the British flora . . . forming an illustrated companion to Mr. Bentham's Handbook.* Sm. 8vo, 328 pp, text figs, Reeve, London 1880; 2ed. 1887; 3ed. 1892; 4ed. 1916; 5ed. 1924. And later revisions. These are illustrations to what is better known as 'Bentham and Hooker'.

1194. [FITTON, Elizabeth & FITTON, Sarah Mary] *Conversations on botany.* 12mo, xvi+213 pp, 20 col. pls, Longman, London 1817; 2ed. 1818; 3ed. 1820; 4ed. 1823; 5ed. 1825; 6ed. 1828; 7ed. 1830; 7ed. 1831; 8ed. 1834; 9ed. 1840. Later printings. Also attributed to *Mrs* Jane Marcet, but not by her.

1195. [FITTON, Sarah Mary] *The four seasons; a short account of the structure of plants.* 16mo, 75 pp, 4 col. pls, text figs, Griffith & Farran, London 1865. By S. M. F.

—— *see* Fitton, *Miss* E.

1196. FITTON, William Henry *A geological sketch of the vicinity of Hastings. List of plants*. 12mo, viii+94 pp, 1 col. pl., text figs, Longman, London 1833.

1197. FITZGERALD, Charles Egerton *Semi-scientific lectures*. 8vo, 281 pp, English, Folkestone 1880.

Fitzerald, Francis *pseudonym. see* Taylor, C.

1198. FITZGERALD, Henry Purefoy *Dictionary of the names of British plants*. 8vo, 90 pp, Baillière, London [1885].

1199. [FITZGIBBON, Edward] *Book of the salmon: in two parts. Part 1. The theory and practice of fly-fishing. Part II. The natural history.* 16mo. xvi+242 pp, 9 pls, Longman, London 1850. By Ephemera *pseudonym*.

1200. FLEMING, George *Animal plagues: their history, nature and prevention*. 8vo, 2 vols: Vol. 1 Chapman & Hall, Vol. 2 Baillière, London 1871, 72.

1201. FLEMING, John *The philosophy of zoology; or, a general view of the structure, functions, and classification of animals*. 8vo, 2 vols, 5 pls, Constable, Edinburgh 1822.

1202. —— *A history of British animals*. 8vo, xxiii+565 pp, Bell & Bradfute, Edinburgh 1828; 2ed. 1842.

1203. —— *Molluscous animals, including shell fish*. 12mo, iv+246 pp, 18 pls, Black, Edinburgh 1837. Reprinted from *Encyclopaedia Britannica* 7ed.

1204. —— *The institutes of natural science, exhibiting the arrangement followed in the lectures delivered by Dr. Fleming*. 8vo, 41 pp, for the author, Edinburgh [1846].

1205. FLINTOFT, J. J. *British mosses in the English Lake District*. 8vo, t.p. + 60 specimens on 50 leaves, Keswick [c.1830].

1206. *Flora Anglica; ou, flore Angloise, contenant les plantes qui croissent naturellement en les Royaumes d'Angleterre*. 8vo, London 1764.

1207. *The flora of Arbroath and its neighbourhood*. Sm. 8vo, viii+63 pp, Horticultural and Natural History Association Arbroath, Arbroath 1882.

*The flora of Liverpool*. 1872. *see* The Liverpool Naturalists Field Club.

*Flora of Stonyhurst and district*. 2ed. 1891. *see* Gerard, J.

1208. *Flora parvula; or, gleanings among favourite flowers*. Sm. 8vo, 87 pp, Wertheim, London 1847.

*Flora Repandunensis. The wild flowers of Repton*. 1866. *see* Wyatt, W. & Thornton, C. G.

*Flora Tunbrigensis*. 1816. *see* Forster, T. F.

*Flora Vectiana*. 1823. *see* Snooke, W. D.

1209. *Flora Wellingtonensis. First list*. 8vo, London [1868]. Second list is by H. Fraser in *Nat. Hist. Rep. Wellington School*, 1868.

*Florae Anglicae specimen . . . anno 1774 inchoatum*. [?1775]. *see* Cullum, *Sir* Thomas G.

1210. FLOWER, John Wickham *On the pleasures and advantages to be derived from the study of natural history*. 8vo, 40 pp, Gray & Warren printed, Croydon 1857.

1211. FLOWER, Thomas Bruges *Flora Thanetensis: or, a catalogue of the plants indigenous to the Isle of Thanet*. 8vo, vi+83 pp, Ramsgate 1847.

1212. [FLOWER, *Sir* William Henry] *Synopsis of the contents of the museum.* 8vo, iv+112 pp, 1 plan, Royal College of Surgeons, London 1862.

1213. —— *editor Recent memoirs on the Cetacea*. Roy. 4to, xii+312 pp, 6 pls, text figs, Ray Society, London 1866. Ray Society Publ. No. 40 for 1866.

1214. —— *An introduction to the osteology of the Mammalia . . . lectures delivered at the Royal College of Surgeons of England*. 8vo, xi+344 pp, text figs, Macmillan, London 1870; 2ed. 1876; 3ed. 1885; reprint of 3ed. 1966. Manuals for Students.

1215. —— *Introductory lecture to the course of comparative anatomy, delivered at the Royal College of Surgeons of England . . . 1870*. 8vo, 43 pp, text figs, Churchill, London 1870.

1216. —— *Catalogue of the specimens illustrating the osteology and dentition of vertebrated animals . . . in the Museum of the Royal College of Surgeons of England. Part 1. Man. (Part 2. Mammalia)*. 8vo, 2 pts, Churchill, London 1879–84; 2ed. Pt. 1 1907. Part 2 is by Garson, J. G. and Flower. For Part 3, *Aves*, *see* R. B. Sharpe.

1217. —— *The horse, a study in natural history*. 8vo, xvi+196 pp, 1 pl., text figs, Kegan Paul, London 1891.

1218. —— *Essays on museums and other subjects connected with natural history*. 8vo, xvi+394 pp, 1 pl., text figs, Macmillan, London 1898.

1219. —— & LYDEKKER, Richard *An introduction to the study of mammals living and extinct*. 8vo, xvi+763 pp, text figs, Black, London 1891.

*Flowering plants, ferns, and fern allies*. [?1865]. *see* Carr, E. D.

1220. *Flowers and their teachings.* 16mo, RTS, London [1874]. By the author of *Sunshine and shadows in Kattern's life.*

1221. *Flowers in their seasons*. Imp. 8vo, Jerrard, London [1854]. At least one copy on F.P. printed in gold.

1222. *The flowers of the year*. 12mo, vi–7–192 pp, 1 pl., RTS, London [c.1860].

1223. FOLKARD, Richard *Plant lore, legends, and lyrics*. 8vo, xxiv+610 pp. 6 pls, text figs, Sampson Low, London 1884; 2ed. 1892.

1224. *Footsteps to the natural history of beasts and birds. Designed for children*. 12mo, 104 pp, 12 pls, Harvey & Darton, London 1803; ed. 1804; ed. 1805; ed. 1822; 4ed. 1825; 5ed. 1836; ed. [?1843].

1225. FORBES, Edward *Malacologia Monensis: a catalogue of the*

*Mollusca inhabiting the Isle of Man and the neighbouring seas*. 12mo, xii+63 pp, 3 pls, Carfrae, Edinburgh 1838.

1226. —— *A history of British starfishes and other animals of the class Echinodermata*. Roy. or demy 8vo, xx+267 pp, text figs, Van Voorst, London [1840–]1841. Natural History of the British Isles Series. Facsimile 1973.

1227. —— *An inaugural lecture on botany considered as a science*. 8vo, 23 pp, Van Voorst, London [1843].

1228. —— *A monograph of the British naked-eyed medusae.* Roy. 4to, viii+104 pp, 13 col. pls, Ray Society, London 1848. Ray Society Publ. No. 12 for 1847.

1229. —— *Literary papers on scientific subjects*. 8vo, xvi+300 pp, 1 portr., Reeve, London 1855.

1230. —— *The natural history of the European seas*. Sm. 8vo, viii+306 pp, 1 map, text figs, Van Voorst, London 1859. Outlines of the Natural History of Europe, Vol. 2. Edited and continued by R. A. C. Godwin-Austen.

—— *see* Latham, R. G.

—— *see* Quiggin, J.

1231. —— & HANLEY, Sylvanus Charles Thorp *A history of British Mollusca and their shells*. Roy. or demy 8vo, 4 vols, 139 col. + 64 plain pls, Van Voorst, London [1848–]1853. Demy copies have all plates plain. Royal copies sometimes have the lettered plates plain. Plates A–I, K–U, W, AA–PP, RR–VV, XX–ZZ, AAA–SSS, I–CXXXIII, CXIVA–F. Natural History of the British Isles Series.

—— & —— *see* British Natural History Society.

1232. [FORD, *Mrs* Laetitia] *The butterfly collector's vade mecum: or, synoptic tables of English butterflies.* 12mo, iv+68 pp, 2 col. pls, Raw, Ipswich 1824; 2ed. 1827; 3ed. 1836. 1ed. signed L.J., 2ed. signed Laetitia Jermyn, 3ed. signed Laetitia Ford.

1233. FORREST, Herbert Edward *The fauna of Shropshire*. Sm. 8vo, 248+vi pp, 25 pls, text figs, Wilding, Shrewsbury 1899.

1234. [FORSTER, Benjamin Meggot] *Peziza cuticulosa*. 12mo, 2 pp, 1 pl., London [1792].

1235. —— *Introduction to our knowledge of funguses*. 8vo, 20 pp, 2 pls, London 1820.

1236. FORSTER, Edward *Catalogus avium in insulis Britannicis habitantium.* 8vo, 48 pp, Valpy, London 1817. Also published in No. 1242.

1237. FORSTER, Johann Reinhold *also known as* John *A*

*catalogue of British insects.* 8vo, 16 pp, Eyres printed, Warrington 1770.

1238. [FORSTER, Thomas Furley] *A list of the rare plants found in the neighbourhood of Tunbridge Wells*. 12mo, [London 1801].

1239. —— *Flora Tonbrigensis: or. a catalogue of the plants growing wild in the neighbourhood of Tunbridge Wells.* 8vo, vii+216 pp, 3 col. pls, Arch, London 1816; ed. Tunbridge Wells 1842.

1240. —— *The botanical pocket book . . . with observations on some of the rarer plants of the neighbourhood of Tunbridge Wells*. 8vo, vi+179 pp, 1 col. pl., Tunbridge Wells 1840.

1241. —— *The pocket encyclopaedia of natural phenomena.* Sm. 8vo, xlviii+440 pp, Nichols, London 1827.

1242. [FORSTER, Thomas Ignatius Maria] *Observations on the brumal retreat of the swallow*. 8vo, 27 pp, for the author, London 1808; 2ed. 1808; 3ed. 1813 signed; 5ed. 1817; 6ed. 1817 title *Observations on the natural history of swallows*; issue 1817 title *Observations on the natural history of the swallow tribe*. 4ed. is in *The Pamphleteer*, Vol. 4, 1814. 1817 editions contain No. 1236. 1ed. is by Philochelidon.

1243. —— *Researches about atmospheric phenomena*. 8vo, xvi+219 pp, Underwood, London 1813; 2ed. 1815; 3ed. 1823 title . . . *to which is added the calendar of nature.*

1244. —— *A synoptic catalogue of British birds*. 8vo, iv+64 pp, 1 col. pl., Nichols, London 1817. Some copies end at p. 37; others continue with *Catalogue and generic division of British birds* to p. 64.

1245. —— *The perennial calendar . . . as connected with . . . botany, natural history*. 8vo, xxvii+804 pp, Harding Mavor & Lepard, London 1824.

1246. FORSYTH, J. S. *The first lines of botany, or primer to the Linnean system.* 12mo, 18+184 pp, 2 col. + 1 plain pls, Bulcock, London 1827.

1247. FORSYTH, William *A botanical nomenclature, containing a systemmatical arrangement . . . of plants as described in the new edition of Linnaeus's Systema naturae by Dr. Gmelin.* 8vo, vii+[355] pp, Cadell London 1794.

Fortescue, Thomas *Baron* Clermont. *see* Clermont, Baron

Foster, Thomas Furley *see* Warner, R.

1248. FOTHERGILL, Charles *Ornithologia Britannica: or, a list of all the British birds. In Latin and English*. Fol., 11 pp, Hick printed, York 1799.

1249. —— *An essay on the philosophy, study, and use of natural history*. 8vo, xxxv+236 pp, White Cochrane, London 1813.

1250. FOTHERGILL, William Edward *Botanical types. Description of structure and life history of ten types*. 8vo, 66 pp, 1 pl., Thin, Edinburgh 1889.

*The four seasons*. 1865. *see* Fitton, *Miss* S. M.

1251. [FOWLER, William Warde] *A year with the birds*. 8vo, ix+144 pp, 1 pl., Blackwell, Oxford 1886; 2ed. 1886 signed; 3ed. London 1889; 1891. By an Oxford Tutor.

1252. —— *Tales of the birds*. Sm. 8vo, xii+209 pp, 8 pls, Macmillan, London 1888; 2ed. 1889; ed. 1901.

1253. —— *The marsh warbler (Acrocephalus palustris) in Oxfordshire and Switzerland*. 8vo, 29 pp, Blackwell, Oxford 1893.

1254. —— *Summer studies of birds and books*. 8vo, x+288 pp, 1 pl., text figs, Macmillan, London 1895.

1255. FOWLER, William Weekes *The Coleoptera of the British Islands*. Roy. or demy 8vo, 6 vols, 5 plain + 200 col. pls, text figs, Reeve, London 1887–91, 1913; suppl. 103 pp, 8 col. pls, London 1931. Vol. 6 is suppl. by Fowler and Donisthorpe, Horace St John Kelly. 1931 suppl. is reprinted from *Ent. mon. Mag*., 1929–31 and is by Donisthorpe alone. Demy copies have only the 5 plain plates.

—— *see* Matthews, Andrew

—— *see* Sharp, D.

1256. FOWNES, G. *An essay on the food of plants.* 8vo, 61 pp, for private distribution, London 1843. *Jl R. Agric. Soc.* reprint.

1257. FOX, George Townshend *editor Synopsis of the Newcastle Museum, late the Allan, formerly the Tunstall or Wycliffe Museum . . . with occasional remarks on the species, by those gentlemen and the editor*. 4to or 8vo, xxii+312 pp, 7 pls, 1 portr., text figs, Charnley, Newcastle-upon-Tyne 1827.

1258. FRANCIS, George William *Catalogue of British flowering plants and ferns.* S.sh., London [1835]; 2ed. [1835]; 5ed. [1840].

1259. —— *An analysis of the British ferns and their allies.* 8vo, iii+68 pp, 6 pls, Simpkin Marshall, London 1837; 2ed. 1842; 3ed. 1849; 4ed. 1858.

1260. —— *The little English flora, or, a botanical and popular account of all our common wild flowers*. 8vo, xiii+174 pp, 14 pls, Simpkin Marshall, London 1839; 2ed. 1842; 3ed. 1847; 1849; 4ed. 1853.

1261. —— *British botany illustrated*. 8vo, 2 pts, 16+32 pp, 6 pls, text figs, [London] 1840. All publ.
1262. —— *Illustrations of the British mosses*. 8vo, 16 pp, London 1840. All publ.
1263. —— *The grammar of botany*. 8vo, 160 pp, 1 col. pl., text figs, Simpkin Marshall, London 1840.
1264. FRANKLAND, Percy Faraday *Our secret friends and foes*. 8vo, viii+167 pp, SPCK, London 1893; 1894; 3ed. 1897; 4ed. 1899. Romance of Science Series.
1265. FRANTZE, Wolfgang *The history of brutes*. 8vo, 256 pp, 1 pl., Francis Haley, London 1670. Translation of *Historia animalium sacra*, Wittenberg 1612; Amsterdam 1643, by N.W.

Franz, Wolfgang *see* Frantze, W.

1266. FRASER, Alexander *Natural history of the salmon, herrings, cod, ling etc.* 8vo, xii+123 pp, Carruthers, Inverness 2ed. 1833.

Fraser, H. *see Flora Wellingtonensis*

1267. FRASER, Louis *Zoologia typica, or figures of new and rare mammals and birds described in the Proceedings, or exhibited in the collections of the Zoological Society of London*. Fol., 70 col. pls with letterpress, by the author, London [1845–] 1846–48 [–1849].
1268. FRASER, Patrick Neill *British ferns and their varieties*. 8vo, 7 pp, Edinburgh 1864; ed. 1868.
1269. FRASER, Robert William *Ebb and flow: curiosities and marvels of the sea-shore*. Sm. 8vo, 251 pp, 13 pls, text figs, Houlston & Wright, London 1860; ed. 1865 unsigned.
1270. —— *Seaside divinity*. 8vo, xii+378 pp, 8 pls, Hogg, London 1861; 2ed. 1861; 3ed. 1868 title *The seaside naturalist.*

Fredol, Alfred *pseudonym. see* Moquin-Tandon, C. H. B. A.

1271. FREEMAN, Strickland *editor Select specimens of British plants*. Fol., 2 pts, 10 col. pls with letterpress, Nicol, London 1797, 1809.
1272. FREKE, Henry *On the origin of species by means of organic affinity*. 8vo, viii+135 pp, 1 tab., Longman, London 1861.

Frey, Johann Friederich Heinrich Konrad *see* Stainton, H. T.

Fric, Antonin Jan *see* Fritsch, A. J.

Friederichs, Hulda *see* Morley, C.

1273. FRIEND, Hilderic *A glossary of Devonshire plant names*. 8vo, vi+63 pp, English Dialect Society, London 1882. Reprinted from *Trans. Devon. Assoc.* 1882.

1274. —— *Flowers and flower law*. 8vo, 2 vols, xvi+704 pp, text figs, Sonnenschein, London 1884; 2ed. 1884; 3ed. 1886; 4ed. 1892.

1275. FRITSCH, Antonin Jan *The birds of Europe*. Fol., 6 11, 61 col. pls, Trübner, London 1877. The plates of Vol. 1 of *Naturgeschichte die Vögel Europas*, Prague [1853–]1870. Author's Polish name is Fric.

Frohawk, Frederick William *see British birds with their nests and eggs*. 1896–98. *see* Butler, A. G.

1276. FROST, Henry *Our common shells and marine birds*. Sm. 8vo, 10 pp, 7 pls, Simpkin Marshall, London [1891].

Frost, James *see* Bingley, T.

1277. FROST, John *Some account of the science of botany*. 8vo, London 1827.

1278. FRY, *Sir* Edward *British mosses*. 8vo, 71 pp, text figs, Witherby, London 1892; 2ed. 1908. Knowledge Series.

1279. —— *The Mycetozoa and some questions which they suggest*. 12mo, viii+82 pp, text figs, Simpkin Marshall, London 1892; 2ed. 1915.

1280. FRYER, Alfred *The potamogetons (pond weeds) of the British Isles*. 4to, x+94 pp, 60 col. pls, Reeve, London [1898–]1915.

1281. FUESSLY, Johann Caspar *Archives of entomology*. 4to, iv+49+xiii+184 pp, 81 col. pls, Johnson, London 1795. Translation of *Archiv der Insekten-gesschichte*, Zurich & Winterthur 1781–86.

1282. FULCHER, *Miss* Florence Anna *Birds of our Islands*. Sm. 8vo, 368 pp, text figs, Melrose, London [1897].

1283. —— *Among the birds*. Sm. 8vo, 253 pp, text figs, SPCK, London 1900.

*A full account of the capture of the monster Royal whale*. 1892. *see* Cordeaux, J.

1284. FULLOM, Stephen Watson *The marvels of science and their testimony to holy writ*. Sm. 8vo, Hurst & Blackett, London 1852. Several later editions.

1285. FURNEAUX, William Samuel *The out-door world, or young collector's handbook*. 8vo, xxviii+411 pp, col. pls, text figs, Longman, London 1893. Many later editions.

1286. —— *Butterflies and moths*. Sm. 8vo, xiv+358 pp, 12 col. pls. Longman, London 1894.

1287. —— *Life in ponds and streams*. 8vo, xix+406 pp, 8 col. pls, text figs, Longman, London 1896; ed. 1911.

1288. FYFE, Andrew *Outlines of comparative anatomy*. 8vo, xxx+352 pp, Black, Edinburgh 1813.

1289. [—— & ROTHERHAM, John] *compilers. A system of anatomy and physiology, with the comparative anatomy of animals*. 8vo, 3 vols, 16 pls, Elliot, Edinburgh 2ed. 1787; ed. 1791; ed. 1795. Contains Monro, A. *first of the name. A system of comparative anatomy*. First edition 1784 is human anatomy alone.

# G

G., G. R. *Notices of insects that are known to form the bases of fungoid parasites*. 1858. *see* Gray, G. R.

1290. GADOW, Hans Friedrich *A classification of the Vertebrata, recent and extinct*. 8vo, xvii+82 pp, Black, London 1898. Printed on one side.

1291. GÄTKE, Heinrich *Heligoland as an ornithological observatory*. 8vo. x+599 pp, 1 pl., 1 portr., text figs, Douglas, Edinburgh 1895. Translation of *Die Vogelarte Helgoland*, Brunswick 1891, by Rudolph Rosenstock. See also Whitlock, F. B.

1292. GALLAUDET, Thomas Hopkins *The youth's book on natural theology*. 12mo, xvii+195 pp, text figs, Fry, London 1832, 2ed. 1835. USA text 1832.

1293. GALLWAY, *Sir* Ralph William Frankland Payne *Bart. The fowler in Ireland, or notes on the haunts and habits of wildfowl and seafowl*. 8vo, xiii+503 pp, 17 pls, text figs, Van Voorst, London 1882.

1294. GALPIN, Francis William *An account of the flowering plants, ferns and allies of Harleston . . . to which are added . . . birds of the district*. 8vo, 157 pp, 1 pl., Bartlett, London 1888.

1295. GALPINE, John *A synoptical compend of British botany (from the Class Monandria to Polygamia inclusive)*. 12mo, viii+194 pp, 1 pl., Salisbury 1806; 2ed. London 1820; 3ed. 1829; 4ed. 8vo 1834. Pagination erratic.

1296. [GALTON, Samuel John] *The natural history of birds; containing a variety of facts . . . intended for the amusement and instruction of children.* 12mo, 6 pts in 3 vols, 116 col. pls, Johnson, London 1786–91; ed. 1791; issue [ ]. There is an issue of the plates alone, dated 1792 or 1793, with title *Plates to the natural history of birds, uncoloured: intended to serve the purpose of a drawing book*. This is also found in two volumes with the text.

[——] *The natural history of quadrupeds*. 1801. *see The natural history of quadrupeds.*

*The garden and menagerie of the Zoological Society delineated*. 1830. *see* Bennett, E. T.

1297. *The garden and wild flowers of the year.* Sm. 8vo, 1 pl., RTS, London 1847. By the author of *A book for the seaside*. Monthly double volume.

1298. GARDINER, William *botanist. Musci Britannici*. 8vo, Glasgow 1836.

1299. —— *Botanical rambles in Braemar in 1844, with an appendix on Forfarshire botany*. 8vo, 51 pp, Dundee 1845.

1300. —— *Twenty lessons on British mosses* [First series]. 8vo, 46 pp, 25 specimens, Dundee 1846; 2ed. Edinburgh 1846; 3ed. 1848; 4ed. 1852.

1301. —— *The Flora of Forfarshire*. 8vo, xxiv+308 pp, 2 pls, Longman, London 1848. Fol. vol. of 150 specimens for subscribers' copies.

1302. —— *Twenty lessons on British mosses. Second series*. 8vo, 58 pp, 25 specimens, London 1849.

1303. —— *Catalogue of British mosses, alphabetically arranged.* S. sh., Dundee [c.1850]; 2ed. [ ]; 3ed. [ ].

1304. GARDINER, William *composer. The music of nature*. 8vo, xii+530 pp, Longman, London 1832.

1305. GARDNER, James *Bird, quadruped and fish preserving: a manual of taxidermy for amateurs*. 8vo, 64 pp, Houlston & Wright, London [1865]; issue Lea, London [1865]; [1868]; ed. 1873.

Gardner, John *see* Robson, J. E.

1306. GARNER, Robert *Figures illustrating the structure of various invertebrate animals (mollusks and Articulata)*. Roy. 8vo, 6 pls, Van Voorst, London 1860.

1307. —— *The natural history of the County of Stafford*. 8vo, vi+551 pp, 7 pls, 1 map, text figs, Van Voorst, London 1844; suppl. 61 pp, 1860.

1308. [——] *Holiday excursions of a naturalist*. Sm. 8vo, viii+345 pp, Hardwicke, London 1867.

1309. GARNETT, Thomas *M.D. Popular lectures on zoonomia or the law of animal life in health and disease*. 4to, portr., London 1804.
1310. GARNETT, Thomas *of Clitheroe. Essays in natural history and agriculture.* 8vo, vi+241 pp, privately printed, London 1883.
1311. —— *Facts on the natural history and habits of the salmon*. 8vo, privately printed, Clitheroe [1867].
1312. GARRATT, G. *Marvels and mysteries of instinct; or, curiosities of animal life*. Sm. 8vo, viii+248 pp, Longman, London 1856; 2ed. 1857; 3ed. 1862.
1313. GARROD, Alfred Henry *The collected scientific papers*. Roy. 8vo, xxv+537 pp, 33 col. pls, 1 portr., London 1881.

Garson, John George *see* Flower, *Sir* William H.

1314. GARTSIDE, M. *Ornamental groups, descriptive of flowers, birds, shells, fruit, insects &c. and illustrative of a new theory of colouring*. Fol., 20 pp, 4 col. + 2 plain pls, Bulmer, London 1808.
1315. GATTY, Margaret, *Mrs* Alfred *British sea-weeds: drawn from Professor Harvey's 'Phycologia Britannica'.* Roy. 8vo, 2 vols, 80 col. pls, Bell & Daldy, London 1863; issue 1872.
1316. —— *Waifs and strays of natural history*. 8vo, 165 pp, London 1871; 1874.
1317. GAUSSEN, Samuel Robert Louis *The world's birthday*. Sm. 8vo, Nelson, London 1860. Several later editions. Translation of *Le premiere chapitre de Genèse exposé*, Toulouse, by J. H.
1318. GAYE, *Miss* Selina *The world's farm*. 8vo, x+365 pp, Seeley, London 1893.
1319. GEDDES, *Sir* Patrick *The rise and aims of modern botany*. 8vo, 23 pp, Dundee 1888.
1320. —— *Chapters in modern botany*. 8vo, xi+201 pp, text figs, Murray, London 1893. University Extension Manuals.
1321. —— & Thomson, *Sir* John Arthur *The evolution of sex*. 8vo, xvi+322 pp, text figs, Walter Scott, London 1889; ed. 1901. Contemporary Science Series.
1322. GEGENBAUR, Carl *Elements of comparative anatomy*. 8vo, xxvi+645 pp, text figs, Macmillan, London 1878. Translation of *Grundriss der vergleichenden Anatomie*, 2ed. Leipzig 1877, by Francis Jeffrey Bell.

*The gem of the peak.* [ ]. *see* Adam, W.

1323. *Gems of nature and art*. 4to, vi+72 pp, 24 col. pls, Groombridge, London [1868].

*A general history of quadrupeds*. 1790. *see* Bewick, T.

*General observations on the advantages which may result from the introduction of the seeds of our best grasses. see* Curtis, W.

1324. GEOGRAPHICAL SOCIETY, The Royal *Hints to travellers*. 8vo, London 4ed. 1878; 5ed. 1883; 6ed. 1889; 7ed. 1893; 8ed. 1901; 9ed. 1906; 11ed. 1935. The first three editions are in. *J. Geogr. Soc*. for 1854 & 1864, and *Proc. Geogr. Soc*. for 1872.

1325. *The geography of plants*. 12mo, vi+192 pp, RTS, London [1850]. Monthly Volume No. 52. See also *Plants of scripture*, 1850.

*Geology, archaeology, and botany . . . of Cheltenham*. 1842. *see* Buckman, J.

1326. [GERARD, John] *Jesuit. A preliminary flora of the Stonyhurst district.* 8vo, 12 pp, Stonyhurst, 1886; 2ed. Clitheroe 1891. title *Flora of the Stonyhurst district*. Reprinted from the *Stonyhurst Magazine*, 1886.

1327. —— *Science and scientists: some papers on natural history*. 8vo, vii+130 pp, Catholic Truth Society, London 1889.

1328. GERARD, John *surgeon. The herball or general historie of plantes*. Fol., 1392 pp, 1 portr., text figs, Bonham & Norton, London 1597; ed. 1633; ed. 1636; abridged ed. 1937. Largely a translation of Dodoens, R. *Stirpium historiae, pemptades sex*, Antwerp 1583.

1329. GERBER, Frederick *Elements of the general and minute anatomy of man and the Mammalia, chiefly after original researches*. 8vo, 2 vols, 34 pls, Baillière, London 1842. Translation of *Handbuch der allgemeine Anatomie*, Bern 1840, by George Gulliver.

1330. GIBBS, John *A lecture on the variation of plants*. 8vo, 24 pp, Fry, Chelmsford [1861].

1331. —— *A first catechism of botany*. 8vo, 60 pp, Chelmsford [1878].

1332. —— *The symmetry of flowers*. 12mo, 19 pp, Dutton, Chelmsford [1890].

1333. GIBSON, George Stacey *The flora of Essex*. 4to or 8vo, 1+469 pp, 1 col. map, Pamplin, London 1862.

1334. GIBSON, John *Science gleanings in many fields. Studies in natural history*. Sm. 8vo, viii+377 pp, 1 pl., text figs, Nelson, London 1884 [=1883].

1335. GIBSON, Robert John Harvey *A textbook of elementary biology*. 8vo, x+362 pp, text figs, Longman, London 1889.

1336. —— & AULD, Helen Park *L.M.B.C. memoirs on typical British marine plants and animals . . . IV. Codium*. 8vo, viii+18 pp, 3 pls, Dobb for the Liverpool Marine Biological Committee, Liverpool 1900.

1337. GIFFORD, Isabel *The marine botanist*. Sm. 8vo, xxvii+141 pp, 6 plain & 6 col. pls, col. text figs, Darton, London [1848]; 2ed. [1848]; 3ed. 12mo, Brighton 1853.

1338. —— *Marine Algae found on the Somerset coast*. 8vo, 7 pp, Minehead 1849. Lithograph.

1339. GILES, George Michael James *A handbook of the gnats, or mosquitoes*. 8vo, viii+374 pp, 8 pls, text figs, Bale, London 1900; 2ed. 1902.

1340. GILPIN, William *Remarks on forest scenery and other woodland views*. 4to or 8vo, 2 vols, 30 pls, 1 map, Blamire, London 1791; 2ed. 1794; 3ed. 1808; ed. Edinburgh 1834; ed. 1879.

1341. GIRAUD, Jane Elizabeth *The flowers of Milton*. Fol., 28 col. pls with letterpress, [Ackermann, London 1846].

1342. —— *The flowers of Shakespeare*. Fol., 29 col. pls with letterpress, [Ackermann, London 1850].

1343. —— *The floral months of England*. 4to, 12 col. pls, London 1850.

1344. GISEKE, Paul Dietrich *Index Linnaeanus in Leonardi Plukenetii . . . Opera botanica . . . Index Linnaeanus in J.J.Dillenii Historia muscorum*. 4to, x+46 pp, Hamburg 1779.

1345. GISSING, Thomas Waller *The ferns and fern allies of Wakefield*. 8vo, x+54 pp, 26 col. pls, Journal and Examiner Office for the author, Wakefield 1862.

1346. —— *Materials for a flora of Wakefield and its neighbourhood*. 8vo, 59 pp, Van Voorst, London 1867.

1347. GLASGOW NATURAL HISTORY SOCIETY *Notes on the fauna and flora of the west of Scotland, with lists*. 8vo, 148 pp, the Society, Glasgow 1876. A British Association Guide Book.

1348. —— *The fauna of Scotland; with special reference to Clydesdale and the western district*. 8vo, 3 pts, the Society, Glasgow 1878–80. ?All publ.

1349. GLEADALL, Eliza Eve *Beauties of flora, with botanic and poetic illustrations*. 4to, 2 vols, 41 col. pls, Sherwood, Wakefield & London [1834–]1837.

1350. *Gleanings from natural history: for the amusement and instruction of youth*. 24mo, 12 pp, text figs, Houlston, Wellington [c.1820].

1351. *Glimpses of natural history*. 12mo, vi+199 pp, text figs, Clark, London [1843]; ed. 16 pls, Harvey & Darton, London 1843.

*Glimpses of the wonderful*. 1844. *see* Gosse, P. H.

Gmelin, Johann Friederich *see* Linnaeus, C.

—— *see* Sibly, E.

Godwin-Austen, Robert Alfred Cloyne *see* Forbes, E.

1352. GOEBEL, Karl Eberhard von *Organography of plants, especially of the Archegonatae and Spermaphyta*. Roy. 8vo, 2 vols, text figs, Clarendon, Oxford 1900–05. Translation of *Organographie der Pflanzen*, Jena 1898–1901, by Isaac Bayley Balfour.

—— *see* Sachs, F. G. J. von.

1353. GOEDART, Joannes *J. Godartius of insects. Done into English and methodised* 4to, 140 pp, 14 col. pls, White for M[artin] L[ister], York 1682; Latin ed. 8vo, London 1685 title *De insectis in methodum cum notularum additione*. Translation of *Metamorphosis et historia naturalis insectorum*. Middleburg 1662–67, by Martin Lister. 150 copies.

1354. GOLDSMITH, Oliver *An history of the earth and animated nature*. 8vo, 8 vols, 101 pls, Nourse, London 1774. ['In eight volumes' on t.p. of Vol. 1 only. Plates signed Isaac Taylor & Elias Martin]; ed. Nourse, London 1774. ['In eight volumes' on t.ps of all volumes. Plates unsigned. ? pirate Edinburgh or Dublin]; 2ed. 1779.

Very many later editions, some of them abridged, including:

1. 8 vols, J. Williams, Dublin 1776–77; 1782–83.
2. 8 vols, F. Wingrave, London 1791; 6 vols 1805; 1806; Wingrave & Collingwood 1816.
3. 4 vols, A. Millar, London, Wilson Spence & Mawman, York 1797; Wilson & Spence, York 1804; T. Wilson, York 1808; 1853.
4. 1 vol., Oriental Press, Vernor Hood & Sharpe, London 1803; 1807; 1810.
5. 1 vol., Scatcherd & Letterman, London 1807. Title *A natural history of birds and beasts* . . . abridged for the use of schools by *Mrs* [Mary] Pilkington.
6. 4 vols, Fowler, Ormskirk 1807.
7. 1 vol., Nuttall Fisher & Dixon, Liverpool 1810.

8. 4 vols, Nuttall Fisher & Dixon, Liverpool 1810; 1811.
9. 4 vols, Mills, Thetford [?1810].
10. 8 vols, Christie, Dublin 1812–1814.
11. 1 vol., R. Edwards, London 1817; 1819.
12. 1 vol., Newman, London [?1820]. Title *A natural history of remarkable birds and fishes*.
13. 6 vols, Richardson, London 1821.
14. 3 vols, Wright, London 1824.
15. 4 vols, Fisher, London 1826.
16. 1 vol., Dove, London 1828.
17. 3 vols, Rivington, London 1831.
18. 4 vols, Fullarton, Glasgow 1832–33.
19. 2 vols, Limberd, London 1834.
20. 1 vol., Whittaker, London 1836.
21. 4 vols, 24mo, Smith & Elder 1838.
22. 3 vols, Dove, London 1838.
23. 1 vol., Nelson, Edinburgh 1839.
24. 2 vols, Fullarton, Edinburgh 1848. Many times reprinted and with imprint sometimes including London and Dublin.
25. 2 vols, Blackie, Glasgow 1852. Many times reprinted and with imprint sometimes including London.
26. 1 vol., Lofts, London 1852.
27. 1 vol., Sprent, Hull 1854.
28. Welsh ed., 4to, Fullarton, Edinburgh [1868]. Title *Hanes y ddaear a'r creaduriaid byw*.

—— *see* Brown, Thomas *Captain*

—— *see* Shaw, G. F.

1355. *The good child's cabinet of natural history.* 18mo, 5 vols, text figs, Wallis, London 1801; 1813.

1356. GOOD, John Mason *The book of nature.* 8vo, 3 vols, Longman, London 1826; 2ed. 1828; 2ed. [c.1830]; 3ed. 1834.

—— *see* Polehampton, E. T. W.

1357. [GOODRICH, Samuel Griswold] *Tales of animals. Comprising quadrupeds, birds, fishes reptiles, and insects.* Sm. 8vo, viii+334 pp, text figs, Tegg, London 1832; 2ed. 1833. Very many later editions. By Peter Parley *pseudonym.* Perhaps not by Goodrich.

*1358. [——] *Tales about plants.* 16mo, xii+500 pp, text figs, Tegg, London [1839]. Later editions. ?By Peter Parley *pseudonym.* Perhaps by *Mrs* Jane Loudon = No. 2247.

Gordon, Alexander *see* Milne, C.

1359. [GORDON, George] *Collecteana for a flora of Moray.* 8vo, 40 pp, privately printed, Elgin 1839.

1360. GORDON, George *botanist. The pinetum.* Roy. 8vo, xxii+353 pp, Bohn, London 1858; suppl. 1862; 2ed. 1875; ed. 1880.

1361. GORDON, Henry Doddridge *The history of Harting . . . and some notice of its fauna and flora.* 8vo, xi+492 pp, 1 pl., text figs, London 1877.

1362. —— *Among the birds of Harting, 1886.* 8vo, 38 pp, 3 pls, Duplock & Woledge, Petersfield [1886]. Reprinted from the *Newcastle Chronicle* and *West Sussex Gazette.*

1363. GORDON, William John *Our country's flowers and how to know them.* Sm. 8vo, vii+152 pp, 33 col. pls, Day, London [1891]; ed. 1896; ed. 1903.

1364. —— *Our country's birds and how to know them.* Sm. 8vo, vii+152 pp, 33 col. pls, Day, London [1892]; suppl. 64 pp, 1905 title *Eggs of the native birds of Britain and list of British birds.*

1365. —— *Our country's butterflies and moths and how to know them.* Sm. 8vo, viii+150 pp, 33 col. pls, Day, London [?1900].

1366. GORING, C. R. & PRITCHARD, Andrew *Microscopic illustrations of a few . . . living objects; with their natural history.* 8vo, 96 pp, 3 col. pls, Whittaker, London 1830; ed. 1833; 1840; 3ed. 1845.

1367. —— & —— *Micrographia: containing practical essays on . . . microscopes.* 8vo, viii+231 pp, 3 pls, text figs, Whittaker, London 1837.

1368. [—— & ——] *Microscopic objects . . . with instructions for preparing and viewing them.* 12mo, vii+210 pp, 1 col. pl., text figs, Whittaker, London 1847. A version of Pritchard's work on the same subject with a posthumous essay by Goring.

1369. GORRIE, Daniel *Summers and winters in the Orkneys.* 8vo, vii+384 pp, Hodder & Stoughton, London 1868; 2ed. [1869].

Gosse, Emily *see* Gosse, P. H. *Sea-side pleasures.* 1853.

1370. GOSSE, Philip Henry *An introduction to zoology.* 12mo, 2 vols, text figs, SPCK, London [1844].

1371. [——] *Glimpses of the wonderful.* 16mo, 115 pp, 31 pls, text figs, Harvey & Darton, London 1844.

1372. —— *The ocean.* 12mo, xii+360 pp, 4 pls, text figs, SPCK,

London 1845; 1846; 1849; 1851; 1854; 1856; ed. (1860). Later undated printings.

1373. —— *Natural history. Mammalia.* Sm. 8vo, vi+302 pp, text figs, SPCK, London 1848; 1862.

1374. —— *Natural history. Birds.* Sm. 8vo, viii+327 pp, text figs, SPCK, London 1849; [ ].

1375. —— *Popular British ornithology.* Roy. 16mo, iv+320 pp, 20 col. pls, Reeve, London 1849; 2ed. 1853. Popular Natural History Series.

1376. —— *Natural history. Reptiles.* Sm. 8vo, iv+296 pp, text figs, SPCK, London 1850.

1377. —— *A text book of zoology for schools.* 12mo, 2+450 pp, SPCK, London 1851.

1378. —— *Natural history. Fishes.* Sm. 8vo, viii+328 pp, text figs, SPCK, London 1851.

1379. —— *A naturalist's rambles on the Devonshire coast.* 12mo, xvi+451 pp, 12 col.+16 plain pls, Van Voorst, London 1853.

1380. [——] *Sea-side pleasures.* Sm. 8vo, 2+86 pp, 8 pls, text figs, SPCK, London 1853. Last chapter is by Emily Gosse.

1381. —— *The aquarium: an unveiling of the wonders of the deep.* 8vo, xiii+278 pp, 6 col.+6 plain pls, Van Voorst, London 1854; 2ed. 1856. See also No. 1383.

1382. —— *Natural history. Mollusca.* Sm. 8vo, viii+328 pp, text figs, SPCK, London 1854.

1383. —— *A handbook to the marine aquarium.* Sm. 8vo, viii+48 pp, 1 pl., Van Voorst, London 1855; 2ed. 1856. This is a new edition of Chapter XI of No. 1381.

1384. —— *A manual of marine zoology for the British Isles.* Sm. 8vo, 2 pts, text figs, Van Voorst, London 1855–56.

1385. —— *Tenby: a sea-side holiday.* 12mo, xviii+400 pp, 20 col.+4 plain pls, Van Voorst, London 1856.

1386. [——] *A tabular view of the orders and families of fishes.* 4to, 4 folded pls, SPCK, London [1856]; issue Stanford [?1868]. Diagrams of Natural History No. VIII.

1387. —— *Life in its lower, intermediate and higher forms.* 8vo, viii+363 pp, 6 pls, text figs, Nisbet, London 1857; 2ed. 1857; 3ed. 1870; 5ed. 1878. Reprinted from articles in *Excelsior.*

1388. —— *Omphalos: an attempt to untie the geological knot,* 12mo, xvii+376 pp, text figs, Van Voorst, London 1857; issue 1857 title *Creation (Omphalos)* . . .

1389. —— *Evenings at the microscope.* 12mo, xii+506 pp, text figs, SPCK, London [1859]; 1874; 1877; 1884; ed. 1895.

1390. —— *Actinologia Britannica. A history of the British sea-anemones and corals.* 8vo, xl+362 pp, 11 col.+1 plain pls, text figs, Van Voorst, London [1858–] 1860. Title of parts . . . *sea-anemones and madrepores.* Pp. 7–10 are cancels. Advertised later as part of Natural History of the British Isles series.

1391. —— *The romance of natural history.* [*First series*]. 8vo, xvi+376 pp, 12 pls, Nisbet, London 1860; 2ed. 1861; 3ed. 1861; 4ed. 1861; 4ed. 1862; 5ed. 1862; 6ed. 1862; 1867; 8ed. 1869; 8ed. 1870; 9ed. 1871; 10ed. 1875; 13ed. 1886. Later undated editions.

1392. —— *The romance of natural history. Second series.* 8vo, xii+396 pp, 10 pls, Nisbet, London 1861; 1862; 1864; 1867; 3ed. 1870; 4ed. 1871; 5ed. 1875. Later undated editions.

1393. —— *A year at the shore.* 8vo, xii+330 pp, 36 col. pls, Strahan, London 1865; 1870; 1873; 1877. Reprinted from *Good Words*, 1864.

1394. —— *Land and sea.* 8vo, ix+425 pp, text figs, Nisbet, London 1865; 4ed. 1874; 1877; 1879. Spine title *Sea and land.*

1395. —— *A marine aquarium.* 8vo, 8 pp, [Herald Press, Birmingham], [1879]. Reprinted from *The Midland Naturalist*, Vol. 2, Jan. 1979.

—— *see* Hudson, C. T.

1396. [GOULD, Charles] *The companion to the microscope.* 12mo, 40 pp, Cary, London 1827; 11ed. 1836; 12ed. 1837; 13ed. 1839; 16ed. 1853.

1397. GOULD, John *The birds of Europe.* Imp. fol., 5 vols, 449 col. pls on 448 ll, Taylor for the author, London [1832–] 1837.

1398. —— *The birds of Great Britain.* Imp. fol., 5 vols, 367 col. pls, Taylor & Francis for the author, London [1862–] 1873.

1399. —— *An introduction to the Birds of Great Britain.* 8vo, 135 pp, for the author, London 1873.

1400. GOULD, William *An account of English ants.* Sm. 8vo, xv+109 pp, Millar, London 1747.

Gower, Charlotte *see* Hogg, R.

1401. GRAHAM, Frederica *Visits to the Zoological Gardens.* 4to, 20 pls with letterpress, Routledge, London [1853]. Books for Young Readers No. 3.

1402. GRAHAM, Henry Davenport *The birds of Iona and Mull.* Sq. 8vo, xv+279 pp, text figs, Douglas, Edinburgh 1890. Vertebrate Fauna of Scotland Series, Vol. 3.

1403. GRAHAM, Peter Anderson *All the year with nature.* 8vo, viii+237 pp, Smith Elder, London 1893.

1404. GRAHAM, Robert *Characters of genera extracted from the British flora of W. J. Hooker.* 8vo, 64 pp, Edinburgh 1830.

1405. GRAHAM, Thomas *Outlines of botany, delivered in a course of lectures.* 8vo, 139 pp, Boone, London [1841]; 2ed. 1848.

1406. GRAHAME, James *The birds of Scotland, with other poems.* 12mo, 248 pp, Blackwood, Edinburgh 1806.

1407. GRANT, Elizabeth *Holiday rambles; or peeps into the book of nature.* 16mo, 6 pls, Binns & Goodwin, Bath 1856.

1408. GRANT, Robert Edmond *An essay on the study of the animal kingdom.* 8vo, 35 pp, Taylor, London 1828; 2ed. 1829.

1409. —— *Outline of a course of lectures on the structure and classification of animals.* 4to, 26 pp, London 1833.

1410. —— *Outlines of comparative anatomy.* 8vo, 2 pts, 656 pp, text figs, Baillière, London [1835–] 1841. All publ.

1411. —— *Syllabus of four lectures, on foraminiferous animalcules, polypiferous animals, insects, and reptiles, to be delivered at Norwich.* 8vo, 12 pp, Taylor & Francis printed, London 1860; ed. Fletcher, Norwich 1860 title *Syllabus of a course of four lectures on zoology.*

1412. —— *Tabular view of the primary divisions of the animal kingdom . . . (Caino-zoology).* 8vo, vi+91 pp, Walton & Maberley, London 1861.

1413. GRANT, William Robert Ogilvie *A hand-book to the game birds.* Sm. 8vo, 2 vols, 47 col. pls, Shaw, London 1895–97; ed. 1896–97. First edition is Allen's Naturalist's Library, second is Lloyd's Natural History.

*Graphic microscopy.* 1856. *see* Draper, E. T.

1414. GRATTANN, W. H. *British marine Algae . . . their collection and preservation.* 8vo, 237 pp, text figs, Bazaar Office, London [1873–74].

1415. GRAVES, George *British ornithology.* Imp. 8vo, 3 vols, 144 col. pls, for the author, London 1811–21; 2ed. 1821. Vol. 3 is the same in both editions.

1416. —— *Ovarium Britannicum; being a correct delineation of the eggs of such birds as are natives of, or domesticated in Great Britain. Part 1.* 8vo, vi pp, 15 col. pls, for the author, London 1816. All publ.

1417. —— *The naturalist's pocket book, or tourist's companion.* 8vo, viii+335 pp, 5 col.+3 plain pls, W. & S. Graves printed for the author, [1817]; 1818; 2ed. 1818; 3ed. 1824 title *The naturalist's companion.*

1418. —— *A monograph of the British grasses.* Roy. 8vo, 5 pts, v pp, 48 col. pls with letterpress, for the author, Peckham 1822. All publ.

1419. GRAY, Asa *Natural selection not inconsistent with natural theology.* 8vo, 55 pp, Trübner, London 1861. USA printed.

1420. —— *The botanical text book, an introduction to scientific botany.* 8vo, xii–520 pp, text figs, Thomas Defl, London 1850; 6ed. 1880; 1887. First edition is New York 1842.

1421. GRAY, George Robert *A list of the genera of birds, with their synonyms, and an indication of the typical species of each genus.* 8vo, viii+ii+80 pp, Taylor & Francis, London 1840; 2ed. 1841; appendix 16 pp, 1842.

1422. —— *Genera of birds.* Fol., 3 vols, 185 col.+149 plain pls, Longman, London 1844–49.

1423. —— *List of specimens of British animals in the collection of the British Museum. Part III. Birds.* 12mo, 16+275 pp, British Museum, London 1850.

1424. [——] *List of specimens of British animals in the collection of the British Museum. Part IX. Eggs of British birds.* 12mo, 143 pp, British Museum, London 1852.

1425. [——] *Notices of insects that are known to form the bases of fungoid parasites.* 4to, 22 pp, 5 pls, privately printed, Hampstead 1858, by G.R.G.

1426. —— *Tabular view of the birds.* 4to, 4 folded pls, SPCK, London [c.1860]; ed. Stanford, London [c.1863]. Diagrams of Natural History No. X.

1427. —— *Catalogue of British birds.* 8vo, xii+247 pp, British Museum, London 1863.

—— *List of British birds.* 1871. *see* Gurney, J. H.

1428. GRAY, John Edward *Gleanings from the menagerie and aviary at Knowsley Hall.* Fol., 17 col. pls with letterpress, privately printed, Knowsley 1846.

1429. —— *List of specimens of British animals in the collection of the British Museum. Part I. Centroniae or radiated animals.* 12mo, xiii+173 pp, British Museum, London 1848.

1430. —— *List of specimens of British animals in the collection of the British Museum. Part II. Sponges.* 12mo, viii+24 pp, British Museum, London 1848.

1431. —— *Hoofed quadrupeds.* Fol., 76 pp, 62 col. pls, privately printed, Knowsley 1850.

1432. [——] *List of specimens of British animals in the collection of the British Museum. Part VII. Mollusca acephala and Brachiopoda.* 12mo, vi+167 pp, British Museum, London 1851.

1433. —— *Handbook of British water-weeds or Algae.* 12mo, iv+123 pp, Hardwicke, London 1864.

—— *see* Gray, S. F.

Gray, Johnnie *pseudonym. see* Speight, H.

1434. GRAY, *Mrs* Maria Emma *Figures of molluscous animals selected from various authors.* 8vo, 5 vols, 381 col. pls, Longman, London [1842–] 1859; ed. 4 vols 1859; ed. 2 vols 1874.

1435. GRAY, Peter & WOODWARD, Bernard Barham *Seaweeds, shells, and fossils.* Sm. 8vo, 94 pp, text figs, Sonnenschein, London [1886]. 2nd ed n.d. given as school prize 1896

—— *see* Holmes, E. M.

1436. GRAY, Robert *The birds of the west of Scotland including the Outer Hebrides.* Sm. 8vo, x+520 pp, 15 pls, Murray, Glasgow 1871.

1437. —— *The birds of Arran.* 8vo, 27 pp, Glasgow 1872. Reprinted from Bryce, J. *Geology of Arran.* 4ed. 1872.

1438. —— *Quadrupeds, birds, and fishes of Loch Lomond and its vicinity.* 8vo, Edinburgh 1877.

1439. —— & ANDERSON, Thomas *The birds of Ayrshire and Wigtownshire.* 8vo, 62 pp, 1 col. pl., Murray, Glasgow 1869. Reprinted from *Proc. nat. Hist. Soc. Glasgow*, 1869.

1440. GRAY, Samuel Frederick *A natural arrangement of British plants.* 8vo, 2 vols, 21 col. pls, Baldwin, London 1821. Largely written by his son John Edward Gray.

1441. Gray, Samuel Octavus *British sea weeds.* Sm. 8vo, xxiii+312 pp, 16 col. pls, Reeve, London 1867. Crown Series.

1442. GREEN, Joseph Reynolds *Manual of botany.* 8vo, 2 vols, text figs, Churchill, London 1895–96; 2ed. 1902; 3ed. Vol. 1 1904.

1443. GREEN, Thomas *The universal herbal, or botanical, medical and agricultural dictionary.* 4to, 2 vols, 109 pls, Liverpool [1816–20].

1444. GREENE, George Clark *Collections and recollections of natural history and sport in the life of a country vicar.* Sm. 8vo, xi+221 pp, text figs, Reeve, London 1886 [=1885]. Crown Series.

1445. GREENE, Joseph *Pupa digging.* 8vo, 23 pp, Newman, London 1857; facsimile ed. Faringdon 1979. Later printings of

1857 ed. are incorporated in No. 1446. Reprinted from *The Zoologist*, Jan. 1857.

1446. —— *The insect hunters' companion; being instructions for collecting and preserving butterflies and moths.* 8vo, 164 pp, text figs, Newman, London [1863]; Van Voorst, London 1863; 2ed. 1870; 3ed. 1880; 4ed. 1892. Later editions.

1447. GREENE, Joseph Reay *Manual of the animal kingdom. I. Protozoa. II. Coelenterata.* Sm. 8vo, 2 pts, text figs, Longman, London 1859–61; ed. 1863; ed. 1869, 71. All publ. The parts occur separately. Galbraith & Haughton's Scientific Manuals Vols 1 & 2.

1448. GREENE, William Thomas *The birds in my garden.* Sm. 8vo, 190 pp, 1 pl., text figs, RTS, London [1889].

1449. —— *Song birds of Great Britain.* 12mo, x+74 pp, text figs, May, London [1891].

1450. —— *Feathered friends, old and new.* 8vo, viii+302 pp, Upcott Gill, London 1896.

1451. GREENWOOD, James *The London vocabulary English and Latin.* 12mo, London c. 1710; 3ed. 1713; 5ed. 1723; 6ed. 1728; 9ed. 1740; 10ed. 1745; 12ed. 1752; 14ed. 1763; 15ed. 1767; 16ed. 1771; 18ed. 1782; 19ed. 1785; 20ed. 1791; 21ed. 1797; 22ed. 1802, 23ed. 1807; 24ed. 1812; 26ed. 1817; new ed. 1828; new ed. 1833; 3 languages ed. Cambridge 1816 title *The London and Paris vocabulary, English Latin and French.* An abridgement of Comenius, J. A. *Orbis sensualium pictus*, 1659. ?No copy of first ed. known to survive.

1452. GREGORY, George *The economy of nature explained and illustrated on the principles of modern philosophy.* 8vo, 3 vols, Johnson, London 1796; 2ed. 1798; 3ed. 1804.

1453. [GREGORY, John] *A comparative view of the state and faculties of man with those of the animal world.* 12mo, iv+203 pp, Dodsley, London 1765; 2ed. 1766; 3ed. 1766; 6ed. 1774 signed; 7ed. 1777; ed. Edinburgh 1788; ed. 1798.

1454. GREIG, John *The world displayed; or, the characteristic features of nature and art exhibited on a new plan.* 12mo, iv+664 pp, 1 pl., Cradock & Joy, London 1810.

1455. GREVILLE, Robert Kaye *Directions for preserving specimens of plants, particularly those of the class Cryptogamia with the least possible trouble.* 32mo, 10 pp, privately printed, Edinburgh 1822.

1456. —— *Scottish cryptogamic flora.* Roy. 8vo, 6 vols, 360 col. pls with letterpress, Maclachlan & Stewart, Edinburgh 1823–28.

1457. —— *Flora Edinensis: or a description of plants growing near Edinburgh.* 8vo, lxxxi+478 pp, 4 pls, Blackwood, Edinburgh 1824.

1458. —— *Algae Britannicae; or, descriptions of the marine and other inarticulated plants of the British Islands belonging to the order Algae.* 8vo, lxxxviii+218 pp, 19 col. pls, Maclachlan & Stewart, Edinburgh 1830.

1459. —— *The collected works of Dr. R. K. Greville on Diatomaceae.* 8vo, 1855–66. A collection of offprints with a printed title leaf.

1460. —— & ARNOTT, George Arnott Walker *Tentamen methodi muscorum. A new arrangement of the genera of mosses. Memoirs 1–3.* 8vo, 127 pp, 4 pls, [Edinburgh 1825].

1461. GREW, Nehemiah *The anatomy of vegetables begun.* 8vo, 198 pp, 3 pls, Hickman, London 1672.

1462. —— *An idea of a phytological history propounded.* 8vo, 144 pp, 7 pls, Chiswell, London 1673.

1463. —— *The comparative anatomy of trunks, together with an account of the vegetation grounded thereupon.* 8vo, 81 pp, 19 pls, Kettilby, London 1675.

1464. —— *Musaeum Regalis Societatis or, a catalogue and description of the natural and artificial rarities belonging to the Royal Society and preserved at Gresham College. Whereunto is subjoined the comparative anatomy of stomachs and guts.* Fol., 2 pts, 386+42 pp, 31 pls, 1 portr., Rawlins for the author, London 1681; 1685; 1686; 2ed. 1694. Errata are usually corrected in ink.

1465. —— *The anatomy of plants. With an idea of a philosophical history of plants.* Fol., 24+304 pp, 83 pls, Rawlins for the author, London 1682. Contains *An idea*, 2ed., *Anatomy of plants*, 2ed., *Anatomy of roots*, 2ed., *Anatomy of trunks*, 2ed., *Anatomy of leaves, flowers, fruits and seeds*, 1ed.

1466. GREY, Henry *Zoo notes for non-zoological visitors.* 12mo, 23 pp, Eden, London 1891.

1467. GRIEVE, Symington *The great auk or garefowl (Alca impennis Linn.).* 4to, x+141+58 pp, 2 col.+2 plain pls, 1 col. map, text figs, Jack, London 1885; additional notes Edinburgh 1898. The notes are reprinted from *Trans. Edinb. Fld. nat. micr. Soc.*

1468. —— *Recent information about the great auk or garefowl (Alca impennis Linn.).* 8vo, 27 pp, 2 pls, Edinburgh 1888. Reprinted from *Trans. Edinb. Fld nat. micr. Soc.*

1469. GRIFFITH, Edward *General and particular description of vertebrated animals . . . Order Quadrumana. (Order Carnivora).* Roy. 8vo, 2 vols, 95 col.+10 plain pls, Baldwin Craddock, London 1821. All publ. L.P.

1470. GRIFFITH, John Edwards *The flora of Anglesey and Caernarvonshire.* 8vo, xx+288 pp, 1 col. map, Nixon & Jarvis, Bangor [1895].

1471. GRIFFITH, John William *An elementary textbook of the microscope.* 12mo, ix+192 pp, 12 col. pls, Van Voorst, London 1864.

1472. —— & HENFREY, Arthur *The micrographic dictionary.* 8vo, xl+696 pp, 41 pls partly coloured, text figs, Van Voorst, London [1854–] 1856; 2ed. 2 vols 1860; 3ed. 1875; 4ed. 1883.

1473. GRIFFITHS, William Handsel *A system of botanical analysis applied to the diagnosis of British natural orders.* 8vo, vi+26 pp, London 1870.

1474. GRINDON, Leopold Hartley *The sexuality of nature.* 8vo, Pitman, London 1853.

1475. —— *Life: its nature, varieties and phenomena. Also times and seasons.* 8vo, viii+328 pp, Whittaker, London 1856; 2ed. 1857; 3ed. 1863; extracts from, 1866.

1476. —— *The little things of nature.* Sm. 8vo, 105 pp, Pitman, London 2ed. 1865; shorthand ed. 1876.

1477. —— *The Manchester flora . . . preceded by an introduction to botany.* 8vo, viii+575 pp, text figs, White, London 1859.

1478. —— *Manchester walks and wildflowers.* 12mo, xi+172 pp, Whittaker, London [1859]; ed. 1882 title *Country rambles and . . .*

1479. —— *A manual of British and foreign plants.* 8vo, xi+271 pp, Pamplin, London 1861.

1480. —— *The phenomena of plant life.* 8vo, 40 pp, Manchester 1864.

1481. —— *British and garden botany.* 8vo, xiii+869 pp, text figs, Routledge, London 1864.

1482. —— *Summer rambles in Cheshire, Derbyshire, Lancashire, and Yorkshire.* 8vo, xi+251 pp, 1 pl., text figs, Palmer & Howe, Manchester 1866.

1483. —— *The trees of old England.* 8vo, 96 pp, text figs, Pitman, London 1868; 2ed. 1870.

1484. —— *Echoes in plant and lower life.* 8vo, v+123 pp, Pitman, London 1869.

1485. —— *The pathway to botany.* 12mo, viii+175 pp, text figs, Pitman, London 1872.

—— *Country rambles and wildflowers.* 1882. *see* No. 1478.

1486. —— *The Shakespere flora.* Sm. 8vo, xii+138 pp, pls, text figs, Palmer & Howe, Manchester 1883; 2ed. 1883.

Groom, Charles Ottley *see* Napier, C. O. G.

1487. GROOM, Percy *Elementary botany.* 8vo, x+252 pp, Bell, London 1898. Many later editions.

1488. GROOS, Karl *The play of animals: a study of animal life and instinct.* 8vo, xxvi+341 pp, Chapman and Hall, London 1898. Translation of *Die Spiele der Tiere*, Jena 1896 by Elizabeth L. Baldwin.

1489. GROVE (*Hon.* Harriet, *Mrs* Edward) *The calendar of nature, or the seasons of England.* Fol. 4 pts, 24 col. pls, Dickinson, London 1849–50.

1490. GROVE, William Bywater *A synopsis of the Bacteria and yeast Fungi and allied species (Schizomycetes and Sacromycetes).* Sm. 8vo, 112 pp, text figs, Chatto, London 1884; 1891.

Grufberg, Isaac Claus *Flora Anglica.* 1754. *see* Linnaeus, C.

1491. GRUNDY, *Sir* Cuthbert Cartwright *Notes on the food of plants.* Sm. 8vo, 142 pp, Simpkin, London 1871.

1492. GÜNTHER, Albert Carl Ludwig Gotthilf *An introduction to the study of fishes.* 8vo, xvi+720 pp, 1 pl., text figs, Black, Edinburgh 1880; reprint 1963.

1493. —— *Instructions for collecting reptiles, batrachians, and fishes.* 8vo, 18 pp, British Museum, London [1891].

*A guide to the Mount's Bay and the Land's End.* 2ed. 1824. *see* Paris, J. A.

1494. GULLIVER, George *A catalogue of plants collected in the neighbourhood of Banbury.* 12mo, 37 pp, London 1841.

1495. —— *Notes on some of the researches in anatomy, physiology and botany.* 8vo, [London 1870]; ed. Canterbury 1880.

1496. GUNN, T. E. *The fish-eating birds of Norfolk.* 8vo, 12 pp, Argus Office, Norwich [ ]. Reprinted from *East Anglian Handbook and Agricultural Annual.*

1497. GURNEY, John Henry *the father* & FISHER, William Richard *An account of the birds found in Norfolk including notices of some of the rarer species which have occurred in the adjoining counties: with remarks on migration.* 8vo, 34 pp, text figs, Newman, London 1846. Reprinted from *The Zoologist.*

1498. GURNEY, John Henry *the son. A summary of the occurrences of the grey phalarope in Great Britain during the autumn of 1866.* 8vo, 24 pp, 1 map, Van Voorst, London 1867.

1499.—— *A list of British birds and doubtful British birds. Gray's arrangement.* Single sheet, privately printed, London 1871.

1500.—— *Rambles of a naturalist in Egypt . . . with an analysis of the claims of certain foreign birds to be considered British.* 8vo, vi+307 pp, Jarrold, London [1876].

1501.—— *A catalogue of the birds of Norfolk.* Sm. 8vo, 47 pp, Wertheimer Lea for R. H. Mason, London 1884. Reprinted from appendix to Mason, R. H. *The history of Norfolk*, 1884.

1502.—— *On the misdeeds of the house sparrow (Passer domesticus)*. Sm. 8vo, 9 pp, Gurney & Jackson, London 1887.

1503.—— and others. *The house sparrow; by an ornithologist, J. H. Gurney junr, by a friend of the farmers, Col. C. Russell; and the English sparrow in America by . . . E. Coues.* 8vo, vi+70 pp, 1 pl., Wesley, London [1885].

Guy's Hospital *see* Smith, P. H. P.

1504. GUYON, George *A list of British species of Geodephaga: intended for marking desiderata and labelling collections.* 8vo, Richmond, Surrey 1855. Based on Dawson, J. F. *Geodephaga Britannica*, 1864.

# H

H., A. *Trees and their nature.* 1856. *see* Harvey, A.

H., A. H. *Prodromus Lepidopterorum Britannicarum.* 1802. *see* Haworth, A. H.

H*, C* *The book of creation unfolded.* 1828. *see* Hoddinott, C.

H., J. *The parlour menagerie.* 1875. *see* Hogg, J.

H., M. M. *Ocean flowers and their teachings.* 1846. *see* Howard, M. M.

H., T. *The dragon-fly.* 1882. *see* Hyatt, T.

H., W. *Phytologia Britannica.* 1650. *see* How, W.

1505. [HACK, *Mrs* Maria] *Stories of animals intended for children between five and seven years old.* [First series.] 16mo, 135 pp, 12 pls, Harvey & Darton, London 1820; 2ed. 1822; 3ed. 1826; 4ed. 1831. Based on William Bingley, *Animal biography*, 1803 and John Church, *Cabinet of quadrupeds*, 1805.

1506. —— *Harry Beaufoy; or, the pupil of nature.* 12mo, vii+194 pp, 1 pl., Harvey & Darton, London 1821; 2ed. 1824; 3ed. 1830; 4ed. 1836. Based on Paley, W. *Natural theology*, 1802.

1507. [——] *Stories of animals intended for children between five and seven years old. Second series.* 16mo, 144 pp, 12 pls, Harvey & Darton, London 1831; 1836.

1508. HAECKEL, Ernst Heinrich Philipp August *The history of creation.* 8vo, 2 vols, 15 pls, text figs, King, London 1876 [=1875]; 2ed. 1876. Later editions. Translation of *Naturliche Schöpfungsgeschichte*, 4ed. Berlin 1874, revised by Edwin Ray Lankester.

1509. —— *The evolution of man.* 8vo, 2 vols, 15 pls, text figs, Kegan Paul, London 1879. Many later editions in various formats. Translation of *Anthropogenie*, 2ed. Leipzig 1874, by Joseph McCabe.

1510. —— *Freedom in science and teaching.* [A reply to R. Virchow.] Sm. 8vo, Kegan Paul, London 1879. Translation of *Freie Wissenschaft und freie Lehre*, Stuttgart 1878, by Thomas Henry Huxley.

1511. —— *The last link. Our present knowledge of the descent of man.* 8vo, 156 pp, Black, London 1898; 2ed. 1899. Later printings. Translation of *Ueber unsere gegenwärtige Kenntniss vom Ursprung des Menschen*, Bonn 1898, with notes by Hans Gadow.

1512. HAILEYBURY NATURAL HISTORY SOCIETY *Fauna and flora of Haileybury. Part 1.* 8vo, 45 pp, for the Society, Hertford 1888; ed. 1902; 3ed. 1926. All publ.

1513. HALES, Stephen *Vegetable staticks.* 8vo vii+376 pp, 14 pls, Innys, London 1727; reprint 1961. Later editions in No. 1514.

1514. —— *Statical essays.* 8vo, 2 vols, 19 pls, Innys, London Vol. 1 2ed. 1731; 3ed. 1738; 4ed. 1769; Vol. 2 1733; 2ed. 1740; 3ed. 1769. First edition of Vol. 1 is No. 1513.

1515. *Half hours in field and forest.* Sm. 8vo, 384 pp, text figs, Nisbet, London 1896. Half Hour Library.

1516. *Half hours in the tiny world: wonders of insect life.* Sm. 8vo, xi+311 pp, Daldy Isbister, London 1875. Later printings. Half Hour Library.

*Half hours with a naturalist. Rambles near the shore.* 1897. *see* Wood, J. G.

1517. *Half hours with the animals.* Sq. 8vo, xiii+239 pp, 16 pls, text figs, Seeley, London 1876.

*Half hours with the microscope.* [1859]. *see* Lankester, E.

1518. HALIDAY, Alexander Henry *Hymenoptera Britannica. Fasc. 1. Oxyura. (Fasc. altera. Alysia).* 8vo, 15+28+4 pp, Baillière, London 1839. All publ.

1519. HALL, *Mrs* S. C., Anna Maria *Animal sagacity.* 8vo, vi+168 pp, Partridge, London 1866.

1520. HALL, John Charles *Interesting facts connected with the animal kingdom.* 8vo, xvi+301 pp, 4 pls, Whittaker, London 1841.

1521. HALL, Robert *Elements of botany, or an introduction to the sexual system of Linnaeus.* 8vo, vii+219 pp, 2 pls, Hodson, London 1802.

1522. HALL, Thomas B. *A flora of Liverpool.* 12mo, viii+186 pp, 1 pl., 1 map, 2 tabs, Whittaker, London [1839].

1523. HALLE, Hughes R. P. Fraser *Letters historical and botanical relating chiefly to places in the Vale of Teign.* 8vo, 158 pp, Houlston, London 1851.

1524. HALSTEAD, Caroline Amelia *The little botanist.* Sm. 16mo, 2 pts, 2 col.+28 plain pls, Harris, London 1835; ed. 1836. The Little Library Vols 16 & 17.

1525. HAMERTON, Philip Gilbert *Chapters on animals.* 420, 252 pp, 20 pls, Seeley, London 1874. Later editions.

1526. HAMILTON, Edward *The river-side naturalist.* Roy. 8vo, vii+401 pp, text figs, Sampson Low, London 1890.

1527. —— *The wild cat of Europe. (Felis cattus).* Roy. 8vo, xxi+99 pp. 9 pls, text figs, Porter, London 1896.

1528. HAMILTON, Gerald Edwin Hamilton Barrett- *Harrow birds.* 16mo, xiii+50 pp, Harrow School Scientific Society, Harrow 1892.

1529. HAMILTON, Robert *The natural history of the ordinary Cetacea or whales.* Sm. 8vo, 264 pp, 28 col.+2 plain pls, 1 portr., Lizars, Edinburgh 1839: ed. [1845–46]; issue 1848. Jardine's Naturalist's Library Vol. XVI, 2ed. Vol. XXVI.

1530. —— *The natural history of the amphibious Carnivora.* Sm.

8vo, 336 pp, 30 col. pls, 1 portr., text figs, Lizars, Edinburgh 1839; ed. [1845–46]; issue 1848. Jardine's Naturalist's Library Vol. XXIII; 2ed. Vol. XXVII [=XV].

1531. —— *The natural history of British fishes.* Sm. 8vo, 2 vols, 68 col. pls, 2 portrs, Lizars, Edinburgh 1843; ed. [1845–46]; issue 1843. Jardine's Naturalist's Library Vols XXXVII & XXXVIII; 2ed. Vols XXVI & XXVII or XXXII & XXXIII.

Hammond, Arthur Rashdall *see* Miall, L. C.

1532. [HAMMOND, Thomas William] *Flowers and plants ... sketched ... from nature.* Fol., 60 col. pls, Mounteney, Nottingham [1879–81]. No text. Plates are signed Émile Favart, a pseudonym.

1533. HANBURY, Frederick Janson *An illustrated monograph of the British Hieracia.* Roy. 4to, 69 pp, 24 col. pls, privately printed, London 1889–98 [–1899]. Pts 1–8 all publ. Pp. 33–36 not published; text of p. 69 on wrapper of Pt. 8.

—— *editor The London catalogue of British plants.* Bed. 1886. *see* Watson, H. C. & Dennes, G. E.

1534. —— & MARSHALL, Edward Shearburn *Flora of Kent.* Sm. 8vo, lxxxiv+444 pp, 2 maps, 1 col., F. J. Hanbury, London 1899.

Hancock, Albany *see* Alder, Joshua.

1535. HANCOCK, John *A fasciculus of eight drawings on stone of groups of birds. The whole being representations of specimens ... contributed ... to the Great Exhibition of 1851.* Fol., 2 11, 8 col. pls, by the author, Newcastle upon Tyne 1853. ?Very rarely coloured.

1536. —— *A catalogue of the birds of Northumberland and Durham.* 8vo, xxvi+174 pp, 14 pls, Williams & Norgate, London 1874. Reprinted from *Trans. nat. Hist. Soc. Northumb. Durham*, Vol. 6, 1873.

1537. *The handbook for Torquay ... with the natural history of the district.* Sm. 8vo, vii+293 pp, Croydon, Torquay [1854].

1538. *Handbook of natural history. An explanation of the natural history diagrams for the use of schools.* 8vo, xvi+234 pp, Bencard, London 1873. This is accompanied by 20 wall charts, 3 with specimens, which are translations of Deyrolle, Émile *Histoire naturelle*, Paris [1872].

Handbooks for children. (C. Tilt). *see* May, W. and Williams, C.

1539. HANHAM, Frederick *Natural illustrations of the British grasses.* Fol., xix+130 pp, 62 specimens, Binns & Goodwin, Bath 1846.

1540. HANLEY, Sylvanus Charles Thorp *The young conchologist's book of species. Univalves.* 12mo, viii+146 pp, 1 col. pl., text figs, Fraser, London 1840; 2ed. 1842 title *The conchologist's book of species.*

1541. —— *An illustrated and descriptive catalogue of recent bivalve shells.* 8vo, xviii+392+24 pp, 16 col. pls, Williams & Norgate, London 1842–56. Title of first part *An illustrated, enlarged and English edition of Lamarck's Species of shells.* This is a companion volume to Hanley's edition of Wood, W. *Index testaceologicus*, 1856. The plates are therefore numbered 9–24.

1542. —— *Ipsa Linnaei conchylia. The shells of Linnaeus described from his manuscript and collection.* 8vo, 556 pp, 5 col. pls, Williams & Norgate, London 1855.

—— *see* Forbes, E.

1543. HANNAFORD, Samuel *Flora Tottoniensis: a catalogue of the flowering plants and ferns . . . of Totnes.* 8vo, Longman, London 1851; suppl. [1852].

1544. HARDCASTLE, Lucy *An introduction to the elements of the Linnaean system of botany.* 4to, 154 pp, text figs, Richardson, Derby 1830; 2ed. London 1833.

1545. HARDING, James Duffield *The park and the forest.* Fol., 26 pls, Maclean, London 1841.

1546. —— *Lessons on trees.* 4to, 30 pls, Bogue, London 1850; ed. 1852.

1547. HARDWICKE, Herbert Junius *Evolution and creation.* 8vo, 10 pts, 52 pls, by the author, Sheffield 1887. The parts have separate paginations.

1548. HARDY, James *A biography of border wild flowers. The ribwort plantain.* 8vo, 11 pp, Alnwick 1864. Reprinted from *The Alnwick Mercury.*

1549. HARMER, Sidney Frederic & SHIPLEY, Arthur Everett *editors The Cambridge natural history.* 8vo, 10 vols, text figs, Macmillan, London 1895–1906; issue 1936; reprint 1958–59.

*Harmonia ruralis; or, an essay towards a natural history of British song birds.* 1824. *see* Bolton, J.

1550. HARPER, John *The sea-side and aquarium.* Sm. 8vo, xi+188 pp, 1 pl., text figs, Nimmo, Edinburgh 1858.

1551. —— *Glimpses of ocean life; or rock pools and the lessons they teach.* 8vo, xiv+379 pp, 12 pls, Nelson, London 1860.

1552. HARRIS, *Miss Coloured drawings of British butterflies from the collection of Mr. W. E. Leach.* Imp. 4to, 4 pts, 12 col. drawings with letterpress, Exeter [1809]. Very few copies.

1553. HARRIS, James *May flowers; a popular and scientific description of the flowers of the month.* 8vo, Chester 1872.

1554. HARRIS, Moses *The aurelian: or natural history of British insects.* Fol., 77 pp, 41 col. pls, for the author, London 1766; issue 1766 title . . . *English moths and butterflies*; ed. 1778; issue 1794; ed. 1840–41.

1555. —— *An essay preceding* [*sic*] a supplement to The aurelian. 4to, 12 pp, 7 col. pls, for the author, London [1767].

1556. —— *The English Lepidoptera: or, the aurelian's pocket companion.* 8vo, xv+66 pp, 1 col. pl., Robson, London 1775.

1557. —— *An exposition of English insects.* Fol., viii+166 pp, 51 col.+2 plain pls, 1 portr., for the author, London 1776 [–1780]; 1781; ed. 1782; ed. 1789; ed. 1791. 1789 ed. is made up from sheets of the earlier eds.

1558. HARRISON, J. *Harrison's natural history of the most remarkable beasts.* 32mo [16]pp, no text, woodcuts, Devizes 1824; 7ed. 1824.

1559. —— *Harrison's natural history of the most remarkable birds.* 32mo, [16]pp, no text, woodcuts, Harrison, Devizes 1824; 7ed. 1824.

—— *Harrison's picture book for children, birds and beasts.* A combination volume of Nos 1558 & 1559.

1560. HART, E. *A catalogue of the birds and animals in Hart's Museum, Christchurch.* 8vo, 38 pp, Southampton 1894.

1561. HART, Henry Chichester *A list of plants found in the islands of Arran, Galway Bay.* 8vo, 32 pp, 1 map, Dublin 1875.

1562. —— *The flora of Howth.* Sm. 8vo, 137 pp, 1 map, Hodges & Figgis. Dublin 1887.

1563. —— *Flora of County Donegal.* 8vo, xxiv+391 pp, 1 col. map, Sealy, Dublin 1898.

1564. HARTING, James Edmund *The birds of Middlesex.* Sm. 8vo, xvi+284 pp, 1 tinted pl., text figs, Van Voorst, London 1866.

1565. —— *The ornithology of Shakespeare critically examined, explained and illustrated.* Sm. 8vo, xxii+321 pp, 1 portr., text figs, Van Voorst, London 1871.

1566. —— *Hints on shore shooting; with a chapter on skinning and preserving birds.* Sm. 8vo, 88 pp, Van Voorst, London 1871.

1567. —— *A handbook of British birds.* 8vo, xxi+198 pp, Van Voorst, London 1872; ed. 1901.

1568. —— *Rambles in search of shells, land and freshwater.* 8vo, viii+110 pp, 10 col. pls, Van Voorst, London 1875.

1569. —— *Our summer migrants.* 8vo, x+336 pp, text figs, Bickers, London 1875; 2ed. 1877; ed. 1889; ed. 1901.

1570. —— *Some British animals extinct within historic times, with some account of the British wild white cattle.* 8vo, vii+258 pp, text figs, Trübner, London 1880.

1571. —— *Essays on sport and natural history.* 8vo, x+485 pp, 1 portr., text figs, Cox, London 1883.

1572. —— *Sketches of bird life from twenty years observation.* 8vo, xii+292 pp, text figs, Allen, London 1883.

1573. —— & ROBERT, Leo Paul Samuel *Glimpses of bird life, portrayed with pen and pencil.* Fol., v pp, 20 col. pls with letterpress, text figs, Sonnenschein, London 1880.

1574. HARTWIG, George Ludwig *The sea and its living wonders.* 8vo, xviii+427 pp, 12 col. pls, 1 map, text figs, Longman, London 1860; 2ed. 1861; 3ed. 1866; 4ed. 1873. Later editions. Translation of *Das Leben des Meeres*, 4ed. Glogau ?1859. See also No. 1577 & 1578.

1575. —— *The harmonies of nature; or, the unity of creation.* 8vo, xix+406 pp, text figs, Longman, London 1866. Translation of *Gott in der Natur*, Wiesbaden 1864.

1576. —— *The aerial world.* 8vo, xviii+566 pp, 8 pls, 1 map, Longman, London 1874; abridged ed. 1888 title *Marvels over our heads.* Translation of *Das Leben des Luftmeeres*, Wiesbaden 1872.

1577. —— *Denizens of the deep. An account of the fishes, molluscs, Crustacea . . . from The sea and its living wonders.* 8vo, 160 pp, text figs, Longman, London 1887. Extracts from No. 1574.

1578. —— *Sea monsters and sea birds . . . from The sea and its living wonders.* 8vo, 159 pp, text figs, Longman, London 1887. Extracts from No. 1574.

—— *Marvels over our heads.* 1888. *see* No. 1576.

1579. [HARVEY, Alexander] *Trees: their nature, and natural longevity and size.* 4to, Aberdeen 1847. Reprinted from *The Family Journal.*

1580. [——] *Trees and their nature; or, the bud and its attributes.* 8vo, xxiii+236 pp, 3 pls, text figs, Nisbet, London 1856. By A. H.

Harvey, Joshua Reuben *see* Cuvierian Society of Cork.

1581. HARVEY, William *Exercitationes de generatione*

*animalium.* 12mo, 568 pp, Elzevier, Amsterdam 1651; ed. 4to, London 1651; ed. 12mo, Janssen, Amsterdam 1651; ed. 12mo, Amsterdam 1622; ed. 12mo, The Hague 1680; English ed. 8vo, London 1653.

1582. HARVEY, William Henry *A manual of the British Algae.* 8vo, lxii+229 pp, Van Voorst, London 1841; 2ed. 1849 title . . . *marine Algae.*

1583. —— *Phycologia Britannica: or, a history of British sea-weeds.* 4to, or Roy. 8vo, 3 or 4 vols, 360 col. pls, Reeve, London 1846–51; issue 1871. 3 vols in order of publication, 4 vols arranged systematically.

1584. —— *The sea-side book; being an introduction to the natural history of the British coasts.* Sm. 8vo, viii+247 pp, text figs, Van Voorst, London 1849; ed. 1849; 3ed. 1854; 4ed. 1857.

1585. [——] *Synopsis of British seaweeds, compiled from Professor Harvey's Phycologia Britannica.* Sm. 8vo, viii+219 pp, atlas vol. Roy. 8vo, 80 col. pls, Reeve, London 1857. Preface signed W.H.R.

1586. —— *An inquiry into the probable origin of the human animal, on the principles of Mr Darwin's theory of natural selection, and in opposition to the Lamarckian notion of a monkey parentage.* 8vo, vi+37 pp, privately printed, Dublin [1860].

—— *see* Gatty, Margaret.

Harvie-Brown, John Alexander *see* Brown, J. A. Harvie-

1587. HASSALL, Arthur Hill *A history of the British freshwater Algae.* 8vo, 2 vols, 103 col. pls, Longman, London 1845; issue 1852; 1857.

1588. HASSELL, Joseph *Common things and elementary science.* Sm. 8vo, vii+384 pp, Blackie, London [1883]. Many later editions.

1589. HASTINGS, *Sir* Charles *Illustrations of the natural history of Worcestershire.* 8vo, iv+184 pp, 1 col. map, Sherwood Gilbert & Piper, London 1834. Publ. of Worcestershire Nat. Hist. Soc.

Haswell, William Aitcheson *see* Parker, T. J.

1590. HATSCHEK, Berthold *The Amphioxus and its development.* Sm. 8vo, 181 pp, 9 pls, Sonnenschein, London 1893. Translation of 'Studien über Entwicklung des Amphioxus' in *Arb. zool. Inst. Wien*, Vol. 4, by James Grove White Tuckey.

1591. HAUGHTON, Samuel *On the forms of the cells made by various wasps, and by the honey bee: with an appendix on the origin of species.* 8vo, 13 pp, Dublin 1863.

1592. —— *The three kingdoms of nature briefly described.* Sm. 8vo, xx+387 pp, text figs, Cassell, London [1869].

1593. —— *Principles of animal mechanics.* 8vo, xiv+495 pp, text figs, Longman, London 1873; 2ed. 1873.

1594. HAVELL, Robert *A selection of British birds from drawings by C. L. E. Perrott . . . of the Worcestershire Natural History Society. Part 1.* Fol., 5 col. pls, London [c.1835]. ?All publ. Very few copies. Facsimile 1980.

1595. HAVILAND, George Darby *Some factors in the evolution of adaptations.* 8vo, 2 pts, Porter, London 1897.

1596. HAWKINS, Benjamin Waterhouse *A comparative view of the human and animal frame.* Fol., 26 pp, 10 pls, Chapman & Hall, London 1860.

1597. [HAWKS, Francis Lister] *Uncle Philip's conversations with children about the habits and mechanical employments of inferior animals.* 12mo, xv+170 pp, text figs, Tegg, London 1834; 3ed. 16mo, London 1837 title . . . *about the tools and trades among the inferior animals*; ed. 16mo, Leeds 1846; 1849; London 1858; 1867. First edition is New York 1834.

1598. [HAWORTH, Adrian Hardy] *Prodromus Lepidopterorum Britannicarum . . . a . . . catalogue of the British lepidopterous insects.* 4to, vi+39+6 pp, Parslee, Holt 1802; facsimile Feltham 1931. By a Fellow of the Linnean Society, A.H.H.

1599. —— *Lepidoptera Britannica.* 8vo, xxxvi+609 pp, Murray, London 1803 [–1828]; errata 1 leaf [ ].

1600. —— *Miscellanea naturalis, sive dissertationes variae ad historiam naturalem spectantes.* 4to, xiv+204 pp, Taylor printed, London 1803.

1601. HAY, Henry Maurice Drummond- *The Grallatores and Natatores of the estuary of the Tay.* 8vo, 24 pp, Dundee Naturalists Society, Dundee 1881.

1602. HAY, William Delisle *An elementary text-book of British Fungi.* 8vo, vi+238 pp, 59 pls, 5 tabs, Sonnenschein, London 1887.

1603. —— *The fungus hunter's guide and field memorandum book.* Sm. 8vo, 157 pp, text figs, Sonnenschein, London 1887; ed. 1906.

1604. HAYES, Charles *The portraits of British birds.* 4to, 120 pp, 120 col. pls, for the author, London 1808.

1605. HAYES, William *A natural history of British birds . . . with their portraits.* Roy. fol., 24 pp, 40 col. pls, Hooper, London [1771–] 1775; ed. pt. 1 1822.

1606. —— *Portraits of rare and curious birds, with their descriptions from the Menagery of Osterley Park.* 4to, 2 vols, 101 col. pls with letterpress, Fauldey for the author, London 1794, 99.
1607. HAYWARD, Jane Mary *Bird notes.* Sm. 8vo, xvii+181 pp, 1 pl., text figs, Longman, London 1895.
1608. HAYWARD, W. R. *The botanist's pocket-book.* Sm. 8vo, xxxvii+198 pp, Bell & Daldy, London 1872; 2ed. 1877; 3ed. 1881; 4ed. 1883; 5ed. 1886; 7ed. 1892; 8ed. 1896; 9ed. 1901; 10ed. 1902; 11ed. 1904; 12ed. 1907; 13ed. 1909; 14ed. 1914; 15ed. 1917; 16ed. 1919; 17ed. 1922; 18ed. 1926; 19ed. 1930. Later reprints.
1609. HEADLEY, Frederick Webb *The structure and life of birds.* 8vo, xx+412 pp, 1 pl., text figs, Macmillan, London 1895.
1610. —— *Problems of evolution.* 8vo, xv+373 pp, Duckworth, London 1900.
1611. HEATH, Francis George *The fern world.* 8vo, xi+459 pp, 17 col. pls, text figs, Sampson Low, London 1877; 2ed. 1877; 3ed. 1877; 4ed. 1878; 5ed. 1878; 6ed. 1879. Later editions.
1612. —— *Our woodland trees.* Sm. 8vo, 8 col. pls, text figs, Sampson Low, London 1878; 2ed. 1878; 3ed. 1878.
1613. —— *Trees and ferns.* 8vo, 138 pp, text figs, Sampson Low, London 1879. Consists of extracts from *The fern world, Our woodland trees* and *The fern paradise.*
1614. —— *Sylvan spring.* 8vo, xvi+446 pp, 12 col. pls, text figs, Sampson Low, London 1880.
1615. —— *Autumnal leaves.* 8vo, xvi+352 pp, 12 col. pls, text figs, Sampson Low, London 1881; 2ed. 1883; 3ed. 1885; 4ed. 1899.
1616. —— *Where to find ferns. With a special chapter on ferns around London.* 8vo, 118 pp, Sampson Low, London 1882. Later editions.
1617. —— *The fern portfolio . . . All the species of British ferns.* Fol., 15 col. pls with letterpress, SPCK, London 1885; ed. 1885; 3ed. 1885.
1618. —— *Tree gossip.* 8vo, viii+176 pp, Field & Tuer, London 1885 [=1884].
1619. —— *Sylvan winter.* 8vo, xvi+354 pp, 17 pls, text figs, Kegan Paul, London 1886.
1620. HEATH, Robert *A natural and historical account of the Islands of Scilly . . . and lastly a general account of Cornwall.* 8vo, xvi+xiii+456 pp, 1 pl., 1 map, text figs, Manby & Cox, London 1750.

1621. HEATHCOTE, John Norman *St Kilda.* 8vo, xiii+229 pp, 3 pls, 1 map, text figs, Longman, London 1900.

Heddle, Robert *see* Baikie, W. B.

1621a. HEHN, Victor *The wanderings of plants and animals from their first home.* 8vo, xiii+523 pp, Sonnenschein, London 1885; issue 1888; ed. 1891 title. *Cultivated plants and domestic animals in their migration from Asia to Europe.* Translation of *Kulturpflanzen und Hausetiere, in ihren Übergang aus Asien*, Berlin 1870, 3ed. 1877, by James Steven Stallybrass.

1622. HEILPRIN, Angelo *The geographical and geological distribution of animals.* Sm. 8vo, xii+435 pp, 1 map, Kegan Paul, London 1887; 3ed. 1907. International Scientific Series Vol. LVIII.

1623. HELE, Nicholas Fenwick *Notes or jottings about Aldeburgh, Suffolk; relating to matters . . . ornithological and entomological.* Sm. 8vo, ii+198 pp, 5 pls, Smith, London 1870; 2ed. Ipswich 1890.

1624. HENFREY, Arthur *Outlines of structural and physiological botany.* Sm. 8vo, xvi+245 pp, 18 pls, Van Voorst, London 1847.

1625. —— *editor Reports and papers on botany.* 8vo, viii+514 pp, 1 col.+2 plain pls, Ray Society, London 1849. Ray Society Publ. No. 16 for 1849.

1626. —— *The rudiments of botany.* 16mo, vi+249 pp, text figs, Van Voorst, London 1849; 2ed. 8vo 1858.

1627. —— *The vegetation of Europe, its conditions and causes.* Sm. 8vo, 387 pp, 1 map, Van Voorst, London 1852. Outlines of the Natural History of Europe, Vol. 1.

1628. —— *editor Botanical and physiological memoirs.* 8vo, xxvi+567 pp, 6 col. pls, Ray Society, London 1853. Ray Society Publ. No. 24 for 1853.

1629. [——] *Tabular view of the vegetable kingdom arranged according to the natural orders.* 4to, 6 folding plates, SPCK, London 1855; ed. Stanford, London [?1860]. Diagrams of Natural History No. IV.

1630. —— *An elementary course of botany, structural, physiological and systematic.* 12mo, xii+702 pp, text figs, Van Voorst, London 1857; 2ed. 1870; 3ed. 1878; 4ed. 1884. The flowerless plants by Bennett, A.

—— *see* Griffith, J. W.

—— *see* Tulk, A.

1631. HENNEDY, Roger *The Clydesdale flora.* 8vo, 8 pls, Glasgow 1865; 2ed. 1869; 3ed. 1874; ed. 1878; 5ed. 1891.
1632. HENRY, Alexander *A glossary of scientific terms for general use.* Sm. 8vo, 202 pp, Walton & Maberly, London 1861 [=1860].
1633. HENSLOW, George *The student's catalogue of British plants.* 8vo, 44 pp, Bateman, London 1879.
1634. —— *Floral dissections, illustrative of typical genera of British natural orders.* Obl. 8vo, iv+20 pp, 8 pls, Stanford, London 1879.
1635. —— *Botany for children. An illustrated elementary textbook.* 8vo, xv+103 pp, text figs, Stanford, London 1880; 3ed. 1881; 4ed. 1881 title *Botany for beginners.* Later editions.
1636. —— *The origin of floral structures through insect and other agencies.* Sm. 8vo, xix+349 pp, text figs, Kegan Paul, London 1888; 2ed. 1893. International Scientific Series. Vol. LXIV.
1637. —— *The making of flowers*. 8vo, vi+168 pp, text figs, SPCK, London 1891. The Romance of Science Series.
1638. —— *The origin of plant structures by self adaptation to the environment*. 8vo, xiii+256 pp, text figs, Kegan Paul, London 1895. International Scientific Series, Vol. LXXVII.
1639. —— *How to study wild flowers*. 8vo, 224 pp, text figs, RTS, London 1896. Later editions.
1640. HENSLOW, John Stevens *Syllabus of a course of lectures on botany*. 8vo, 49 pp, Cambridge 1828; 2ed. 1848; 3ed. 1853.
1641. —— *A catalogue of British plants arranged according to the natural system*. 8vo, 40 pp, Cambridge 1829; 2ed. 1835.
1642. —— *The principles of physiological and descriptive botany*. 8vo, vii+322 pp, text figs, Longman, London 1835; ed. 1844 title *The principles of descriptive and physiological botany*. Lardner's Cabinet Cyclopaedia.
1643. —— *British plants growing wild . . . in the parish of Hitcham, Suffolk*. Fol., s. sh., Ipswich 1851; ed. [1855]; ed. 1859.
1644. —— *Illustrations to be employed in practical lessons on botany, adapted to beginners*. 8vo, Chapman & Hall, London 1858.
1645. —— *A dictionary of botanical terms*. Sm. 8vo, 218 pp, text figs, Groombridge, London [1858].
—— *see* Oliver, D.
1646. —— & SKEPPER, Edmund *Flora of Suffolk*. 12mo, x+140 pp, Simpkin & Marshall, London [1860].

*The herb of the field.* 1853. *see* Yonge, *Miss* C. M.

1647. HERDMAN, John *Essays on the causes and phenomena of animal life*. 8vo, xii+236 pp, Creech, Edinburgh 1795.

1648. HERDMAN, William Abbott *A phylogenetic classification of animals*. 8vo, 76 pp, 1 pl., text figs, Macmillan, London 1885.

1649. —— *Handbook to Liverpool and the neighbourhood . . . Appendix, The Isle of Man*. 8vo, v+191 pp, 5 col. maps, British Association, Liverpool 1898.

1650. —— *L.M.B.C. memoirs on typical marine plants and animals . . . I. Ascidia*. 8vo, v+52 pp, 5 pls, Dobb for the Liverpool Marine Biological Committee, Liverpool 1899.

—— *see* Leslie, G.

1651. HERON, Robert *Elegant extracts of natural history*. 8vo, 2 vols, Balfour, Edinburgh 1792.

1652. HERRICK, *Mrs* Sophia Bledsoe *The wonders of plant life under the microscope*. 8vo, v+248 pp, text figs, Allen, London 1884. First ed. 12mo, New York 1883.

1653. HERSCHEL, *Sir* John Frederick William *editor A manual of scientific inquiry; prepared for the use of Her Majesty's Navy and adapted for travellers in general*. 8vo, xi+488 pp, 2 charts, text figs, Murray, London 1849; 2ed. 1851; 3ed. 1859; 4ed. 1871; 5ed. 1886; erratum leaf 1906.

1654. HETT, Charles Louis *A dictionary of bird notes, to which is appended a glossary of popular, local and old-fashioned synonyms of British birds*. 8vo, 136 pp, Jacksons, Brigg 1898; eds of the glossary alone [1899]; London 1902.

1655. HEURCK, Henri Ferdinand van *The microscope; its construction and management*. Roy. 8vo, xv+382 pp, 3 pls, 1 portr., text figs, Crosby Lockwood, London 1893. Translated from *La microscope*, Brussels 1865, 4ed. 1891, by Wynne Edwin Baxter.

1656. —— *A treatise on the Diatomaceae*. Imp. 8vo, xx+558 pp, 36 pls, text figs, Wesley, London 1896; reprint 1962. Translation of *Traité des diatomacées*, Anvers 1889, by Wynne Edwin Baxter.

1657. HEWETSON, Henry Bendelack *Nature cared for and uncared for . . . A lecture on ornithology*. 8vo, vi–36, 3 pls, West Newman, London 1879.

1658. HEWETT, George Mottram Andrews *Bug hunting specially round about Winchester*. 8vo, 16 pp, Wells, Winchester [c.1880].

1659. HEWITSON, William Chapman *British oology*. Roy. 8vo, 2 vols, 15 pp, 155 col. pls with letterpress, Vol. 1 Empson, Vol. 2 Currie and Bowman for the author, Newcastle upon Tyne [1831–38]; suppl. 14 col. pls, Van Voorst, London 1842; ed. Roy. or demy 8vo, London 1846 title *Coloured illustrations of the eggs of British birds*; 2ed. London 1856. Pls 87 & 119 are cancels. First ed. is sometimes found in 3 vols. London eds are in Natural History of the British Islands Series.

Hewitt, *Miss* Mary *see* No. 40.

1660. HIBBERD, James Shirley *Brambles and bay leaves*. 8vo, vi+312 pp, 3 col. pls, text figs, Groombridge, London 1855; 2ed. 1862; 3ed. 1872.

1661. —— *The book of the water cabinet*. 12mo, [48]pp, text figs, Groombridge, London 1856.

1662. —— *The book of the fresh-water aquarium*. 8vo, [50] pp, text figs, Groombridge, London 1856. Later editions.

1663. —— *The book of the marine aquarium*. 8vo, [48] pp, Groombridge, London 1856. Later editions.

—— *The book of the aquarium*. 8vo, 3 pts, [50+48+48] pp, text figs, Groombridge, London [1856]. A combination volume of Nos 1661–1663; ed. 148 pp, 1856. Later editions contain Nos 1662 & 1663 only.

1664. —— *Rustic adornments for homes of taste*. Cr. 8vo, 7 col. pls, text figs, Groombridge, London 1856; 2ed. 1857; ed. 4to 1870; ed. 1895; ed. 4to [c.1895].

1665. —— *Field flowers . . . the outdoor study of British plants*. 8vo, iv+156 pp, 8 col. pls, text figs, Groomridge, London 1870; 1878; ed. [1894]; ed. 1931 title *Wild flowers through the seasons*.

1666. —— *The seaweed collector; a handy guide to the marine botanist . . . in the study of the British Algae, and the British sponges*. 8vo, vii+152 pp, 8 col. pls, text figs, Groombridge, London [1872].

1667. —— *The ivy*. 4to, 115 pp, 3 col. pls, text figs, Groombridge, London 1872; 2ed. 1893.

Hicks, John Braxton *see* Samuelson, James.

1668. HICKSON, Sydney John *The fauna of the deep sea*. Sm. 8vo, xvi+169 pp, 1 pl., text figs, Kegan Paul, London 1894. Modern Science Series Vol. 6.

1669. —— *The story of life in the seas*. Sm. 8vo, 182 pp, 1 pl., text figs, Newnes, London 1898. Library of Useful Stories.

1670. HIGGINS, Henry Hugh *Synopsis of an arrangement of*

*invertebrate animals in the Free Public Museum of Liverpool*. 8vo, xix+104 pp, 1 pl., Liverpool 1874; 2ed. 1880.

1671. HIGHMORE, Nathaniel *The history of generation*. Sm. 8vo, 141 pp, 2 pls, Martin, London 1651.

1672. HILL, John *calling himself Sir A general natural history . . . of the animals, vegetables, and minerals*. Fol., 3 vols, 56 col. pls, 1 tab., Osborne, London 1748–51–52; ed. 1773. L.P. copies only are coloured.

1673. —— *A review of the works of the Royal Society of London; containing animadversions on such of the papers as deserve particular observation. In eight parts: under the several heads of arts, antiquities, medicine, miracles, zoophytes, animals, vegetables, minerals*, 4to, viii+265 pp, London 1751.

1674. —— *Essays in natural history and philosophy, containing a series of discoveries by the assistance of microscopes*. 8vo, 415 pp, 4 pls, Whiston, London 1752.

1675. —— *The British herbal: an history of plants and trees, natives of Britain, cultivated for use, or raised for beauty*. Fol., 533 pp, 75 col. pls, Osborne & Shipton, London 1756. L.P.

1676. —— *A system of botany with the latest improvements, in which the Genera and Species plantarum of Linnaeus are connected in one work*. Fol., iv, 5–12 pp, 2 pls, Osborne & Shipton, London 1757. ?Not published. First part only.

1677. —— *The sleep of plants, and cause of motion in the sensitive plant explained*. 12mo, 57 pp, Baldwin, London 1757; 2ed. 8vo 1762.

1678. —— *An account of a stone . . . which on being watered produces excellent mushrooms*. 4to, viii+38 pp, 2 pls, Baldwin, London 1758.

1679. —— *Outlines of a system of vegetable generation*. 8vo, 46 pp, 6 pls, the author, London 1758.

1680. —— *Botanical tracts*. 8vo, Baldwin, London 1758–62. Contains 7 tracts, with own pagination, previously published separately.

1681. —— *The origin and production of proliferous flowers*. 4to, 38 pp, 7 pls, the author, London 1759.

1682. —— *Flora Britanica* [sic]: *sive synopsis methodica stirpium Britanicarum*. 8vo, 672 pp, 30 pls, Waugh, London 1759; issue 1760; ed. 1763.

1683. —— *The vegetable system*. Fol., 26 vols, 1549 pls on 1547 leaves, the author, London 1759–75; index 1775; 2ed. vols 1 & 2 only 1775, 71; ed. vol. 1 only 1762. L.P.

1684. —— *Herbarium Britannicum, exhibens plantae Britanniae indigens*. 8vo, 2 vols, 296 pp, 199 pls, for the author, London 1769–70. Pls 1–195, 89*, 151*, 170*, [176]. All publ.

1685. —— *The construction of timber from its earliest growth, explained by the microscope*. 8vo, 170 pp, 43 pls, Baldwin for the author, London 1770; 2ed. fol. 1774.

1686. —— *A decade of curious insects . . . shewn in their natural size; and as they appear enlarg'd before the lucernal microscope*. 4to, 20 pp, 10 col. pls, for the author, London 1773.

1687. HILLER, H. Croft *Against dogma and free-will*, 8vo, viii+168 pp, Williams & Norgate, London 1892; 2ed. 1893 title . . . *and for Weismannism*.

1688. HILLHOUSE, William *Contributions towards a new flora of Bedfordshire*. 8vo, 20 pp, 1 map, [Bedford 1876].

1689. —— *Botanical divisions of Bedfordshire*. 8vo, 4 pp, [Bedford] 1876.

1690. —— *Hints for plant collectors*. 8vo, 8 pp, [Bedford 1877].

1691. —— *Bedfordshire plant list for 1876*. 8vo, [Bedford 1877].

1692. HILLMAN, William Augustus *Catalogue of the specimens of Mammalia and birds in store bottles*. 4to, 32 pp, Royal College of Surgeons, London 1841.

1693. HINCKS, Thomas *A catalogue of the zoophytes of South Devon and South Cornwall*. 8vo, 6 pls, Van Voorst, London 1868. Reprinted from *Ann. Mag. nat. Hist*., Vols 8–10, 1861–62.

1694. —— *A history of British hydroid zoophytes*. Roy. or demy 8vo, lxviii+338 pp, 1 pl., text figs, atlas vol. 67 pls, Van Voorst, London 1868. Natural History of the British Islands Series.

1695. —— *A history of British marine Polyzoa*. Roy. or demy 8vo, cxli + 601 pp, atlas vol. 1 col. + 82 plain pls, Van Voorst, London 1880. Natural History of the British Islands Series.

1696. HIND, William Marsden *The flora of Suffolk*. 8vo, xxxiv+508 pp, 1 col. map, Gurney & Jackson, London 1889.

1697. HINTON, James *Life in nature*. Sm. 8vo, xvi+258 pp, text figs, Smith & Elder, London 1862; ed. 1932.

1698. *Hints for the formation of a freshwater aquarium*. 12mo, 132 pp, 2 col. pls, text figs, SPCK [1857].

*Hints to travellers*. 4ed. 1878. *see* Royal Geographical Society.

1699. *The historical pocket library. Vol. VI. Natural history*. 16mo, 1 pl., text figs, Hazard, Bath, for G. Riley, London 1790. A set of 6 volumes of which only this one is relevant.

1700. *History of beasts*. 32mo, 16 pp, text figs, Davis, London [c.1825]. No. 32 in Davis's Chapbook Series.

1701. *History of birds*. 32mo, 16 pp, text figs, Davis, London [c.1825]. No. 33 in Davis's Chapbook Series.

1702. *History of birds, (Beasts)*. 8vo, 2 pts, Bysh, London [1861].

1703. *The history of birds in verse*. 32mo, 8 pp, text figs, Davison, Alnwick [c.1820].

1704. *The history of birds. With a familier* [sic] *description of each in verse and prose*. 12mo, text figs, London 1791.

1705. *History of British birds*. 8vo, 180 pp, 12 col. pls, Phillips, London 1807.

*History of British birds. Land birds. (Water birds). see* Bewick, T.

1706. *History of domestic and rapacious birds, containing an interesting and familiar account of their modes of life*. 32mo, 72 pp, text figs, Dutton, London 1806.

1707. *A history of esculent fish . . . and an essay on the breeding of fish*. 4to, iv+80 pp, 18 col. pls, London 1794. The plates are by the Albins (Eleazar, Elizabeth & Firmin); the essay a late edition of Roger North, but the *History* is anonymous.

1708. *History of fishes in miniature adapted to the juvenile capacity*. Sq. 16mo, [46] pp, 23 pls, Langley, London [c.1795].

1709. *The history of insects*. 16mo, xii+192 pp, 16 col. pls, text figs, RTS, London 1839; ed. 1842.

*A history of the Mammalia*. 1849. *see Sketches in natural history*.

1710. *The history of Ripon . . . and lists of the rarer indigenous plants found in the neighbourhood*. 8vo, 304 pp, 2 tinted pls, text figs, Farrer, Ripon 1806.

*History of singing birds*. 1791. *see* Albin, E.

1711. *A history of wonderful animals; containing an account of the most remarkable fishes, beasts and birds*. 12mo, 180 pp, text figs, Courtney printed, Dublin 2ed. 1818; 2ed. Wetton, London [?1818].

1712. *A history of wonderful fishes*. 12mo, vi+181 pp, text figs, Bentham printed, Dublin 1820.

1713. *A history of wonderful fishes and monsters of the ocean, to which is added an account of the lion;—the tiger;—the elephant; the rhinoceros, and the hippopotamus*. 12mo, 141 pp, text figs, Graisberry & Campbell, Dublin 1816.

1714. *The history of wonderful fishes, and other animals*. 12mo, text figs, Whittaker, London 1818.

1715. HITCHINS, Fortescue *The history of Cornwall*. 4to, 2 vols, 12 pls, tabs, Penaluna, Helston 1824.

1716. [HOARE, Sarah] *A poem on the pleasures and advantages of botanical pursuits, with notes*. 12mo, xi+134 pp, Rose printed, Bristol [?1825]. Previously published, signed, in Priscilla Wakefield's *An introduction to botany*. 8ed. 1818; 9ed. 1823.

1717. —— *Poems on conchology and botany, with plates and notes*. 12mo, iv+106 pp, 5 col. pls, Simpkin Marshall, London 1831.

1718. HOBKIRK, Charles Coddrington Pressick *Huddersfield: its history and natural history*. 8vo, viii+164 pp, 5 pls, Tindall, Huddersfield 1859; 2ed. 1868.

1719. —— *A synopsis of British mosses*. Sm. 8vo, vii+196 pp, Reeve, London 1873; 2ed. 1884. 2ed. is Crown Series.

1720. —— & BOSWELL, Henry *The London catalogue of British mosses*. 8vo, 11 pp, Botanical Record Club, Welwyn 1877; 2ed. London 1881 title . . . *and hepatics*.

1721. HOBLYN, Richard Dennis *A dictionary of scientific terms*. 12mo, 386 pp, Whittaker, London 1849.

1722. —— *A treatise on botany*. 12mo, 11 pp, Scott, London 1851.

1723. —— *British plants*. 12mo, Scott, London 1851.

1724. [HOBSON, Edward] *Catalogue of British insects*. 8vo, 23 pp, n.p., [c.1825]. Printed on one side only.

1725. [HODDINOT, C.] *The book of creation unfolded: or, the creator as seen in his works*. 12mo, 129 pp, 1 pl., Wellington (Salop) 1828; 3ed. signed, London [ ]. By the author of *The voice of truth*. Preface signed C*H*.

1726. HODGSON, William *Flora of Cumberland*. 8vo, xxxvi+398 pp, 2 maps, Meale, Carlisle 1898.

1727. HOEVEN, Jan van der *Handbook of zoology*. 8vo, 2 vols, 24 pls, University Press, Cambridge 1856–58. Translation of *Handbook der Dierkunde*, 2ed. Rotterdam 1828–33, by William Clark.

1728. HOFMANN, Ernst *The young beetle-collector's handbook*. Sm. 8vo, viii+178 pp, text figs, Sonnenschein, London 1897; ed. 1902. Young Collector Series.

1729. HOFMEISTER, Wilhelm *On the germination, development, and fructification of the higher Cryptogamia, and on the fructification of the Coniferae*. 8vo, xviii+506 pp, 65 pls, Ray Society, London 1862, Ray Society Publ. No. 35 for 1861. Translation of *Vergleichende Untersuchungen der Heimung, Entfaltung und Fruchtbildung höherer Kryptogamen*, Leipzig 1851, by Frederick Currey.

1730. HOGAN, Edmund Ignatius, HOGAN, John & MACEARIEAN, John C. *Luibleabran: Irish and Scottish gaelic names of herbs, plants, trees*. 8vo, xii+137 pp, Gill, Dublin 1900.

Hogan, John *see* Hogan, E. I.

1731. HOGG, Jabez *The microscope: its history, construction and applications*. 8vo, xvii+440 pp, 15 pls, text figs, Illustrated London Library, London 1854; 2ed. 1855; 1856; 3ed. 1858; 4ed. 1859; 6ed. 1867; 1870; 8ed. 1871; 1882; 10ed. 1883; 11ed. 1886; 12ed. 1887; 15ed. 1898; 1911.

1732. HOGG, John *The natural history of the vicinity of Stockton-on-Tees*, 8vo, 94 pp, 1 col. map, Stockton-on-Tees 1827. Preprint from Brewster, John *The parochial history of Stockton-on-Tees*, 1829.

1733. —— *Observations on the Spongilla fluviatilis*. 8vo, 372 pp, London 1840. Reprinted from *Trans. Linn. Soc. Lond*., Vol. 18.

1734. —— *A catalogue of birds observed in south-eastern Durham*. 8vo, 50 pp, Newman, London 1845. Reprinted from *The Zoologist*, 1845.

1735. —— *On the classification of birds, and particularly the genera of European birds*. 8vo, 22 pp, Neill printed, Edinburgh 1846. Reprinted from *Edinburgh new phil. J*. 1846.

1736. —— *On the distinction of a plant and an animal, and on a fourth kingdom of nature*. 8vo, 12 pp, Edinburgh 1860. Reprinted from *Edinburgh new phil. J*., 1860.

1737. [——] *editor The parlour menagerie. Wherein are exhibited . . . the habits, the resources, and the mysterious instincts of the . . . animal creation*. 8vo, 520 pp, text figs, Hogg, London 1875; 2ed. 1878; 3ed. 1879; ed. 1882. Signed J.H.

1738. HOGG, Robert *The vegetable kingdom and its products; serving as an introduction to the natural system of botany*. 8vo, vi+882 pp, Kent, London [1857–]1858.

1739. —— & JOHNSON, George William *The wild flowers of Great Britain*. 4to, 11 vols, 924 col. pls, Cottage Gardeners Office, London [1860–] 1863–80. Vol. 2 by Gower, Charlotte & Smith, William G. Vols 3–11 by Smith, W. G. Also available in 4 or 9 vols, according to the natural system.

1740. HOLDER, Charles Frederick *Marvels of animal life*. 8vo, x+240 pp, 31 pls, Sampson Low, London 1885; ed. [ ]. First ed. is New York 1885.

1741. —— *Living lights*. 8vo, ix+187 pp, 25 pls, Sampson Low, London 1887; ed. [1893]. First ed. is New York 1887.

Holdsworth, Edmund William Hunt *see* Zoological Society of London.

*Holiday excursions of a naturalist*. 1867. *see* Garner, R.

1742. *A holiday stroll in the Gardens of the London Zoological Society; describing the various animals*. 12mo, 48 pp, text figs, Wallis, London [c.1830].

Holland, Robert *see* Britten, James.

1743. HOLLOWAY, William & BRANCH, John *The British Museum; or elegant repository of natural history*. 12mo, 4 vols, 146 col. pls, Badcock, London 1803–04; issue 1806. 1806 issue bears Holloway's name only.

1744. HOLMES, Edward Morell *Holmes's botanical notebook or a practical guide to a knowledge of botany*. 8vo, iv+186 pp, text figs, Christy, London 1878.

1745. —— & BATTERS, Edward Arthur Lionel *A revised list of the British marine Algae*. Roy. 8vo, Frowde, London 1892. Reprinted from *Ann. Bot*., Vol. 5, 1890–91.

1746. —— & BRENT, Francis *Scale mosses, liverworts, and lichens of Devon and Cornwall*. 8vo, 71 pp, Plymouth [1865].

1747. —— & GRAY, Peter *British Fungi, lichens, and mosses*. 8vo, 32+32+32 pp, text figs, Sonnenschein, London 1886. Later printings. Young Collector Series.

Holmesdale Natural History Club *see* Saunders, G. S. & Saunders, E. O.

Holt, Sestertius *pseudonym see* Cooper, W. W.

1748. *Home and foreign birds. A book for young children*. Sm. 8vo, 94 pp, text figs, Routledge, London 1856. Picture reading books. Printed on linen.

1749. HOME, *Sir* Everard *Lectures on comparative anatomy . . . (to which is subjoined Synopsis systematis regni animalia)*. 4to, 4 vols, 371 pls, 1 portr., Longman, London 1814–23; suppl. 2 vols 1828. L.P.

1750. —— & BLIZARD, *Sir* William *Summary of the arrangement of the Hunterian Collection in the Museum*. 8vo, 14 pp, Royal College of Surgeons, London 1813.

1751. HOOKE, Robert *Micrographia: or, some physiological descriptions of minute bodies made by magnifying glasses*. Fol., [xxxiv+246] pp, 34 pls, Martyn & Allestry, London 1665; ed. 1667; ed. 8vo Oxford 1938. 1938 ed. is Vol. 13 of

Gunther, Robert T. W. *Early science in Oxford. See also The wonders of the microscope.*

—— *Micrographia restaurata; or, the copper plates of Dr. Hooke's . . . discoveries by the microscope*. Fol., 65 pp, 33 pls, Bowles London 1745; issue 1780 title *Microscopic observations . . .*

—— [Extracts from *Micrographia*]. Alembic Club, Edinburgh 1902. Alembic Club Reprints No. 5; Oxford 1926. Old Ashmolean Reprints No. VI.

1752. HOOKER, *Sir* Joseph Dalton *The student's flora of the British Islands.* Sm. 8vo, xx+504 pp, Macmillan, London 1870; 2ed. 1878; 3ed. 1884. Later reprintings.

1753. —— *Botany*. Sm. 8vo, viii+117 pp, text figs, Macmillan, London 1876; ed. 1878; 1881; 3ed. 1884. Later reprintings. Science Primers.

—— *see* Bentham, G.

1754. HOOKER, *Sir* William Jackson *British Jungermanniae*. Fol. or 4to, 88 col. pls with letterpress, Longman, London [1812–]1816; issue 1846. Plates 1–84, supplementary 1–4.

1755. —— *Flora Scotica; or a description of Scottish plants*. 8vo, 2 pts, Taylor, London 1821.

1756. —— *Botanical illustrations*. Obl. fol., 3 pts, 21 pls with letterpress, Constable, Edinburgh 1821–22; ed. Glasgow 1837. All publ.

1757. —— *Directions for collecting and preserving plants in foreign countries*. 4to, 3 pp, Curll printed, Glasgow 1823.

1758. —— *The British flora*. 12mo, x+480 pp, Longman, London 1830; 2ed. 1831; 3ed. 2 vols 1835; 4ed. 1838; 5ed. 1842; 6ed. 1850; 7ed. 1855; 8ed. 1860. Vol. 2 of eds 4 & 5 are the same printing as that of ed. 3. Parts of Vol. 2 are by Berkeley, Miles Joseph, and this volume is also Vol. 5 of Smith, *Sir* James Edward *English flora,* 2ed. 1833. 6ed. and later are by Hooker and Arnott, George Arnott Walker.

1759. —— *A catalogue of the British flowering plants included in Dr. Hooker's British flora*. Fol., s. sh., London [1835].

1760. —— *The British ferns*. Roy. 8vo, 66 col. pls with letterpress, Reeve, London 1861[–1862].

—— *see* Curtis, W.

1761. —— & BAKER, John Gilbert *Synopsis filicum; or, a synopsis of all known ferns*. 8vo, 482 pp, 9 col. pls, Hardwicke, London [1865–] 1868; 2ed. 1874; 1883.

1762. —— & —— *An alphabetical index of all known ferns*. 8vo, 33 pp, Hardwicke, London 1874. A separate printing of the index to No. 1761.

1763. —— & TAYLOR, Thomas *Muscologia Britannica: containing the mosses of Great Britain and Ireland*. 8vo, xxxv+152 pp, 31 pls, Longman, London 1818; 2ed. 1827; 3ed. 1855; ed. 1863 title *Bryologia Britannica* . . .; 3ed. revised by Wilson, W. is almost a new work.

1764. HOOLE, Charles *A plain and easy primer for children, both the pictures of beasts and birds for each letter in the alphabet for their easie remembrance and delight in learning*. ?12mo, William London, London ?c.1650. ?Any surviving copy, t.p. in British Library.

1765. HOOPER, Benjamin *The cruise of the Nelly, summer wanderings of a marine zoologist on the coasts of the British Isles*. 8vo, 192 pp, 3 pls, Tubbs & Brook, Manchester [?1871].

1766. HOOPER, Robert *Observations on the structure and economy of plants*. 8vo, vi+129 pp, Fletcher, Oxford 1797.

1767. HOPE, Frederick William *The coleopterist's manual*. 8vo, 3 pts, 11 col. pls, Bohn, London 1837–40. All publ.

1768. [HOPE, John] *Characteres genera plantarum: ex Systemate vegetabilium Linnaei, et Horto Kewensi praecipue excerpti. In usus academices*. 8vo, 2+138 pp, 1 tab., University Press, Edinburgh 1793. Partly from Linnaeus, C. *Systema vegetabilium*, 3ed. Gottingen 1774.

—— *see* Linnaeus, C.

1769. HOPE, Thomas Charles *Tentamen inaugurale, quaedam de plantarum motibus et vita complectens*. 8vo, 37 pp, Balfour & Smellie, Edinburgh 1787.

1770. HOPKINSON, John *A sketch of the geology and climate of Hertfordshire with notes on its botanical districts*. 4to, 32 pp, Austin printed, Hertford 1887. Reprinted from Pryor, A. R. *Flora of Hertfordshire*.

1771. HOPKIRK, Thomas *Flora Glottiana. A catalogue of the indigenous plants on the banks of the river Clyde, and in the neighbourhood of the City of Glasgow*. 8vo, 170 pp, Smith, Glasgow 1813.

1772. —— *Flora anomoia. A general view of the anomalies in the vegetable kingdom*. 4to or 8vo, 198 pp, 14 or 11 pls, text figs, Smith, Glasgow 1817. Nature printed. The 4to has 14 pls.

1773. HOBLEY, Catherine Cooper *Snakes; curiosities and wonders of serpent life*. 8vo, viii+614 pp, 2 col. pls, text figs, Griffith & Farren, London 1882.

1774. —— *British reptiles and batrachians*. Sm. 8vo, 94 pp, text figs, Sonnenschein, London 1888. Young Collector Series.

1775. HORN, George *An entire new treaty on leeches*. 8vo, 29 pp, London 1798.

1776. HORNADAY, William Temple *Taxidermy and zoological collecting*. 8vo, xix+362 pp, 23 pls, text figs, Kegan Paul, London 1891. First ed. is New York 1891.

*Hortas sanitatis see The noble lyfe & natures* . . . [?1521].

1777. HOUGHTON, William *Country walks of a naturalist with his children*. Sm. 8vo, vi+154 pp, 8 col. pls, text figs, Groombridge, London 1869; 2ed. 1870. Several later editions. See also No. 1779.

1778. —— *Sea-side walks of a naturalist with his children*. Sm. 8vo, vi+154 pp, 8 col. pls, text figs, Groombridge, London 1870. Later editions. See also No. 1779.

1779. —— *Walks of a naturalist with his children*. Sm. 8vo, vi+154+vi+154 pp, 16 col. pls, Groombridge, London [1872]. Combination volume of Nos 1777 & 1778.

1780. —— *The microscope and some of the wonders it reveals*. Sm. 8vo, text figs, Cassell, London [1872].

1781. —— *Sketches of British insects. A handbook for beginners in the study of entomology.* Sm. 8vo, vi+161 pp, 6 col. pls, text figs, Groombridge, London 1875; 2ed. 1877. Later editions.

1782. —— *British fresh-water fishes.* Roy. 4to, 2 vols, xxxvi+204 pp, 41 col. pls [of fish] text figs, Mackenzie, London [1879]; ed. demy 4to, 24 pls [of views], Drane, London 1879; ed. 1884; 2ed. 8vo, Hull 1895; 3ed. 1900. First ed. also occurs as one vol.

1783. —— *Gleanings from the natural history of the ancients.* 8vo, 252 pp, text figs, Cassell, London [1879].

1784. HOUSMAN, Henry *The story of our museum: showing how we formed it, and what it taught us*. Sm. 8vo, xvi+280 pp, text figs, SPCK, London 1881; [ ].

1785. HOUSSAY, Frederic *The industries of animals*. 8vo, ix+258 pp, text figs, Scott, London 1893. Contemporary Science Series. Translation of *Les industries des animaux*, Paris, 1889, by Henry Havelock Ellis.

1786. HOUSTON, David *Practical botany for elementary students. Introduction to the systematic botany of flowering plants*. 8vo, 128 pp, text figs, Stewart, London [1880]. Stewart's Educational Series.

1787. HOVENDEN, Frederick *What is life? Or, where are we? What are we?* 8vo, xvi+329 pp, text figs, Chapman & Hall, London 1897; 2ed. 1899; 3ed. 1909.

*How to collect mosquitoes.* [1898]. *see* Austen, E. E.

1788. [HOW, William] *Phytologia Britannica, natales exhibens indigenarum stirpium sponte emergentium.* 8vo, [xvi]+133 pp, Pulleyn, London 1650. By W.H. L.P.

1789. HOWARD, John Elliot *An examination of the Belfast address of the British Association, 1874, from a scientific point of view.* 8vo, 25 pp, Hardwicke, London 1875. The address by Tyndall, J.

1790. [HOWARD, Mary Matilda] *Wildflowers and their teaching.* 8vo, 36 specimens, Binns & Goodwin, Bath 1845; 2ed. 1848. Captions are extracts from Hooker's *British flora.*

1791. [——] *Ocean flowers and their teachings.* 8vo, iv+146 pp, 39 specimens, Binns & Goodwin, Bath 1846; 1847. Introduction to Part 3 signed M.M.H.

Howard, R. J. *see* Mitchell, F. S.

1792. HOWE, W. E. *The ferns of Derbyshire, illustrated from nature.* 8vo, x pp, 24 col. pls with letterpress, Wertheim & Macintosh, London 1861; 2ed. 1861; ed. 1863; 1865; 6ed. 1867; 1869; [1877]; ed. 1880 unsigned.

1793. HOWES, George Bond *An atlas of practical elementary biology.* 4to, viii+116 pp, 24 pls, Macmillan, London 1885; 2ed. of zoological part only 1885 title *Atlas of practical elementary zootomy*; ed. 1902.

1794. HOWIE, Charles *The moss flora of Fife and Kinross.* Sm. 8vo, 116 pp, Cupar [1889].

1795. —— *Musci Fifenses: specimens of the mosses of Fifeshire.* 8vo, vii+112 pp, specimens, St Andrews [c.1890].

1796. HOWITT, Godfrey *The Nottinghamshire flora.* 12mo, 124 pp, Hamilton Adams, London 1839.

1797. —— & VALENTINE, W. *Muscologia Nottinghamiensis.* 8vo, specimens with letterpress, Nottingham 1833.

1798. HOWITT, Mary *Sketches of natural history.* 12 mo, vi+167 pp, 23 pls, Effingham Wilson, London 1834; 7ed. [1851].

1799. —— *Birds and flowers and other country things.* Sm. 8vo, ix+208 pp, text figs, Darton, London 1837. Many later editions. The Christmas Library Vol. 1.

1800. —— *With the birds.* Sm. 8vo, viii+128 pp, 1 pl., text figs, London [?1871].

1801. —— *Birds and their nests*. Sm. 4to, vi+124 pp, 23 pls, text figs, Partridge, London 1872. Later editions.
—— *see* Aikin, J.

1802. HOWITT, William *The book of the seasons, or the calendar of nature*. 8vo, Bentley, London 1831; 2ed. 1831. Later editions.

1803. —— *The boy's country book . . . written by himself*. 12mo, xi+308 pp, text figs, Longman, London 1839; 2ed. 1841. Later editions.

1804. HUBER, Jean Pierre *The natural history of ants*. 12mo, xlv+398 pp, 1 col. + 1 plain pls, Johnson, London 1820. Translation of *Recherches sur les moeurs des fourmis*, Paris 1810, by James Rawlins Johnson.

1805. HUBERT, *alias* Forges, Robert *A catalogue of those rarities collected in thirty years time with a great deal of pains and industry, by one of His Majestie's sworn servants*. 12mo, London [1664]; 2ed. 1665 title *A catalogue of many natural rarities . . . collected by Robert Hubert*.

Huddersfield Economic & Educational Museum *see* Burr, M.
—— *see* Swanton, E. W.

1806. HUDSON, Charles Thomas *assisted by* GOSSE, Philip Henry. *The Rotifera; or wheel-animalcules*. Imp. 8vo, 2 vols, 30 col. + 4 plain pls, Longman, London 1886; suppl. vi+64 pp, 4 pls, 1889.

1807. HUDSON, John *A treatise on the medicinal leech*. 8vo, 106 pp, 4 pls, for the author, Hull 1841.

1808. HUDSON, John *A complete guide to the Lakes . . . botanical notices by T. Gough*. 8vo, x+vii+259 pp, 7 pls, maps, text figs, Hudson, Kendall 2ed. 1843; 4ed. 1853; 5ed. 1859; another ed. 1859. This John Hudson is not the same as the author of No. 1807.

1809. HUDSON, William *Guiliemi Hudsoni . . . Flora Anglica, exhibens plantas per regnum Angliae sponte crescentes*. 8vo, xv+506 pp, for the author, London 1762; 2ed. 2 vols 1778; issue 1798.

1810. HUDSON, William Henry *Birds in a village*. Sm. 8vo, x+232+8 pp, Chapman & Hall, London 1893; ed. [1920].

1811. —— *Lost British birds*. 8vo, 32 pp, text figs, Society for the Protection of Birds, London [1894]; ed. 1894; ed. 1923 title *Rare, vanishing and lost British birds*. Society for the Protection of Birds Pamphlet No. 14.

1812. —— *British birds; with a chapter on their structure and classification by Frank E. Beddard*. Sm. 8vo, xxii+363 pp, 8 col. + 11 plain pls, text figs, Longman, London 1895. Later editions.

1813. —— *Pipits*. 4to, Society for the Protection of Birds, London [1897]. Educational Series No. 21.

1814. —— *Birds in London*. 8vo, xvi+339 pp, 15 col. + 1 plain pls, 1 map, text figs, Longman, London 1898. Later editions.

1815. —— *Nature in downland*. 8vo, xii+307 pp, 12 pls, text figs, Longman, London 1900. Later editions.

Huebner, Jacob *see* Stephens, J. F.

1816. HUGHES, William R. *On the principles and management of the marine aquarium*. 8vo, 28 pp, 3 pls, Birmingham 1875; ed. 56 pp, Van Voorst London 1875.

1817. HUISH, Robert *The wonders of the animal kingdom in the various menageries of this country*. 4to, ii+122 pp, 63 pls, Kelly, London [1829–]1830.

1818. HULL, John *The British flora*. 8vo, 2 pts, vii+449 pp, Dean printed, Manchester 1799; 2ed. Vol. 1 1808.

1819. —— *Elements of botany*. 4to, 2 vols, 16 pls, Dean, Manchester 1800.

1820. HULME, Frederick Edward *A series of sketches from nature of plant form*. Imp. 8vo, 54 pp, 100 col. pls, London [1867–]1868.

1821. —— *Plants, their natural growth and ornamental treatment*. Fol., Ward, London 1874.

1822. —— *Familiar wild flowers*. Sm. 8vo, 5 vols, 200 col. pls, Cassell, London [1877–85]; ed. 7 vols 1902; ed. 8 vols 1906; ed. 9 vols 1906. Later reprints.

1823. —— *Wayside sketches*. Sm. 8vo, viii+340 pp, text figs, SPCK, London [1889].

1824. —— *Natural history lore and legend*. 8vo, viii+350 pp, text figs, Quaritch, London 1895.

1825. —— *The wild flowers collecting book*. 4to, 6 pts, Cassell, London [1895].

1826. HULME, J. R. *The Scarborough Algae*. 4to, iv+103 pp, 25 specimens, Scarborough 1842.

1827. HUMBOLDT, Friedrich Heinrich Alexander von, *Baron. Kosmos; a general survey of the physical phenomena of the Universe*. 8vo, 2 vols, Baillière, London 1845, 48. Translated by Pritchard, A. See also Nos 1828 & 1829.

1828. —— *Cosmos: sketch of the physical description of the Universe* 12mo, 4 vols, Longman, London 1846–58; 2ed. Vol. 1

1847; 4ed. 1848; 6ed. 1849; 8ed. Vols 1 & 2 1850. Vol. 4 is pt. 1 only, all publ. Translated by Sabine *Mrs* Elizabeth Juliana & Sabine, *Sir* Edward. See also Nos 1827 & 1829.

1829. —— *Cosmos: a sketch of a physical description of the Universe*. Sm. 8vo, 5 vols, portr., Bohn, London 1849–58. Translated by Otté, Elise C., Paul, Benjamin Horatio & Dallas, William Sweetland. Nos 1827–28 are translations, in whole or in part of *Kosmos* 5 vols, Stuttgart 1845–62.

1830. —— *Aspects of nature in different lands and different climates*. Sm. 8vo, 2 vols, Longman, London 1849; ed. 1850. Translated by Sabine, *Mrs* Elizabeth Juliana. See also No. 1831.

1831. —— *Views of nature; or, contemplation on the sublime phenomena of creation*. 8vo, xxi+452 pp, 1 pl., Bohn, London 1850. Nos 1830 & 1831 are translations of *Ansichten der Natur*, 3ed., Stuttgart 1849. No. 1831 translated by Otté, Elise C. & Bohn, Henry G.

1832. [HUMPHREYS, Henry Noel] *Insect changes*. 16mo, 32 pp, col. text figs, Grant & Griffith, London 1847; 1851.

1833. —— *Ocean gardens: the history of a marine aquarium*. Sm. 8vo, viii+112 pp, 12 col. pls, text figs, Sampson Low, London 1857. See also No. 1834.

1834. —— *River gardens . . . cultivating fresh-water plants in aquaria*. Sm. 8vo, vii+108 pp, 8 col. pls, text figs, Sampson Low, London 1857. This and No. 1833 were also issued together with collective title page *Ocean and river gardens*, 1857; 2ed. 1857.

1835. [——] *Butterflies in their floral homes*. Fol, 9 11, Jerrard, London [1858]. By a dreamer in the woods.

1836. —— *The butterfly vivarium, or insect home*. Sm. 4to, xii+288 pp, 8 col. pls, Lay, London 1858.

1837. —— *The genera and species of British butterflies*. Imp. 8vo, xii+66+4 pp, 33 col. pls, Jerrard, London [1859].

1838. —— *The genera of British moths*. Imp. 8vo, vi+206 pp, 62 col. pls, Jerrard, London [1860].

—— *see* Westwood, J. O.

Humphreys, John D. *see* Cuvierian Society of Cork.

1839. HUMPHRY, *Sir* George Murray *Observations on the limbs of vertebrate animals: the plan of their construction: their homology*. 4to, 44 pp, 3 pls, Macmillan, Cambridge 1860.

1840. —— *Observations in myology, including the myology of Cryptobranchus . . . and the disposition of muscles in vertebrate animals*. 8vo, 192 pp, 9 pls, Macmillan, Cambridge 1872.

1841. HUNT, John *British ornithology; containing portraits of all the British birds*. Sm. 8vo, 2 vols, 180 col. + 12 plain pls, Bacon for the proprietor, Norwich 1815–22. The plates are from *The natural history of birds*, Bungay 1815, and their number varies from copy to copy. All publ.

1842. HUNTER, John *Observations on certain parts of animal economy*. 4to, 18 pls, London 1786; 2ed. 1792; ed. 8vo 1837. Reprints of paper in *Phil. Trans. Roy. Soc.*, with new title.

1843. —— *Memoranda on vegetation*. 4to, vi+34 pp, London 1860.

1844. —— *Essays and observations on natural history, anatomy, physiology . . . posthumous papers*. Roy. 8vo, 2 vols, 2 pls, Van Voorst, London 1861.

—— *see* Abernethy, J.

Hurst, Charles Herbert *see* Marshall, A. M.

1845. HUSSEY, *Mrs* Thomas John *Illustrations of British mycology* [First series]. Roy. 4to, viii pp, 90 col. pls with letterpress, Reeve, London 1847; issue 1849.

1846. —— *Illustrations of British mycology. Second series*. Roy. 4to, 50 col. pls with letterpress, Reeve, London 1855.

1847. HUTCHINS, John *The history of the County of Dorset . . . some remarkable particulars of natural history*. Fol., 2 vols, pls, 1 map, Bowyer, London 1774; 2ed. 4 vols 1798–1815; 3ed. 1861–73.

1848. HUTCHINSON, Peter Orlando *The ferns of Sidmouth*. 8vo, 40 pp, 19 pls, Harvey, Sidmouth 1862. Nature printed.

1849. HUTCHINSON, William *Handbook of grasses*. Sm. 8vo, 92 pp, text figs, Sonnenschein, London 1895. Young Collector Series.

1850. HUTTON, Frederick Wollaston *Darwinism and Lamarckism old and new*. Sm. 8vo, x+169 pp, Duckworth, London 1899.

1851. HUTTON, William *The book of nature laid open . . . and the appearance of nature during each month of the year*. 12mo, xii+279 pp, Newman, London 1820; 2ed. 1821; 3ed. 1821; 4ed. 1821; 6ed. 1828; abridged ed. Penrith [ ].

1852. —— *The wonders of nature and art, containing an account of the most remarkable and curious animals, and mineral and vegetable productions in the world*. 12mo, iv+306 pp, 1 pl., Newman, London 1823.

1853. HUXLEY, Thomas Henry *On the educational value of the natural history sciences*. 8vo, 32 pp, Van Voorst, London 1854.

1854. —— *The oceanic Hydrozoa*. Roy. 4to, x+143+24 pp, 12 pls, text figs, Ray Society, London 1859. Ray Society Publ. No. 30 for 1838.

1855. —— *Evidence as to man's place in nature*. 8vo, 159 pp, 1 pl., text figs, Williams & Norgate, London 1862; 1863; 1864. Later printings, some with title *Man's place in nature.*

1856. —— *On our knowledge of the causes of the phenomena of organic nature*. Sm. 8vo, 157+[2] pp, text figs, Hardwicke, London 1862 [= 1863]; issue 156+[4] pp, 1863. First issued as 6 separate pamphlets dated 1862.

1857. —— *An elementary atlas of comparative osteology in twelve plates*. Roy. 40, 12 pls with letterpress, Williams & Norgate, London 1864.

1858. —— *Lectures on the elements of comparative anatomy . . . on the classification of animals . . . Vol. 1* 8vo, xi+303 pp, text figs, Churchill, London 1864. All publ. See also No. 1859.

1859. —— *An introduction to the classification of animals*. 8vo, 147 pp, text figs, Churchill, London 1869. A new ed. of pt. 1 of No. 1858.

1860. —— *Lay sermons, addresses and reviews*. 8vo, xi+378 pp, Macmillan, London 1870; 2ed. 1871; ed. 1880; 3ed. 1887. Later editions.

1861. —— *Essays selected from Lay sermons, addresses and reviews.* Sm. 8vo, Macmillan, London 1871. Later editions. Selections from No. 1860.

1862. —— *A manual of the anatomy of vertebrated animals*. 8vo, vi+510 pp, text figs, Churchill, London 1871. Later editions.

1863. —— *Critiques and addresses*. 8vo, xiv+350 pp, Macmillan, London 1873. Later editions.

1864. —— *American addresses, with a lecture on the study of biology*. 8vo, xi+378 pp, Macmillan, London 1877. Later editions.

1865. —— *A manual of the anatomy of invertebrated animals*. 8vo, viii+698 pp, text figs, Churchill, London 1877; ed. 1879.

1866. —— *Physiography: an introduction to the study of nature*. 8vo, xix+384 pp, 5 col. pls, text figs, Macmillan, London 1877. Later editions.

1867. —— *The crayfish. An introduction to the study of zoology*. Demy or cr. 8vo, xiv+371 pp, 1 pl., text figs, Kegan Paul, London 1880. Later editions. International Scientific Series Vol. XXVIII. L.P. 250 copies, not I.S.S. Facsimile 1974.

1868. —— *Science primers: introductory*. 16mo, 94 pp, Macmillan, London 1880. Later editions.

1869. —— *Science and culture and other essays*. 8vo, ix+346 pp, text figs, Macmillan, London 1881. Later editions.

1870. —— *Essays upon some controverted questions*. 8vo, vii+625 pp, Macmillan, London 1892.

1871. —— *Evolution and ethics*. 8vo, 57 pp, Macmillan, London 1893. Later editions.

1872. —— *Collected essays*. 8vo, 9 vols, Macmillan, London 1893–94. Later reprints. Eversley Series. Almost all the contents had been published before.

1873. —— *The scientific memoirs of Thomas Henry Huxley*. Roy. 8vo, 5 vols, pls, portrs, text figs, London 1898–1903. Vol. 5 is suppl. Reprints of papers in serials from 1845 onwards.

1874. —— *assisted by* MARTIN, Henry Newell *A course of practical instruction in elementary biology*. Sm. 8vo, xii+268 pp, text figs, Macmillan, London 1875; 2ed. 1876 title *A course of elementary instruction in practical biology*. Many later editions.

# I

I., A. *see* P., W.

1875. I., M. H. H. *The instructive picture book. Lessons from the geographical distribution of animals*. Fol., 60 col. pls, Edmonston & Douglas, Edinburgh 1860; 2ed. 1867; 4ed. London 1878.

1876. IBBOTSON, Henry *A catalogue of the phaenogamous plants of Great Britain*. 4to, 206 pp, London [1846–]1848; ed. 1855.

1877. —— *The ferns of York*. 8vo, 22 pp, Sessions, York 1884.

Illiger, Johann Carl Wilhelm *see* Bowdich, T. E.

*An illustrated handbook of British dragonflies*. 1890. *see* Bath, W. H.

An illustrated monograph of the British Hieracia. 1889–98. *see* Hanbury, F. J.

1878. *Illustrated sketches of natural history*. Sm. 8vo, 2 vols, 2 col. pls, text figs, SPCK, London [1864–66].

*Illustrations of British hawk moths and their larvae*. 1874–76. *see* Johnson, Theodore.

*Illustrations of natural history and mineralogy*. [?1880]. *see* Wilson, A.

1879. *Illustrations of natural history, birds*. 8vo, no text, 18 col. pls, Kitton, Norwich [c.1815].

1880. *Illustrations of natural history: embracing a series of engravings and descriptive account . . . of the animal world*. 8vo, 384+vi pp, 50 pls, Longman, London [1829–30]. Signatures state Vol. 1; all publ.

1881. *Illustrations of the British flora*. Sm. 8vo, pls, Reeve, London 1880.

1882. *Illustrations of zoology. The engravings by J. W. Lowry and T. Landseer . . . with descriptive letterpress . . . according to Cuvier*. Imp. 4to, xii+163 pp, 88 pls, text figs, Griffin, London 1851. Reprinted from *Encyclopaedia Metropolitana.*

1883. IM THURN, *Sir* Everard Ferdinand *The birds of Marlborough.* Sm. 8vo, viii+117 pp, Perkins, Marlborough 1870.

1884. *The infant's library. Book 5* [Quadrupeds]. 32mo, 32 pp, full page text figs, John Marshall, London [c.1800]. Various issues.

1885. INGEN-HEUSZ, Jan *Experiments upon vegetables, discovering their power of purifying the common air*. 8vo, xlviii+302 pp, 1 pl., Elmsley, London 1779.

1886. INGPEN, Abel *Instructions for collecting . . . insects: also for collecting . . . Crustacea and shells*. 12mo, xv+92 pp, 1 col. pl., Bulcock, London 1827; 2ed. 1839; 3ed. 1843; ed. 1849 title *Manual of the butterfly collector or . . .*

*Insect architecture*. 1830. *see* Rennie, J.

*Insect changes, forming a first lesson in entomology*. 1847. *see* Humphreys, H. N.

*The insect hunters.* [1857]. *see* Newman, E.

1887. *Insect manufactures*. 16mo, iv+162 pp, text figs, SPCK, London 1847.

*Insect miscellanies*. 1831. *see* Rennie, J.

*Insect transformations*. 1830. *see* Rennie, J.

1888. *The insect world: or, a brief outline of the classification, structure, and economy of insects*. Sm. 8vo, xi+276 pp, 1 col. + 2 plain pls, Tilt & Bogue, London 1843.

1889. *Insects and reptiles, with their uses to man*. 8vo, 10 col. pls with letterpress, Darton, London [1844].

1890. *Insects and their habitations*. 12mo, iv+96 pp, text figs, Parker, London 1839.

Insecta Britannica. (Reeve). *see*
Stainton, H. T. *Lepidoptera Tineina.* 1854.
Walker, E. *Diptera*. 1851–56.
Wilkinson, S. J. *British Tertrices*. 1859 (Van Voorst). This is shown as part of the series in the *Catalogue of the British Museum (Natural History)*.

*Instructions for collecting and preserving insects, particularly moths and butterflies*. 1771. *see* Curtis, W.

*The instructive picture book, lessons from the geographical distribution of animals. see* I., M.H.H.

*The instructive picture book: or, instructive lessons from the natural history of animals and plants. see* White, A.

*The instructive picture book; or, lessons from the vegetable world*. 1857. *see* Yonge, *Miss* C. M.

1891. *The intellectual observer . . . also the student . . . a collection of eighty plates*. 8vo, title page and 80 col. or tinted pls, London [?1871]. Plates from *The intellectual observer,* Vol. 2 onwards, and its successor *The student*, which ceased in 1871.

1892. *Interesting walks of Henry and his tutor; or, rational conversations on natural history*. 18mo, Harvey & Darton, London 1827; 1828.

International Botanical Congress. 1867. *see* Candolle, A. L. P. P. de

International Scientific Series. (King to 1876, later Kegan Paul) *see*
Benenden, P. J. van *Animal parasites*. 1876.
Binet, A. & Féré, Charles *Animal magnetism*. 1887.
Cooke, M. C. *Fungi*. 1875.
—— *Fresh-water Algae*. 1890.
Demoor, J. *et al. Evolution by atrophy*. 1899.
Heilprin, A. *Geographical distribution of animals*. 1887.
Henslow, G. *Origin of floral structures*. 1888.
—— *Origin of plant structures*. 1895.
Huxley, T. H. *The crayfish*. 1880.
Kew, H. W. *Dispersal of shells*. 1893.
Lubbock, *Sir* John *Bart. Ants, bees, and wasps*. 1882.
—— *Seedlings*. 1892.
—— *Buds & stipules*. 1899.
Marey, E. J. *Animal mechanism*. 1874.
Pettigrew, J. B. *Animal locomotion*. 1873.

Phillips, W. *British Discomycetes*. 1887.
Poulton, E. B. *Colours of animals*. 1890.
Romanes, G. J. *Animal intelligence*. 1882.
—— *Jelly fish, star fish and sea urchins*. 1885.
Schmidt, E. O. *The doctrine of descent and darwinism*. 1875.
—— *The Mammalia*. 1885.
Semper, C. G. *The natural conditions of existence*. 1881.
Stebbing, T. R. R. *A history of Crustacea*. 1893.
Trouessart, E. L. *Microbes, ferments, and moulds*. 1886.

*In the green leaf and the sere*. 1896. *see* Jordan, D.

1893. *In the temperate regions; or, nature and natural history in the temperate zones.* Sm. 8vo, viii+249 pp, 1 pl., text figs, Nelson, London 1882.

*Introduction to the catalogue of the Ashmolean Museum.* [1826]. *see* Duncan, P. B.

1894. *An introduction to the natural history of curious and interesting birds*. 12mo, 3 vols, text figs, Vol. 1 Christopher & Jennett, Vols 2 & 3 Appleton, Stockton [c.1810].

*An introduction to the study of birds; or elements of ornithology on scientific principles.* [1835]. *see* Martin, W. C. L.

1895. *Introduction to vegetable physiology*. 12mo, 144 pp, Pickering, London 1845. Small Books on Great Subjects, No. IX.

*Invitation to surgeons and naturalists for donations to the Museum*. 1826. *see* Royal College of Surgeons.

1896. INWARDS, Richard *Weather lore; a collection of proverbs, sayings, and rules concerning the weather*. Sm. 8vo, 91 pp, Tweedie, London 1869 [= 1868]; issue 1869 title *Weather wisdom . . .*; ed. 1893; 3ed. 1898; 4ed. 1950.

1897. IRBY, Leonard Howard Lloyd *British birds: key list*. 8vo, 58 pp, Porter, London 188; 2ed. 1892.

Ireland, National Museum of *see* More, A. G.

*The Irish flora*. 1833. *see* Bailey, *Miss* C.

1898. IRVINE, Alexander *The London flora*. 8vo, xiii+340 pp, Smith Elder, London 1838; ed. 1846.

1899. —— *The illustrated handbook of British plants*. 8vo, xxiv+840 pp, text figs, Nelson, London 1858.

1900. —— *Introduction to the science of botany*. 8vo, 152 pp, London [1858].
—— *see* P., W.

1901. IRVING, Christopher *A catechism of botany*. 12mo, 1 pl., Souter, London [1821]. Many later editions. Souter's Catechisms.

1902. —— *Outlines of the kingdoms of nature*. 18mo, Darton, London [1841].

Isle of Wight Philosophical and Scientific Society. *see* More, A. G.

# J

J., C. A. *Sea weeds.* [c.1855]. *see* Johns, C. A.

J., J. *An account of the rare fish, Regalecus glesne*. 1849 *see* Jacob, J.

J., L. *The butterfly collector's vade-mecum*. 1824. *see* Ford, *Mrs* L.

J., M. E. *Botanical lectures*. 1804. *see* Jackson, *Mrs* M. E.

—— *Sketches of the physiology of vegetable life* 1811. *see* Jackson, *Mrs* M. E.

J., R. *The gamekeeper at home*. 1878. *see* Jefferies, R.

J., R. *Wild life in a southern county*. 1879. *see* Jefferies, R.

J., R. *Juvenile rambles through the paths of nature*. 1786. *see* Johnson, R.

J., S. *A natural history of fishes, and of reptiles, insects . . . vegetables.* 1795. *see* Jones, S.

1903. JACKSON, Benjamin Daydon *A glossary of botanic terms, with their derivation and accent*. 8vo, xi+327 pp, Duckworth, London 1900; 2ed. 1905; 3ed. 1916; 4ed. 1928.

1904. JACKSON, Julian *known as* John Richard *What to observe; or, the traveller's remembrancer*. 12mo, xx+577 pp, 20 tab., Madden, London 1841; 2ed. 1845; 3ed. 1861. Translation of *Guide du voyageur*, Paris 1822.

1905. JACKSON, *Miss* M. A. *The pictorial flora, or British botany delineated.* 8vo, text figs, Longman, London 1840.

1906. —— *Botanical terms illustrated. Series I*. 8vo, 4 pls, Birmingham [?1842]. All publ.

1907. [JACKSON, *Mrs* Maria Elizabeth] *Botanical dialogues, between Hortensia and her four children*. 8vo, xlvi+335 pp, 11 pls, Johnson, London 1797. By a lady.

1908. [——] *Botanical lectures . . . altered from botanical dialogues*. 8vo, xix+243 pp, Johnson, London 1804. By a lady, M. E. J.

1909. [——] *Sketches of the physiology of vegetable life*. 8vo, xi+172 pp, 14 pls, Hatchard, London 1811. By the authoress of *Botanical dialogues*, M. E. J.

1910. JACKSON, Thomas *Our dumb neighbours*. 4to, Partridge, London [1870].

1911. —— *Our feathered companions*. 4to, viii+148 pp, Partridge, London [1870].

1912. —— *Stories about animals*. 4to, Cassell, London [1874].

1913. JACOB, Edward *Plantae Favershamienses. A catalogue of the more perfect plants growing spontaneously about Faversham*. 12mo, 146+liii pp, 1 pl., 1 portr., March for the author, London 1777.

1914. [JACOB, J.] *An account of the rare fish, Regalecus glesne, caught off Cullercoats, on the 26th of March, 1849*. 8vo, 16 pp, 1 pl., text figs, London 1849. By J. J.

1915. JAEGER, Gustav *Problems of nature. Researches and discoveries*. 8vo, ii+261 pp, text figs, 1 map, Williams & Norgate, London 1897. Translations of selections of his work, by Heinrich G. Schlichter.

1916. JAMES, Ivor *Handbook for Cardiff and district, prepared for the use of the British Association*. 8vo, 244pp, 2 pls, Owen, Cardiff 1891.

1917. JAMES, William Powell *Guesses at purpose in nature with special reference to plants*. Sm. 8vo, 192 pp, SPCK, London [1882].

1918. JAMESON, Hampden Gurney *Key to the genera and species of British mosses*. 8vo, 32 pp, 1 pl., London 1891. Reprint from *J. Bot*., 1891.

1919. —— *Illustrated guide to British mosses*. 8vo, 75 pp, 59 pls, the author, Eastbourne [1893]. Enlarged from No. 1918.

1920. JAPP, Alexander Hay *Hours in my garden and other nature sketches*. Sm. 8vo, 340 pp, text figs, Hogg, London [1894].

1921. —— *Our cuckoo, and other cuckoos and parasitical birds*. Sm. 8vo, xi+292 pp, 12 pls, text figs, Burleigh, London 1899.

Jardine, *Sir* William *Bart*, *editor The naturalist's library*. Sm. 8vo, 40 vols, col. pls, 40 portrs, Lizars, Edinburgh 1833–43; issue [1845–46]; issue London 1848. First ed. may have all title pages dated 1843; these were issued with the last vol. There are several later reissues. Relevant titles are entered under author. Apart from those by the editor, *see*

Bushnan, J. S. *Fishes*. 1840.

Duncan, J. *Coleopterous insects*. 1835.

—— *British butterflies*. 1835.

—— *British moths*. 1836.

—— *Introduction to entomology*. 1840.

Hamilton, R. *Whales*. 1839.
—— *Amphibious Carnivora*. 1839.
—— *British fishes*. 1843.
Macgillivray, W. *British quadrupeds*. 1838.
Selby, P. J. *Pigeons*. 1835.
Smith, C. H. *Introduction to Mammalia*. 1842.
Swainson, W. *Flycatchers*. 1838.

1922. JARDINE, *Sir* William *Bart The natural history of lions tigers etc.* Sm. 8vo, 276 pp, 34 col. pls, 1 portr., Lizars, Edinburgh 1834; issue [1845–46]; issue London 1848; 1860. Naturalist's Library Vol. IV 2ed. Vol. XVI.

1923. —— *The natural history of gallinaceous birds. Vol. 1*. Sm. 8vo, 255 pp, 29 col. pls, 1 portr., Lizars, Edinburgh 1834; issue [1845–46]; issue London 1848; 1860. Naturalist's Library Vol. V; 2ed. Vol. XIV.

1924. —— *The natural history of game birds. Gallinaceous birds Vol. 2*. Sm. 8vo, 197 pp, 30 col. pls, 1 portr., Lizars, Edinburgh 1834; issue [1845–46]; issue London 1848. Naturalist's Library Vol. VI, 2ed. Vol. VIII.

1925. —— *The natural history of fishes of the perch family*. Sm. 8vo, 177 pp, 34 col. pls, portr., Lizars, Edinburgh 1835; issue [1845–46]; issue London 1848; 1861. Naturalist's Library Vol. VII; 2ed. Vol. XXXVIII.

1926. —— *The natural history of deer, antelopes, camels etc.* Sm. 8vo, 229 pp, 33 col. pls, 1 portr., Lizars, Edinburgh 1835; issue [1845–46]; ed. London 1848; 1855. Naturalist's Library Vol. XI, 2ed. Vol. XXI.

1927. —— *The natural history of goats, sheep, oxen etc.* Sm. 8vo, 264 pp, 32 col. pls, 1 portr., Lizars, Edinburgh 1835; issue [1845–46]; issue London 1848. Naturalist's Library Vol. XII, 2ed. Vol. XXII.

1928. —— *The natural history of the birds of Great Britain and Ireland*. Sm. 8vo, 4 pts, 129 col. pls, 4 portr., Lizars, Edinburgh 1838–43; issue London 1848; 1852–55; 1866; [1876]. Naturalist's Library Vol. XL; 2ed. Vol. IV.

1929. —— *British Salmonidae*. Fol., 2 pts, 12 col. pls with letterpress, for the author, [Edinburgh 1839, 41]. All publ. Usually in a portfolio; no t.p. issued.

1930. —— *The natural history of bees etc.* Sm. 8vo, 301 pp, 30 col. pls, 1 portr., Lizars, Edinburgh 1840; issue [1845–46]; issue London 1848; 1858; 1866. Naturalist's Library Vol. XXVX; 2ed. Vol. XXIV.

1931. —— *Illustrations of the duck tribe*. 4to, 9 col. pls, for the author, Lockerby [c.1840]. Very few copies, usually uncoloured. Plates from Jardine & Selby, *Illustrations of ornithology*.

1932. —— editor *Leaves from the books of nature, containing the principal illustrations in Jardine's Naturalist's Library*. Fol., 117 11 of colour plates, Lizars, Edinburgh [c.1843].

Jarrett, Emily Elizabeth *see* Kirby, W. & Spence, W.

1933. JAUFFRET, Louis Francois *Visits to the menagerie and the botanical garden at Paris*. 12mo, 2 vols, pls, Phillips, London 1801; ed. 1804; ed. 1 vol. 1811. French ed., Rowell, Rugby 1832. Translation of *Voyage au Jardin des Plantes*. Paris An VI [1798].

1934. JAYNE, Horace *Mammalian anatomy: a preparation for human and comparative anatomy. Part 1. The skeleton of the cat . . . compared with the skeleton of man.* Roy. 8vo, xix+816 pp, text figs, Lippincott, London 1898. USA printed.

1935. [JEFFERIES, John Richard] *The gamekeeper at home; sketches of natural history and rural life*. 8vo, xi+216 pp, Smith Elder, London 1878. By R. J. Many later reprints.

1936. [——] *Wild life in a southern county*. 8vo, xii+387 pp, Smith & Elder, London 1879. By the author of *The gamekeeper at home*. R. J. Later reprints.

1937. —— *Round about a great estate.* 8vo, vii+204 pp, Smith Elder, London 1880.

1938. —— *Nature near London.* 8vo, vi+242 pp, Chatto & Windus, London 1883. Later reprints. Reprinted from *The Standard.*

1939. —— *The life of the fields.* 8vo, viii+262 pp, Chatto & Windus, London 1884. Later reprints.

1940. —— *Red deer.* 8vo, 207 pp, Longmans, London 1884. Later reprints.

1941. —— *The open air*, 8vo, 270 pp, Chatto & Windus, London 1885. Later reprints.

1942. —— *After London, or wild England.* 8vo, vii+442 pp, Cassell, London 1885.

1943. —— *Field and hedgerow.* 8vo, viii+331 pp, Longman, London 1889; issue L.P. 1 portr., 1889; ed. 1895.

—— *see* Toplis, G.

1944. JEFFREYS, Charles *A complete catalogue of British Mollusca. Compiled from* [*John Gwyn*] *Jeffrey's 'British conchology'*. Sm. 8vo, 53 11, Marsden, Gloucester 1887.

1945. —— *A catalogue of land and fresh water Mollusca.* Sm. 8vo, Marsden, Gloucester 1887. Reprint of part of No. 1944.

1946. JEFFREYS, John Gwyn *British conchology.* 12mo, 5 vols, 107 col.+40 plain pls, Van Voorst, London 1862–69; ed. Vol. 1 1904 title *Land and freshwater shells of Great Britain.* See also Nos 1944 & 1945. The nudibranchs by Joshua Alder.

1947. —— *Gleanings in British conchology. (More gleanings. Further gleanings)*. 8vo, 3 pts, 4 pls, Taylor & Francis, London 1858–59. Reprinted from *Ann. Mag. nat. Hist.*, Vols 1–3, 1858–59.

1948. —— *Address delivered at the anniversary meeting of the Hertfordshire Natural History Society . . . 1880. On evolution.* 8vo, 13 pp, Austin, Hertford 1880. Reprinted from *Trans. Herts. nat. Hist. Soc.*, 1880.

1949. —— *Deep-sea exploration. A lecture.* 8vo, 20 pp, Van Voorst, London 1881.

—— *see* Jeffreys, C.

—— *see* Jordan, H. K.

1950. JENNER, Edward *Observations on the natural history of the cuckoo.* 4to, 19 pp, London [1788]. Reprinted from *Phil. Trans. Roy. Soc.* Vol. 78, 1788. 50 copies.

1951. —— *Some observations on the migration of birds.* 4to, 36 pp, Nicol printed, London 1824; ed. 8vo, Chipping Sodbury 1882. Reprinted from *Phil. Trans. Roy. Soc.*, Vol. 114, 1824.

1952. JENNER, Edward *A.L.S. The flora of Tunbridge Wells.* 8vo, xx+260 pp, 2 col. pls, Colbran, Tunbridge Wells [1845]. pp. xvii–xx are supplementary and not always present.

1953. JENNER, J. H. A. *A list of the land and freshwater Mollusca of East Sussex.* 8vo, 13 pp, Chronicle Office, Eastbourne [1835]. Reprinted from *Proc. Eastbourne nat. Hist. Soc.*, 1884–85.

1954. JENNINGS, C. *The eggs of British birds.* 8vo, xxx+200 pp, 7 col. pls, Binns & Goodwin, Bath [1853]; 2ed. 1853.

1955. JENNINGS, James *Ornithologia, or the birds: a poem in two parts; with an introduction to their natural history and copious notes.* 12mo, 2 pts, xxix+468 pp, 2 pls, Poole & Edwards for the author, London 1828; 2ed. 1829.

1956. *Jenny and the insects; or, little toilers and their industries.* Sm. 8vo, 267 pp, text figs, Nelson, London 1880.

1957. JENYNS, Leonard *later* Blomefield *A monograph of the British species of Cyclas and Pisidium.* 4to, 24 pp, 3 pls, Smith printed, Cambridge 1832. Reprinted from *Trans. Cambridge phil. Soc.* Vol. 4.

1958. —— *A manual of British vertebrate animals.* 8vo, xxxii+559 pp, Deighton, Cambridge 1835.

1959. —— *A systematic catalogue of British vertebrate animals.* 8vo, iv+[20] pp, Deighton & Stevenson, Cambridge 1835.

1960. —— *A catalogue of the collection of British quadrupeds and birds in the Museum of the Cambridge Philosophical Society.* 8vo, 47 pp, Cambridge 1836.

1961. —— *Observations in natural history . . . also a calendar of periodic phenomena.* 8vo, xvi+440 pp, Van Voorst, London 1846; ed. Cambridge 1903 of calendar only, title *A naturalist's calendar kept at Swaffham Bulbeck*; ed. 1922 of Calendar only.

Jermyn, Laetitia *later* Ford *see* Ford, *Mrs* L.

1962. JESSE, Edward *Gleanings in natural history.* 8vo, 3 series, 2 pls, text figs, Murray, London 1832, 34, 35; ed. Vol. 1 1832; ed. 2 vols 1838; 5ed. 1843; 6ed. 1845; 7ed. 1849; 8ed. 1854; 9ed. 1856; 10ed. 1861, 11ed. 1869; 1870; 12ed. 1877; 13ed. 1887. Eds 6–13 in one vol.

1963. —— *Scenes and tales of country life; with recollections of natural history.* 12mo, vi+395 pp, 4 pls, text figs, Murray, London 1844; ed. 1844; ed. 1853 title *Scenes and occupations . . .*

1964. —— *Favourite haunts and rural studies.* 8vo, vi+365 pp, 1 pl., text figs, Murray, London 1847.

1965. —— *Lectures on natural history . . . delivered at the Fisherman's Home Brighton.* 8vo, 48 pp, Murray, London 1860; 1861; 2ed. 1863. First edition is lectures 1–6 only; 1861 is lectures 1–12.

1966. —— *Lectures on natural history.* Sm. 8vo, viii+157 pp, Booth, London 1861; 2ed. 1863.

1967. JOHNS, Charles Alexander *Flora sacra; or a knowledge of the works of nature.* Sm. 8vo, 47 pp, 23 specimens [of Mosses], Parker, London 1840.

1968. —— *Botanical rambles.* Sm. 8vo, 187 pp, 1 pl., text figs, SPCK, London 1846; [ ].

1969. —— *Botanical rambles. Spring. (Summer, Autumn, Winter).* Sm. 8vo, 4 pts, 4 pls, text figs, SPCK, London 1847–52.

1970. —— *The forest trees of Britain.* 12mo, 2 vols, SPCK, London 1847; 1849; 1860; 1869; ed. 1 vol. [1873]; 1882, 1886; 1889; 1892; 1893; 8ed. 1899; 9ed. 1903; 10ed. 1912; 11ed.

1919; 12ed. 1921; [c.1925]; 1929; another ed. [1911] title *British trees.*

1971. —— *A ramble in spring.* 16mo, 110 pp, 1 pl., text figs, SPCK, London [1847].

1972. —— *A week at the Lizard.* 16mo, viii+322 pp, 1 pl., 1 map, text figs, SPCK, London 1848; ed. [1863]; ed. [1874]; ed. [?1880].

1973. —— *First steps to botany.* 16mo, National Society for Promoting the Education of the Poor, London 1853.

1974. —— *Flowers of the field.* 16mo, 2 vols, text figs, SPCK, London [1853]; 2ed.–12ed. [ ]; 13ed. 1878; 14ed. 1878; 15ed. 1878; 17ed. 1880; 18ed. 1881; 19ed. 1882; 20ed. 1884; 21ed. 1885; 24ed. 1886; 25ed. 1888; 26ed. 1889; 27ed. 1890; 28ed. 1894; 29ed. 1896; 1899; 30ed. 1902; 31ed. 1905; 32ed. 1910; 33ed. 1911; 34ed. 1919; 35ed. 1925; another set of editions 1–16, 1907–49. For appendix on grasses see No. 1982.

1975. [——] *Birds' nests and eggs.* 16mo, vii+248 pp, 22 col. pls, SPCK, London [1854]; ed. [1865].

1976. [——] *Sea-weeds.* 16mo, 24 pp, 12 col. pls, SPCK, London [c. 1855]. Text signed C.A.J.

1977. —— editor. *Birds of the wood and field.* [*Series 1.*] *(Series 2; Series 3.)* 12mo, 3 vols, text figs, SPCK, London 1859, 60, [62].

1978. —— editor. *Monthly gleanings from the field and garden.* 8vo, 24 col. pls, SPCK, London [1859].

1979. —— editor. *Monthly wild flowers.* 12mo, SPCK, London [1860].

1980. —— *British birds in their haunts.* Sm. 8vo, xxxii+626 pp, text figs, SPCK, London 1862; 2ed. 1867; 1870; 1879; 1882; 1887; 1888; 1907; 1909; 1910; 12ed. 1911; 14ed. 1918; 15ed. 1920; 25ed. 1948; another set of editions 1–7, 1909–22; abridged ed. 2 vols, Edinburgh 1905, 07. Title *I go a walking*.

1981. —— *Home walks and holiday rambles.* Sm. 8vo, vi+273 pp, 4 pls, text figs, Longman, London 1863.

1982. JOHNS, Charles Henry *Grasses . . . an appendix to the late Rev. C. H. Johns* [i.e. C. A. Johns's] *Flowers of the field.* 8vo, 96 pp, SPCK, London 1892.

1983. JOHNS, William *Practical botany; an improved arrangement of the generic characters of British plants.* 8vo, xi+156 pp, Longman, London 1826; ed. 1840.

1984. JOHNSON, Charles *The ferns of Great Britain.* Roy. 8vo, 87 pp, 49 col. pls, Sowerby, London 1855; ed. 1859.

1985. —— *British poisonous plants.* 8vo, iv+59 pp, 28 col. pls, Sowerby, London 1856; 2ed. 1861.

1986. —— *The fern allies; a supplement to The ferns of Great Britain.* Roy. 8vo, 52 pp, 31 col. pls, Sowerby, London 1856; ed. 1859.

1987. —— *The grasses of Great Britain.* Roy. 8vo, xxxii+192 pp, 144 col. pls, Sowerby, London [1857–] 1858–61; ed. 1861.

1988. JOHNSON, Charles Pierpoint *British wild flowers illustrated.* 8vo, xlix+168 pp, 81 col.+2 plain pls, Sowerby, Lambeth [1858–] 1860; ed. with suppl. London 1863; 1876; 1882; 1890; 1894; 1902; 1914. Suppl. by Salter, John William and Sowerby, John Edward

—— *see* Sowerby, John Edward.

1989. JOHNSON, George William *The British ferns popularly described.* Sm. 8vo, 284 pp, text figs, Cottage Gardener Office, London 1857; 3ed. 1859; 4ed. 1861; 5ed. 1867.

—— *see* Hogg, H.

1990. JOHNSON, James Rawlins *Disputatio physica inauguralis, quaedam de hirudine complectens.* 8vo, 45 pp, 1 pl., Stewart, Edinburgh 1814.

1991. —— *A treatise on the medicinal leech; including its medical and natural history.* 8vo, xii+147 pp, 2 pls, Longman, London 1816.

1992. —— *Further observations on the medicinal leech.* 8vo, xv+112 pp, 4 col. pls, Longman, London 1825.

1993. [JOHNSON, Richard] *Juvenile rambles through the paths of nature; in which many parts of the . . . creation . . . are made familiar.* 16mo, 128 pp, 1 pl., text figs, Newbery, London 1786; eds. [ ]; eds Harris, London [ ]; ed. 12mo, Swaffham 1830. Attribution doubtful; dedication signed R.J. 1830 ed. by the author of *The toyshop.*

*1994. JOHNSON, Theo *Illustrations of the larvae and pupae of British moths.* Roy. 8vo, 56 col. drawings with letterpress, privately printed, London 1867.

1995. —— *Illustrations of British butterflies and their larvae.* Roy. 8vo, 63 col. drawings with letterpress, privately printed, London 1871.

1996. —— *Illustrations of British hawk moths, and their larvae*

* All Theo Johnson's works have hand drawn and coloured plates and very few copies were produced. Their bibliography is complex.

*(including the Sesidae).* Roy. 8vo, 46 pp, 36 col. drawings on 35 leaves (3 & 4 on same leaf) in some copies, privately printed, London 1874–76; 1888; addenda 1 leaf, 2 col. drawings 1888.

1997. —— *Illustrations of the larvae and pupae of British Lepidoptera.* Roy. 8vo, 53 col. drawings with letterpress, privately printed, London 1878; partial ed. 1906 title *The caterpillars of familiar butterflies and moths*, 20 pp, drawings Nos 34–53.

1998. —— *Wild animals, birds.* Fol., 25 col. drawings with letterpress, London [c.1880].

1999. —— *Zoological sketches, or typical mammals drawn and described.* Fol., 2 vols, 50 col. drawings with letterpress, privately printed, London [1887].

2000. —— *Familiar British insects, selected from the principle orders.* 8vo, 42 col. drawings with letterpress, privately printed, London 1887.

2001. —— *Zoological lectures. Vol. 1.* 8vo, 22 col. & 40 plain drawings, London 1889. T.p. & introduction printed; rest of text in mss. ?All publ.

2002. —— *Illustrations of British Lepidoptera.* Roy. 8vo, 25 col., drawings with letterpress, privately printed, London 1900.

2003. —— *Illustrations of the larvae of British Lepidoptera, the butterflies and hawk moths.* Roy. 8vo, 32 col. drawings with letterpress, privately printed, London 1900.

2004. JOHNSON, Thomas *Iter plantarum investigationis ergo susceptum a decem sociis in agrum Cantianum ... Ericetum Hamstedianum sive plantarum ibi crescentium observatio.* 4to, [13] pp, [London 1629]; ed. with No. 2005, 1972. See also No. 2008.

2005. —— *Descriptio itineris plantarum investigationis ergo suscepti in agrum Cantianum ... Enumeratio plantarum in Ericeto Hamstediano locisq. vicinis crescentium.* 8vo, 48 pp, 2 pls, Cotes, London 1632; ed. 1849; ed. with No. 2004, 1972. See also No. 2008. Ed. 1849 is from sheets of No. 2008.

2006. —— *Mercurius botanicus; sive plantarum gratia suscepti itineris, anno MDCXXXIV descriptio.* 4to, 78+19 pp, 2 pls, text figs, Cotes, London 1634.

2007. —— *Mercurii botanici pars altera, sive plantarum gratia suscepti itineris in Cambriam, sive Walliam descriptio.* 4to, 37 pp, Cotes, London 1641.

2008. —— *Opuscula omnia botanica Thomae Johnsoni.* 4to, 5 pts, 3 pls, Pamplin, London 1847. A collection of Nos 2004–2007 with one other. The pagination is as that of the originals.

2009. JOHNSTON, Alexander Keith *The physical atlas of natural phenomena.* Fol., 122 pp, 25 col. maps, Edinburgh 1848; 2ed. 1850.

2010. JOHNSTON, George *A flora of Berwick-on-Tweed.* Sm. 8vo, 2 vols, 2 col. & 6 plain pls, Carfrae, Edinburgh 1829–31.

2011. —— *Some account of a whale stranded near Berwick-on-Tweed.* 4to, Newcastle-on-Tyne 1830.

2012. —— *A history of British zoophytes.* 8vo, xii+341 pp, 44 tinted pls, text figs, Lizars, Edinburgh 1838; 2ed. 2 vols, London 1847. 2ed. is Natural History of the British Isles Series.

2013. —— *A history of British sponges and lithophytes.* 8vo, xii+264 pp, 25 col. pls, text figs, Lizars, Edinburgh 1842.

2014. —— *An introduction to conchology.* Roy. 8vo, xvi+614 pp, text figs, Van Voorst, London 1850.

2015. —— *Terra Lindisfarnensis. The natural history of the eastern borders . . . Vol. 1. The botany.* 8vo, xii+336 pp, 1 col.+13 plain pls, text figs, Van Voorst, London 1853. All publ.

2016. —— *A catalogue of the British non-parasitical worms in the collection of the British Museum.* 8vo, 365 pp, 24 pls, text figs, British Museum, London 1865.

2017. JOHNSTONE, Alexander *Botany, a concise manual.* 8vo, xiv+260 pp, text figs, Pentland, Edinburgh 1891.

2018. [JOHNSTONE, *Mrs* Christian Isobel] *Scenes of industry displayed in the bee-hive and the ant-hill.* 12mo, xii+212 pp, 6 pls, Harris, London [?1829]; 2ed. 1830. By the author of *The wars of the Jews* etc.

2019. JOHNSTONE, James *L.M.B.C. memoirs on typical marine plants and animals . . . II. Cardium.* 8vo, vii+84 pp, 7 pls, 1 map, Dobb for the Liverpool Marine Biology Committee, Liverpool 1899.

2020. JOHNSTONE, William Grosart & CROALL, Alexander *The nature-printed British sea weeds.* 4to or 8vo, 4 vols, 221 col.+1 plain pls, Bradbury & Evans, London 1859–60.

Jones, Greville *see* Sibly, E.

Jones, John *see* Sibly, E.

2021. [JONES, John Pike] *A botanical tour through various parts of the Counties of Devon and Cornwall.* Sm. 8vo, viii+74 pp, Treadwin, Exeter 1820.

2022. —— & KINGSTON, J. F. *Flora Devoniensis: or, a descriptive catalogue of plants growing wild in the County of Devon.* 8vo, 2 pts, Longman, London 1829.

2023. [JONES, Stephen] *The natural history of beasts compiled from the best authorities.* 12mo, xii+204 pp, 38 pls, Newbery, London 1793; ed. 2 vols 1798.

2024. [——] *The natural history of birds.* 12mo, xii+204 pp, 38 pls, Newbery, London 1793; ed. 2 vols 1798.

2025. [——] *A natural history of fishes, and of reptiles, insects . . . vegetables.* 12mo, 208 pp, text figs, Newbery, London 1793; ed. 1795. Advertisement signed S.J., attribution doubtful.

2026. JONES, Thomas Rymer *A general outline of the animal kingdom, and manual of comparative anatomy.* 4to, Imp., Roy. or demy 8vo, xiv+732 pp, text figs, Van Voorst, London [1838–] 1841; 2ed. 1855; 3ed. 1861; 4ed. 1871. Companion vol. to Natural History of the British Isles Series.

2027. —— *The natural history of animals.* 12mo, 2 vols, text figs, Van Voorst, London 1845–52.

2028. —— *The aquarian naturalist. A manual for the sea-side.* Sm. 8vo, xviii+524 pp, 8 col. pls, Van Voorst, London 1858.

2029. —— *The animal creation; a popular introduction to zoology.* 12mo, xvi+594 pp, text figs, SPCK, London 1865. Later editions.

2030. —— *The natural history of birds.* 12mo, xx+452 pp, text figs, SPCK, London 1867. Later editions.

2031. —— *Mammalia; a popular introduction to natural history.* 8vo, xx+428 pp, text figs, Warne, London [1873].

2032. JONES, Thomas Wharton *Evolution of the human race from apes, and of apes from lower animals, a doctrine unsanctioned by science.* Sm. 8vo, xviii+69 pp, Smith Elder, London 1876.

2033. —— *Glimpses of animal life. A naturalist's observations on the habits and intelligence of animals.* Sm. 8vo, x+229 pp, Elliot Stock, London 1889.

2034. JONES, William *The religious use of botanical philosophy . . . a sermon . . . on the wisdom of God, as displayed in the vegetable creation.* 4to, iii+18 pp, Robinson, London 1784; ed. 1786.

2035. JONSTON, Johann *An history of the wonderful things of nature.* Fol., 354 pp, Streater printed, London 1657. Translation of *Thaumatographia naturalis*, 2ed. Amsterdam 1633, by Joannes Jonstonus.

2036. —— *A description of the nature of four-footed beasts.* Fol.,

119 pp, 80 pls, Widow of J. J. Schipper & S. Swart, Amsterdam 1678. Translation of *Historiae naturalis de quadrupedibus*, Amsterdam 1657.

2037. [JORDAN, Denham] *Woodland, moor and stream, being notes of a naturalist.* 8vo, viii+224 pp, Smith Elder, London 1889; 3ed. 1896. By a son of the marshes.

2038. [——] *On Surrey hills.* Sm. 8vo, vi+301 pp, Blackwood, Edinburgh 1891; 2ed. 1892; 4ed. 1893; ed. 1894. By a son of the marshes.

2039. [——] *Within an hour of London town, among wild birds and their haunts.* Sm. 8vo, 314 pp, Blackwood, London 1892; 3ed. 1894. By a son of the marshes.

2040. [——] *Forest tithes, and other studies from nature.* Sm. 8vo, viii+208 pp, Smith Elder, London 1893. By a son of the marshes.

2041. [——] *With the woodlanders and by the tide.* 8vo, 305 pp, Blackwood, Edinburgh 1893; 2ed. 1893. By a son of the marshes.

2042. [——] *From spring to fall, or when life stirs.* Sm. 8vo, 244 pp, Blackwood, London 1894; 2ed. 1894. By a son of the marshes.

2043. [——] *In the green leaf and the sere.* 8vo, 228 pp, text figs, Kegan Paul, London 1896. By a son of the marshes.

2044. [——] *The wild-fowl and sea-fowl of Great Britain.* 8vo, viii+326 pp, 12 pls, Chapman & Hall, London 1896. By a son of the marshes.

2045. [——] *Drift from the longshore.* Sm. 8vo, 261 pp, 1 pl., Hutchinson, London 1898. By a son of the marshes.

2046. JORDAN, Henry Keys *A complete catalogue of the British Mollusca: compiled from British conchology – Jeffreys.* Sm. 8vo, 2 pts, 62 11, Mardon, son, & Hall, Bristol 1870.

*The journal of a naturalist.* 1829. *see* Knapp, J.L.

2047. JOYCE, Jeremiah *Dialogues on the microscope.* 12mo, 2 vols, 10 pls, Johnson, London 1812.

2048. JUSSIEU, Adrien de *The elements of botany.* 8vo, xv+750 pp, text figs, Van Voorst, London 1849; ed. 1858. Translation of *Cours élémentaire d'histoire naturelle; botanique*, Paris 1842; 4ed. 1845, by James Hewetson Wilson.

*Juvenile conversations on natural history . . . The moss and the lichen.* 1850. *see* Macnab, Catherine M.

2050. *Juvenile history of birds, or the power and wisdom of the supreme being in the works of creation attempted to be unfolded to young minds.* 12mo, 25 pp, text figs, Dean & Munday, London [c.1820].

Juvenile Library. *see Longlands Castle.*

2051. *The juvenile naturalist, enigmatically arranged; consisting of beasts, birds.* 12mo, London 1811.

*Juvenile rambles through the paths of nature.* 1830. *see* Johnson, R.

2052. *Juvenile visit to the menagerie.* 32mo, pp 73–140, text figs, Fairburn, London [c.1825]. The relevant part of Fairburn's Juvenile Library.

# K

K. *A list of birds found in the neighbourhood of Sudbury.* 1838. *see* King, W. D.

K., *Miss* E. *Sylvan sketches.* 1825. *see* Kent, *Miss* E.

2053. KALM, Pehr *Kalm's account of his visit to England on his way to America in 1748.* 8vo, xvi+480 pp, 2 maps, text figs, Macmillan, London 1892. Translation of the relevant parts of *En Resa til Norra America*, Stockholm 1753–61. First English ed. of whole is Warrington 1770–71.

Kane, Katherine, *Lady. see* Bailey, *Miss* K.

2054. KANE, William Francis de Vismes *European butterflies.* Sm. 8vo, xxxi+184 pp, 15 pls, text figs, Macmillan, London 1885.

2055. KAPPEL, August Wilhelm & KIRBY, William Egmont *Beetles, butterflies and moths, and other insects.* 4to, 182 pp, 12 col. pls, Cassell, London 1892.

2056. —— & —— *British and European butterflies and moths; (Macrolepidoptera).* Imp. 8vo, xvi+273 pp, 30 col. pls, text figs, Nister, London 1895.

2057. KARR, Jean Baptiste Alphonse *A tour round my garden.* Sm. 8vo, xii+332 pp, text figs, Warne, London 1855; [ ]. Translation of *Voyage autour de mon jardin*, Paris 1845, by John George Wood.

2058. KEARSLEY, George *Links in the chain, or popular chapters in the curiosities of animal life.* Sm. 8vo, xii+295 pp, 7 pls, Hogg, London [1862].

2059. —— *Curiosities and wonders of animal life.* Sm. 8vo, text figs, Ward & Lock, London [1872].

2060. —— *Wonders and curiosities of animal life.* Sm. 8vo, 1 col. pl., Ward & Lock, London [1878].

2061. KEARTON, Richard *Birds' nests, eggs, and egg-collecting.* Sm. 8vo, 71 pp, 16 col. pls, Cassell, London 1890. Later editions.

2062. —— *British birds' nests, how, where, and when to find them and identify them.* 8vo, xx+368 pp, text figs, Cassell, London 1895; 1898; 2ed. 1907; 1913.

2063. —— *With nature and a camera.* Roy. 8vo, xvi+368 pp, text figs, Cassell, London 1897. Later editions.

2064. —— *Wild life at home, how to study and photograph it.* 8vo, xv+188 pp, 1 pl., text figs, Cassell, London 1898. Later editions.

2065. —— *Our rarer British breeding birds.* 8vo, xvi+149 pp, text figs, Cassell, London 1899. Later editions.

2066. —— *Our bird friends.* 8vo, xvi+215 pp, text figs, Cassell, London 1900; ed. 1923.

2067. KEEPING, Walter *A popular handbook to the natural history collection.* 8vo, 63 pp, 1 map, 1 pl., York Museum, York 1881.

2068. KEITH, Patrick *A system of physiological botany.* 8vo, 2 vols, 9 pls, Baldwin, London 1816.

2069. —— *A botanical lexicon.* 8vo, vi+416 pp, Orr, London 1837.

2070. [KENDALL, E. A.] *Little stories of pretty little birds.* 32mo, 58 pp, 4 pls, text figs, Dean, London [c.1845].

2071. KENNEDY, Alexander William Maxwell Clark *The birds of Berkshire and Buckinghamshire.* Sm. 8vo, xiv+232 pp, 4 col. pls, Ingalton & Drake, Eton 1868. The plates are hand coloured photographs.

2072. [KENT, *Miss* Elizabeth] *Sylvan sketches.* 8vo, xliv+408 pp, 1 pl., Taylor & Hessey, London 1825; ed. 1831. By *Miss* E. K.

2073. KENT, William Saville- *A manual of the Infusoria.* Roy. 8vo, 2 vols & atlas, 1 col.+52 plain pls, Bogue, London 1880–82. Plates 1–51, 48a.

2074. —— *Handbook of the marine and freshwater fishes of the British Islands.* 8vo, viii+129 pp, text figs, Clowes, London 1883. Great International Fisheries Exhibition Handbook, Vol. 1.

2075. —— *Official guide to the Brighton Aquarium.* 8vo, Brighton [c.1885].

2076. KENTISH, Richard *An essay on the method of studying natural history; being an oration delivered to the Societas Naturae Studiosum at Edinburgh.* 8vo, viii+110 pp, Elmsley & Johnson, London 1787.

2077. KEOGH, John *Botanologia universalis Hibernica ; or, a general Irish herbal.* 4to, vi+145 pp, Harrison, Cork 1735.

2078. —— *Zoologia medicinalis Hibernica; or, a treatise of the birds, beasts, fishes, reptiles, or insects, which are commonly found in this Kingdom.* 8vo, xiv+167 pp, Kelburn, Dublin 1739; ed. London 1744. Ed. of 1744 has name of Bernard de Mandeville as author, but is a reissue with new title page.

2079. KERMODE, Sidney Alfred Pizey *The Isle of Man. List of flowering plants.* 8vo, 21 pp, Brown printed, Douglas 1900. Reprinted from *Yn Lioar Manninagh*, Vol. 3, pp, 273–91.

2080. KERNER VON MARILAUN, Anton *Flowers and their unbidden guests.* 8vo, xvi+164 pp, 3 pls, Kegan Paul, London 1878. Translation of *Die Schütsmittel der Blüthen gegen unberufene Gäste*, Innsbruck 1876, by William Ogle.

2081. —— *The natural history of plants.* Imp. 8vo, 2 vols, 16 col. pls, portr, text figs, Blackie, London [1894–95]. Several later editions in 2, 4 or 6 vols, some without col. pls. Translation of *Pflanzenleben*, Leipzig [1887–] 1888 [–1891], by Francis Wall Oliver.

2082. KERR, Richard *Hidden beauties of nature.* 8vo, xii+256 pp, 1 pl., text figs, RTS, London 1895; 2ed. 1900; 3ed. 1905; 4ed. [ ].

2083. KEW, Henry Wallis *The dispersal of shells.* 8vo, xiv+291 pp, Kegan Paul, London 1893. International Scientific Series, Vol. LXXV.

2084. KEYS, Isaiah Waterloo Nicholson *Flora of Devon and Cornwall.* 8vo, 5 pts, Plymouth 1865–71. Reprinted from *Trans. Plymouth Instn,* 1865–66.

2085. [KEYS, J. H.] *Catalogue of British Tenthridinidae, Siricidae, Cynipidae and British Orthoptera.* 8vo, 13 pp, Plymouth 1892. Printed on one side only.

2086. —— *Exchange list of British Hemiptera-Heteroptera.* 8vo, 1896. Based on Saunders, E. *A catalogue of the British Hemiptera-Heteroptera.* 1876.

2087. KIDD, John *Introductory lecture to a course in comparative anatomy, illustrative of Paley's Natural theology.* 8vo, 72 pp, University Press for the author, Oxford 1824.

2088. KIDD, Walter *Design in nature's story.* Sm. 8vo, 166 pp, Nisbet, London 1900.

2089. KIDD, William *The book of British song-birds, and aviary companion; including interesting notes on the warblers.* Sm. 8vo, 7 pts, Groombridge, London [1856]. The parts also issued separately.

2090. KINAHAN, John Robert *A review of the genera of terrestrial Isopoda (Oniscoidea) with descriptions of all the known British species and genera.* 8vo, 38 pp, 4 pls, privately printed, Dublin 1857. Reprinted from *Nat. Hist. Rev.*, Vol. 4.

*Kindness to animals.* [1844]. *see* Phelan, *Mrs* C. E.

2091. KING, James J. Francis Xavier *A contribution towards a catalogue of the neuropterous fauna of Ireland.* 8vo, 38 pp, Natural History Society of Glasgow, Glasgow 1889. Reprinted from *Trans. nat. Hist. Soc. Glasgow.* Vol. 11, 1887–88.

2092. [KING, William Doubleday] *A list of the birds found in the neighbourhood of Sudbury*, 4to, 3 pp, Sudbury 1838. By K.

2093. —— *A paper on the study of natural history.* 8vo, v+28 pp, Wright, Sudbury 1849.

2094. KINGSLEY, Charles *Glaucus; or, the wonders of the shore.* Sm. 8vo, 165 pp, 1 pl., Macmillan, Cambridge 1855; 2ed. 1855; 3ed. 1856 [without col. pls]; 3ed. 1858 [with col. pls]; 4ed. 1859 with 12 col. pls; 5ed. 1873. Many later editions.

2095. —— & [SOWERBY, George Brettingham, *the second*] *Glaucus; or, the wonders of the shore. With an appendix containing coloured illustrations of the objects mentioned in the work.* Sm. 8vo, 3ed. 168+48 pp, 12 col. pls, Macmillan, Cambridge 1858. Combination vol. of No. 2094 with Sowerby's *Companion to Mr. Kingsley's Glaucus*, 1858.

Kingston, J. F. *see* Jones, J. P.

—— *see* Turton, W.

2096. KINGSTON, William Henry Giles *Stories of animal sagacity.* 8vo, text figs, Nelson, London 1874.

2097. —— *Wonders of the ocean.* Sm. 8vo, 128 pp, 1 pl., text figs, Gall & Inglis, London [1880]. Later editions, some with title *The ocean and its wonders.*

Kirby, Elizabeth *see* Kirby, Mary.

2098. KIRBY, *Miss* Mary *later* Gregg *A flora of Leicestershire.* Sm. 8vo, ix+183 pp, Leicester 1848; ed. London 1850. With notes by her sister, i.e. Kirby, E.

2099. —— & KIRBY, *Miss* Elizabeth *Plants of the land and water.* 12mo, xi+347 pp, text figs, Jarrold, London [1857]; 2ed. [ ]. Observing Eye Series.

2100. —— & —— *The world at home; or pictures and scenes from far-off lands.* 4to, xii+296 pp, text figs, Nelson, London 1869; ed. 1876; 1892.

2101. —— & —— *Caterpillars, butterflies and moths.* 12mo, 158 pp, Jarrold, London 1860. Observing Eye Series.

2102. —— & —— *Things in the forest.* 8vo, text figs, London 1861; ed. 1874.

2103. —— & —— *The sea and its wonders.* 12mo, xii+384 pp, text figs, Nelson, London 1871 [=1870]. Later editions.

2104. —— & —— *Chapters on trees.* 8vo, 320 pp, pls, text figs, Cassell, London 1873; 2ed. 1876 title *Talks about trees.*

2105. —— & —— *Stories about birds of land and water.* 4to, text figs, Cassell, London [1873].

2106. —— & —— *Sketches of insect life.* 8vo, RTS, London 1874.

2107. KIRBY, William *Monographia apum Angliae; or . . . such species of the Linnean genus Apis as have been discovered in England.* 8vo, 2 vols, 4 col.+14 plain pls, J. Raw for the author, Ipswich 1802.

2108. —— *On the power, wisdom and goodness of God as manifested in the creation of animals, and in their history, habits and instincts.* 8vo, 2 vols, 17 pls, Pickering, London 1835; ed. 1847; ed. 1853. Bridgewater Treatise No. VII.

2109. —— & SPENCE, William *An introduction to entomology; or elements of the natural history of insects.* 8vo, 4 vols, 6 col.+24 plain pls, Longman, London 1815–26; 2ed. Vol. 1 1816, Vol. 2 1818; 3ed. Vol. 1 1818, Vol. 2 1823; 4ed. Vol. 1 1822 or 1826, Vol. 2 1826; 5ed. 1828; 6ed. 2 vols 1843; 7ed. 1 vol. 1856, 1857; 1858; 1865; 1876; index [1876] by *Miss* E. E. Jarrett. In 6ed. Vols 3 & 4 were not reprinted.

2110. KIRBY, William Egmont *Insects: friends and foes.* 16mo, x+138 pp, pls, Partridge, London 1898.

—— *see* Kappel, A. W.

2111. KIRBY, William Forsell *A manual of European butterflies on the plan of Stainton's Manual of British butterflies and moths.* 12mo, xii+153 pp, Williams & Norgate, London 1862.

2112. —— *European butterflies and moths.* 4to, lvi+427 pp, 62 col. pls, Cassell, London 1882. Based on Berge, Friederich *Schmetterlingsbuch*, Stuttgart 1842; 5ed. 1876.

2113. —— *The young collector's handbook of beetles.* Sm. 8vo, 32 pp, 1 pl., text figs, Sonnenschein, London [1883]. Young Collector's Handbook Series.

2114. —— *The young collector's handbook of butterflies and moths.* Sm. 8vo, 32 pp, text figs, Sonnenschein, London [1883]. Young Collector's Handbook Series.

2115. —— *The young collector's handbook of the orders of insects.* Sm. 8vo, 32 pp, text figs, Sonnenschein, London [1883]. Young Collector's Handbook Series.

2116. —— *Elementary text book of entomology.* 8vo, viii+240 pp, 57 pls, Sonnenschein, London 1885; 2ed. 1892.

2117. —— *British butterflies, moths and beetles.* Sm. 8vo, 93 pp, text figs, Sonnenschein, London 1885. Young Collector Series.

2118. —— *A hand-book to the order Lepidoptera.* Sm. 8vo, 5 vols, 156 col.+2 plain pls, text figs, Shaw, London 1894–97; issue 1896–97. First issue is Allen's Naturalists' Library; second is Lloyd's Natural History.

2119. —— *Marvels of ant life.* 8vo, 174 pp, 1 pl., text figs, Partridge, London 1898.

2120. KITCHENER, Frances Anna *A year's botany, adapted to home and school use.* 8vo, xi+264 pp, text figs, Rivington, London 1874; 2ed. 1878; 4ed. 1894; ed. 1907.

2121. KITCHENER, Francis Elliott *Naked-eyed botany.* 8vo, xii+182 pp, text figs, Percival, London 1892; 2ed. 1894; 3ed. 1898.

Klasing, Otto *see* Coultas, H.

2122. KLEIN, Hermann Joseph *Land, sea and sky; or, the wonders of life and nature.* 8vo, viii+832 pp, text figs, Ward Lock, London 1881–82; ed. 2 pts [ ] title *God's glorious creation.* Translation of *Die Erde und ihr organisches Leben*, Stuttgart [1880–81] by Janet Minshull.

2123. KNAGGS, Henry Guard *The lepidopterist's guide.* 8vo, 122 pp, Van Voorst, London 1869; ed. 1884; 2ed. [ ]; 3ed. [1901].

2124. —— *A list of Macro-lepidoptera occurring in the neighbourhood of Folkestone.* 8vo, 24 pp, Folkestone Natural History Society, Folkestone 1870.

2125. —— *A cabinet list of the Lepidoptera of Great Britain and Ireland.* 8vo, London 1870.

2126. KNAPP, F. H. *The botanical chart of British flowering plants and ferns.* 8vo, x+90 pp, Binns & Goodwin, Bath 1846.

2127. KNAPP, John Leonard *Gramina Britannica; or representa-*

*tions of the British grasses.* 4to, 119 col. pls with letterpress, White for the author, London 1804; 2ed. 1842.

2128. [——] *The journal of a naturalist.* 8vo, xiii+403 pp, 11 pls, Murray, London 1829; 2ed. 1829; 3ed. 1830; ed. 1831; ed. 1835 title *Country rambles in England.* Pls 1–7, but 2, 4, 6 & 7 on two leaves.

2129. KNIGHT, Charles *The menageries. Quadrupeds, described and drawn from living subjects.* 12mo, 2 vols, text figs, Knight, London 1829–31. Library of Entertaining Knowledge.

2130. —— *The pictorial museum of animated nature and companion for the Zoological Gardens.* Fol., 2 vols, text figs, Printing and Publishing Co., London [1856–58].

2131. KNIGHT, D. *Morphology of the Vertebrata; dog-fish, cod, pigeon and rabbit.* 8vo, 48 pp, 4 pls, Pentland, Edinburgh 1883.

2132. KNIGHT, Francis A. *Idylls of the field.* Sm. 8vo, vi+182 pp, 1 pl., Elliot Stock, London 1889.

2133. —— *By leafy ways; brief studies from the book of nature.* 4to, or Sm. 8vo, viii+197 pp, 1 pl., text figs, Elliot Stock, London 1889; 2ed. 1889; 3ed. 1889.

2134. —— *By moorland and sea.* Sm. 8vo, x+215 pp, 1 pl., text figs, Elliot Stock, London 1893.

2135. KNIGHT, Richard Payne *A selection from the physiological and horticultural papers.* 8vo, xii+379 pp, portr., 7 pls, Longman, London 1841.

2136. KNIGHT, William *Outlines of botany.* 8vo, 32 pp, Chalmers, Aberdeen 1813; ed. 1828.

2137. KNOX, Alexander *A history of the County of Down; including an account of its . . . natural history.* 8vo, viii+724 pp, 1 pl., 1 map, Dublin 1875.

2138. KNOX, Arthur Edward *Ornithological rambles in Sussex.* Sm. 8vo, vi+250 pp, 4 pls, Van Voorst, London 1849; 2ed. 1850; 3ed. 1855.

2139. —— *Game birds and wild fowl; their friends and their foes.* Sm. 8vo, x+264 pp, 4 pls, Van Voorst, London 1850.

2140. —— *Autumns on the Spey.* 8vo, 171 pp, 4 pls, text figs, Van Voorst, London 1872.

2141. KNOX, Robert *Fish and fishing in the lone glens of Scotland; with a history of the propagation, growth, and metamorphosis of the salmon.* 8vo, 144 pp, text figs, Routledge, London 1854.

2142. KOENIG, Carl Dietrich Eberhard *Tracts relative to botany, translated from different languages.* 8vo, 277 pp, 9 pls, Phillips & Fardon printed, London 1805. Translations of ten separate tracts.

2143. [KOLB, C.] *Animal life in Europe.* 4to, 14 col. pls, text figs, RTS.

Komensky, Jan Amos *see* Comenius, J. A., London [1870].

# L

L, A. E. *The wild flowers of Clevedon.* 1877. *see* Lainson, A. E.

L, H. M. *Rough notes on natural history in Norfolk.* [1890]. *see* Leathes, H. M.

2144. [LÁCÉPÈDE, Bernard Germaine Etiene de la Ville sur Illon, *Conte de* & CUVIER, George Leopold Chretien Frederic Dagobert, *Baron de*] *The wonders of animated nature; consisting of . . . the principal animals and birds in the Royal Menageries of London and Paris.* 12mo, 198 pp, 15 pls, Phillips, London 1810. Based on *La Ménagerie du Muséum National*, Paris 1801.

—— *see* Buffon, G. L. L., *Conte de.*

2145. [LAINSON, A. E.] *The wild flowers of Clevedon.* Sm. 8vo, 67 pp, Caple, Clevedon 1877.

2146. LAISHLEY, Richard *A popular history of British birds' eggs.* Roy. 16mo, xi+313 pp, 20 col. pls, Reeve, London 1858. Popular Natural History Series.

Lamarck, Jean-Baptiste Pierre Antoine de Monet, *Chevalier de. see* Crouch, E. A.

—— *see* Dubois, C.

—— *see* Hanley, S. C. T.

2147. LAMB, Henry *The flora of Maidstone.* 8vo, 15 pp, Maidstone 1889.

2148. LAMOUROUX, Jean Vincent Felix *Corallina; or, a classical arrangement of flexible coralline polypidoms, selected from the French.* 8vo, xxvi+284 pp, 19 pls, Sherwood, London 1824. Translation of parts of *Histoire des polypiers coralligenes flexibles*, Caen 1816.

2149. LANDSBOROUGH, David *the father. Arran: a poem in six cantos; and, excursions to Arran, with reference to the natural history of the island.* 16mo, Edinburgh 1847; ed. 8vo 1852 title *Excursions to Arran, Ailsa Craig and the two Cumbraes*; ed. 8vo Ardrossan 1875 title *Arran; its topography, natural history and antiquities.* 1875 ed. is by The Landsboroughs, i.e. the father and David the son.

2150. [——] *Treasures of the deep: or, specimens of Scottish seaweeds.* 4to, t.p., 39 leaves of specimens with printed names, Bryce, Glasgow 1847. Later issues. Attribution doubtful.

2151. —— *A popular history of British sea-weeds.* Roy. 16mo, xx+368 pp, 20 col. pls, Reeve, London 1849; 2ed. 1851; 3ed. 1857. Popular Natural History Series.

2152. —— *A popular history of British zoophytes or corallines.* Roy. 16mo, xii+404 pp, 20 col. pls, Reeve, London 1852. Popular Natural History Series.

Landsborough, David *the son. see* Landsborough, David *the father*, No. 2149.

Landseer, Thomas *see* Barrow, J. H.

—— *see* Lowry, J. W.

2153. LANG, Arnold *Text-book of comparative anatomy.* 8vo, 2 pts, text figs, Macmillan, London 1891–96. Translation of *Lehrbuch der vergeleichenden Anatomie der wirbellosen Thiere*, Jena 1890–94, by Henry Meyners Bernard & Matilda Bernard.

2154. LANG, Henry Charles *Rhopalocera Europae descripta et delineata. The butterflies of Europe.* Roy. 8vo, 2 vols, 82 col. pls, Reeve, London [1881–] 1884.

2155. LANKESTER, Edwin *An account of Askern and its mineral springs; together with a sketch of the natural history.* Sm. 8vo, 151 pp, 1 col. pl., Churchill, London 1842.

2156. —— *The natural history of creation.* 16mo, Young Men's Christian Association, London 1848; ed. 8vo. 1856; ed. 1884.

2157. —— *Vegetable physiology, in a series of easy lessons.* 8vo, 80 pp, Cassel, London [1869].

2158. —— *The aquavivarium; fresh and marine.* Sm. 8vo, 71 pp, 4 col. pls, text figs, Hardwicke, London [1856].

2159. [——] *Half-hours with the microscope.* Sm. 8vo, xx+92 pp, 8 pls, Hardwicke, London [1859]; 2ed. signed 1860. Very many later editions.

2160. LANKESTER, *Sir* Edwin Ray *On comparative longevity in man and the lower animals.* 8vo, ix+135 pp, Macmillan, London 1870.

2161. —— *Notes on embryology and classification for the use of students.* 8vo, 58 pp, 1 pl., Macmillan, London 1870.

2162. —— *Degeneration. A chapter in darwinism.* 8vo, 75 pp, Macmillan, London 1879; ed. 1880. Nature Series.

2163. —— *Observations and reflections on the appendages and on the nervous system of Apus cancriformis.* 8vo, 36 pp, text figs, Adlard, London 1881. Reprinted from *Quart. Jl micr. Sci.*

2164. —— *Studies on Apus, Limulus and Scorpio.* 8vo, 36+76 pp, 3 pls, text figs, Churchill, London 1887. Reprinted from *Quart. Jl micr. Sci.*

2165. —— *Spolia maris. Contributions to the knowledge of Rhabdopleura and Amphioxus.* Roy. 4to, 70 pp, 5 col.+5 plain pls, Churchill, London 1889. Reprinted from *Quart. Jl micr. Sci.*

2166. —— *The advancement of science.* 8vo, vi+387 pp, text figs, Macmillan, London 1890.

2167. —— *Zoological articles contributed to the Encyclopaedia Britannica.* 4to, viii+195 pp, text figs, Black, London 1891. Reprinted from *Encyclopaedia Britannica*, 9ed. 1875.

2168. —— *editor. A treatise on zoology.* 8vo, 7 vols in 8, text figs, Black, London 1900–1909; reprint 1964. All publ.

2169. LANKESTER, Phebe, *Mrs* Edwin *A plain and easy account of British ferns.* 8vo, xv+108 pp, 9 col. pls, Hardwicke, London 1860; 6ed. 1876; [1877]; ed. 1881 title *British ferns.* For earlier editions see Bosanquet, E. The first *Mrs* Lankester edition is the third. Many later editions.

2170. —— *Wild flowers worth notice.* Sm. 8vo, xiv+15+138, 7 col. pls, Hardwicke, London 1861. Later editions.

2171. —— *Botany for elementary schools.* 16mo, Allman, London 1872. Allman's Penny School Series.

2172. —— *Talks about plants; or, early lessons in botany.* 8vo, 252 pp, 6 col. pls, text figs, Griffith, London 1879. Later editions.

2173. LARDNER, Dionysius *The museum of science and art.* Sm. 8vo, 12 vols, text figs, Walton & Maberley, London 1854–56. Issued in 156 parts; some parts were reissued up to 1859. A number of articles were also issued as separate works by the same publisher.

2174. LARTER, C. Ethelinda *Manual of the flora of Torquay.* 8vo, 83 pp, Torquay 1900.

2175. LATHAM, John *A general synopsis of birds.* 4to, 3 vols, 106 col. pls, Vol. 1 White, Vols 2 & 3 Leigh & Sotheby, London 1781–85; suppl. iii+298 pp, 14 col. pls, 1787; 2nd suppl. 376 pp, 23 col. pls, 1801; 2nd suppl. issue 1802; ed. 11 vols 8vo Winchester 1821–28 title *A general history of birds.* Plates 1–140, 136*, 138*.

2176. —— *Index ornithologicus, sive systema ornithologiae.* 4to, 2 vols, xviii+790 pp, for the author, London 1790; suppl. lxxiv pp, 1801; ed. 8vo Paris 1809. Index to No. 2175.

2177. LATHAM, Robert Gordon & FORBES, Edward *The natural history department of the Crystal Palace described. Ethnology, Zoology and Botany.* Sm. 8vo, 92 pp, text figs, Crystal Palace Library, London 1854.

—— *see* Ansted, D. T.

2178. [LAURIE, Mrs J. Werner] *Natural history for the young.* 16mo, Edinburgh [1869]. By the author of *Home and its duties.*

2179. LAVER, Henry *The mammals, reptiles, and fishes of Essex.* 8vo, viii+138 pp, 8 pls, text figs, Chelmsford 1898. *Essex Field Club, Spec. Mem.* No. III.

2180. LAWRENCE, *Sir* William *Bart. An introduction to comparative anatomy and physiology, being the two introductory lectures delivered at the Royal College of Surgeons.* 8vo, vi+179 pp, Callow, London 1816.

2181. —— *Lectures on physiology, zoology, and the natural history of man.* 8vo, xxiii+579 pp, 12 pls, Callow, London 1819; ed. 1822; 2ed. 1822; ed. 2 vols 1822; 3ed. 1823; 6ed. 1834; 7ed. 1838; 8ed. 1840 title *Lectures on comparative anatomy . . .*; 9ed. 1844; 9ed. 1848. First ed. was withdrawn by the author; the rest are pirates.

2182. LAWSON, Roderick *Ailsa Craig; its history and natural history.* Sm. 8vo, 80 pp, 1 map, text figs, Parlane, Paisley 1888; ed. 1895.

2183. LEACH, F. S. *Plain instructions for the management of the aquarium.* 8vo, 46 pp, Dean, London 1856.

2184. —— *Handbook of plain instructions for the construction and management of fresh water aquaria.* 8vo, 72 pp, Dean, London [1856].

2185. LEACH, William Elford *The zoological miscellany.* Roy. 4to or 8vo, 3 vols, 150 col. pls, Nodder, London 1814–17. Plates

1–149, 135; pp. 27–28 in Vol. 2 not published. Continuation of Shaw, G. & Nodder, F. R. *The naturalist's miscellany*, 1790–1813.

2186. —— *Malacostraca podophthalmata Britanniae; or, descriptions of such British species of the Linnean genus Cancer as have their eyes elevated on footstalks.* Roy. Fol. or Roy. 4to, 19 pts, 54 col. pls with letterpress, London 1815 [–1820], 1875. Parts 18 & 19 first published by Quaritch, with text by G. B. Sowerby II, and some other parts reissued at the same time. Plates 1–45, 9a, 9b, 21b, 22b, 22c, 24a, 28b, 37b, 37c. Title of parts issue . . . *or descriptions of the British species of crabs, lobsters, prawns and other Malacostraca with pedunculated eyes.* Parts 18/19 was printed on Imp. 4to not Fol. L.P. 30 copies.

2187. —— *Systematic catalogue of the . . . indigenous Mammalia and birds that are preserved in the British Museum.* Roy. 8vo, 42 pp, British Museum, London 1816; ed. 1822. Ed. of 1822 is Willughby Society Reprint.

2188. —— *Molluscorum Britanniae synopsis. A synopsis of the Mollusca of Great Britain.* Sm. 8vo, xvi+376 pp, 13 pls, Van Voorst, London 1852.

2189. [LEATHES, Hill Mussenden] *Rough notes on natural history in Norfolk and the eastern counties.* Sm. 8vo, 141 pp, text figs, Jarrold, London [1890]. By H. M. L.

2190. LEBOUR, George Alexander *Visit of the British Association to Newcastle-upon-Tyne, 1889 . . . Geology and natural history.* 8vo, vi+219 pp, 5 pls, text figs, Lambert, Newcastle-upon-Tyne 1889. British Association Handbook. Made up of *Outlines of the geology of Northumberland and Durham*, 1886 to p. 147, the rest new.

Leclerc, George Louis *Conte de* Buffon *see* Buffon, *Conte de*

2191. LEE, Arthur Bolles *The microtomist's vade-mecum.* 8vo, xiv+424 pp, Churchill, London 1885; 2ed. 1890; 3ed. 1893; 4ed. 1896; 5ed. 1900; 6ed. 1905; 7ed. 1913; 8ed. 1921; 9ed. 1928; 10ed. 1937; 11ed. 1905.

2192. LEE, Henry *Aquarium notes. The octopus; or devil fish of fiction and of fact.* Sm. 8vo, xv+114 pp, 1 pl., text figs, Chapman & Hall, London 1875.

2193. —— *Sea fables explained.* 8vo, x+122 pp, text figs, Clowes, London 1883. *Great International Fisheries Exhibition Handbook*, Vol. III.

2194. —— *Sea monsters unmasked.* 8vo, vii+103 pp, 1 pl., text figs,

Clowes, London 1883. *Great International Fisheries Exhibition Handbook*, Vol. III.

2195. LEE, James *An introduction to botany*. 8vo, xvi+320 pp, 12 pls, Tonson, London 1760; 2ed. 1765; 3ed. 1776; 4ed. 1788; 5ed. 1794; 4ed. 1796; 6ed. Edinburgh 1796; ed. Edinburgh 1799; ed. Edinburgh 1806; 4ed. 1810 title . . . *to the science of botany*; ed. Edinburgh 1811. Based on Linnaeus, C. *Philosophia botanica*, Stockholm 1751.

2196. —— *Coloured specimens to illustrate the natural history of butterflies. From the collection of Mr. Lee, Hammersmith*. Fol., vii pp, 20 col. pls, Miller, London 1806. ?50 copies.

2197. —— *Termini botanici. A dictionary of botanical terms*. 8vo, [88] pp, Jennings, London 1813. A separate printing of a work which first appeared in No. 2195, 2ed. 1765.

2198. LEE, John Henry *British Pyralides, including the Pterophoridae*. 8vo, viii+121 pp, 7 col.+11 plain pls, London 1886.

2199. LEE, Oswin A. J. *Among British birds in their nesting haunts*. Fol., 4 vols, 160 pls, text figs, Douglas, Edinburgh 1896–99. All publ.

Lee, Sarah, *Mrs* Robert *see* Bowdich, Sarah, *Mrs* T. E.

2200. LEES, Edwin *The botanical looker-out among the wild flowers*. Sm. 8vo, xvi+376 pp, 1 pl., text figs, Tilt & Bogue, London 1842; 2ed. 1851; 3ed. 1868.

2201. —— *The botany of the Malvern hills*. 8vo, viii+64 pp, Tilt & Bogue, London [1843]; 2ed. [1852]; 3ed. [1868].

2202. —— *Pictures of nature in the Silurian region around the Malvern hills and the Vale of Severn*. Sm. 8vo, xii+336 pp, 1 pl., text figs, Lamb, Malvern 1856.

2203. —— *The botany of Worcestershire*. 8vo, xci+147+48 pp, 3 pls, 1 map, text figs, Worcestershire Naturalists Club, Worcester 1867.

2204. LEES, Frederic Arnold *A summary of comital plant-distribution additional to that detailed in Topographical botany* [by H. C. Watson]. 8vo, 74 pp, Wellwyn 1878; ed. 54 pp, West Newman, London 1878.

2205. —— *The flora of West Yorkshire*. 8vo, x+843 pp, 1 col. map, York 1888; suppl. v+134 pp, London [1941]. *Trans. Yorks. Nat. Union.*

2206. —— *Botany and outline flora of Lincolnshire*. 8vo, 29 pp, Sheffield 1892. Reprinted from White's *History of the County*.

—— *see* Davis, J. W.

2207. LEEUWENHOEK, Anthony van *The select works of Anthony van Leeuwenhoek, containing his microscopical discoveries in many of the works of nature.* 4to, 2 vols (3 pts), 1 col.+22 plain pls, for the translator S. Hoole, London 1798–1807; ed. 1 vol. 1800. The col. pl., of a lily, is nearly always absent. There is a 1797 issue of Nos 1 & 2 only. Pt. 2 1798 usually lacks pp. i–viii and the plate of liverfluke which is present in 1800 ed. Translation of parts of dutch and latin editions with suppressions, by Samuel Hoole.

2208. LEFEVRE, August Alfred. *An atlas of the eggs of the birds of Europe.* Roy. 8vo, 136 col. pls, Paris 1844 [–?1845]; ed. 1848.

2209. [LEGG, John] *A discourse on the emigration of British birds; or this question at last solv'd; whence came the stork and the turtle, the crane and the swallow, when they know and observe the appointed time of their coming.* 8vo, ix+45 pp, for the author, Salisbury 1780; ed. London 1781; ed. 1795; ed. 1814. By a naturalist. The ed. of 1814 is attributed to G. Edwards on title page. For another edition see Lupton, T.

2210. LEGRAND, Anthony *Philosophia veterum e mente Renati Descartes.* 12mo, J. Martyn, London 1671; ed. 8vo 1672 title *Philosophiae secundum principia R. Descartes*; 3ed. 1675; ed. 1678; 4ed. 4to 1680; English ed. Fol., 1694 title *An entire body of philosophy according to the principles of the famous Renate des Cartes.* Translated by Richard Blome.

Leicester Literary and Philosophical Society. *see* Mott, F. T.

2211. LEIGH, Charles *The natural history of Lancashire, Cheshire, and the Peak in Derbyshire.* Fol., 3 books, 196+97+112 pp, 24 pls, 1 portr., 1 col. map, for the author, Oxford 1700.

2212. LEIGH, *Hon.* Mary Cordelia Emily & THEOBALD, Frederick Vincent *An account of British flies (Diptera). Part 1.* 8vo, London 1891. All publ. Later withdrawn and republished by Theobald alone, 1892 q.v.

2213. LEIGHTON, William Allport *Catalogue of the Cellulares, or flowerless plants of Great Britain.* 4to, 15 pp, London [1837].

2214. —— *A flora of Shropshire.* 8vo, xii+573 pp, 20 pls, Van Voorst, London 1841.

2215. —— *The British species of angiocarpous lichens.* 8vo, v+101 pp, 30 col. pls, Ray Society, London 1851. Ray Society Publ. No. 20 for 1851.

2216. —— *The lichen flora of Great Britain, Ireland and the Channel*

*Islands*. 12mo, xlvii+467 pp, for the author, Shrewsbury 1871; 2ed. 1872; 3ed. 8vo 1879.

Le Keux, J. *Illustrations of natural history*. [1829–30]. *see Illustrations of natural history*.

2217. LE LIEVRE, F. *Le Lievre's guide to Guernsey, containing the topography, history (natural and political) . . . with notices of Sark and Alderney*. 8vo, viii+200 pp, 1 map, Guernsey 1863.

2218. LE MAOUT, Emmanuel & DECAISNE, Joseph *A general system of botany, descriptive and analytical*. Roy. 8vo, 2 pts, xii+1066 pp, text figs, Longman, London 1873; ed. 1876. Translation of *Traité générale de botanique*, Paris 1868 by *Mrs* Joseph Dalton (Frances) Hooker.

2219. LE MOYNE DE MORGUES, Jacques *La clef des champs pour trouver plusiers animaux, tant besstes qu'oyseaux, avec plusiers fleurs & fruitz*. Obl. fol., 50 11, col. text figs, for the author, London 1586.

2220. LEMPRIERE, William *Popular lectures on the study of natural history and the sciences*. 8vo, Wightman, London 1827; 2ed. 1830.

2221. LESLIE, George & HERDMAN, William Abbott *The invertebrate fauna of the Firth of Forth*. 8vo, 106 pp, M'Farland and Erskine, Edinburgh 1881. Reprinted from *Proc. R. phys. Soc. Edinb.*

2222. LESLIE, George Dunlop *Letters to Marco*. Sm. 8vo, xviii+259 pp, pls, text figs, Macmillan, London 1893.

2223. —— *Riverside letters*. Sm. 8vo, xvi+256 pp, 12 pls, text figs, Macmillan, London 1896.

2224. LESSER, Friedrich Christian *Insecto-theology*. 8vo, xvi+439 pp, Cadell, London 1799. Translation of *Theologie des insectes*. The Hague 1742, itself a translation of *Insecto-theologia*, Frankfurt 1738. Very few copies.

2225. *Lessons derived from the animal world*. Sm. 8vo, 2 ser., text figs, SPCK, London 1847, 51.

*Lessons on shells*. 1832. *see* Mayo, Elizabeth.

2226. *Lessons on the geographical distribution of animals*. Fol., 60 col. pls with letterpress, London 3ed. 1877; 4ed. [c.1880].

2227. LETOURNEAU, Charles Jean Marie *Biology*. 8vo, xi+480 pp, text figs, Chapman & Hall, London 1878; ed. 1890. Translation of *La biologie*, Paris 1876, by William Maccall.

*The letters of Rusticus on the natural history of Godalming*. 1849. *see* Newman, E. *editor*

2228. *Letters on entomology, intended for the amusement and instruction of young persons*. 12mo, 160 pp, 3 col. pls, Whittaker, London 1825.

2229. [LETTSON, John Coakley] *The naturalist's and traveller's companion*. 8vo, 69 pp, 1 col. pl., for the author, London 1772; 2ed. 1774 signed; 3ed. 1799.

2230. LEUTEMANN, Gottlob Heinrich *Coloured lithographs of animals after Leutemann's Zoological atlas for schools*. Obl. fol., 36 col. pls without text, SPCK, London 1881. Captions translated from *Zoologisches Atlas für den Schulgebrauch*, Leipzig 1876, by Arabella Burton Buckley.

2231. —— *Animals from the life ... coloured illustrations with descriptive letterpress*. 4to, 47 pp, 24 col. pls, Stanford, London 1887. Translation of *Bilder und Studien aus dem Tierreich*, Leipzig 1882, by Arabella Burton Buckley.

2231a LEVERIAN MUSEUM *A companion to the Museum, (late Sir Ashton Lever's) removed to Albion Street, the Surrey end of Black Friars Bridge*. 4to, 2 parts, 122 pp, 1 col. & 2 plain pls, London 1790.

—— *see* Fenn, *Lady* Eleanor; Martyn, W. F.; Shaw, George

2232. LEWES, George Henry *Sea-side studies at Ilfracombe, Tenby, the Scilly Isles, and Jersey*. Sm. 8vo, ix+414 pp, 7 pls, Blackwood, Edinburgh 1858; 2ed. 1860. Reprinted from *Blackwood's Magazine*.

2233. —— *The physiology of common life*. 8vo, 2 vols, Blackwood, Edinburgh 1859, 60.

2234. —— *Studies in animal life*. Sm. 8vo, viii+196 pp, 1 col. pl., text figs, Smith Elder, London 1862. Reprinted from *The Cornhill Magazine*.

2235. LEWIN, William *The birds of Great Britain with their eggs*. Imp. or Roy. 4to, 7 vols, 323 col. pls, for the author, London 1789 [–1794]; 2ed. 8 vols 1795–1801 title *The birds of Great Britain systematically arranged*; issue Vol. 1 1812; ed. 1838. First edition ?50 or 66 copies; plates are original water colours.

2236. —— *British birds' eggs, painted from the Portman Museum*. 4to, 90 col. pls, London 1795. Very few copies.

2237. —— *Insects of Great Britain ... Vol. 1*. 4to, 97 pp, 46 col. pls, Johnson, London 1795; another issue 1795 title *The papilios of Great Britain*. All publ.

2238. LEWIS, William Arnold *A discussion of the law of priority in entomological nomenclature*. 8vo, 86 pp, Williams & Norgate, London 1872.

Ley, Augustin *see* Purchas, William H.

2239. *Life beneath the waves and a description of the Brighton Aquarium*. Sm. 8vo, 95 pp, 22 pls on 20 leaves, Tinsley, London 1871.

*Life in the walls*. [1871]. *see* Chambers, *Miss* A. C.

2240. *The life of a plant*. 12mo, London [1895]. Science for the Household Series.

2241. *The life of an insect*. 16mo, 2 vols, 16 pls, text figs, SPCK, London [1857]. Vol. 1 not so called. Last page of each vol. signed R.E.

2242. LIGHTFOOT, John *Flora Scotica; or a systematic arrangement . . . of the native plants of Scotland and the Hebrides*. 8vo, 2 vols, xli+1151 pp, 35 pls, White, London 1777; 2ed. 1789; 2ed. 1792. Contains Thomas Pennant, 'A sketch of Caledonian zoology', 66 pp, 5 pls.

2243. LILFORD, Thomas Littleton Powys *Baron*. *Notes on the birds of Northamptonshire*. Roy. 8vo, 2 vols, 24 pls, 1 map, text figs, privately printed, Northampton 1880–93. Reprinted from *J. Northants. nat. Hist. Soc.*, 4to issue 100 copies.

2244. —— *Coloured figures of the birds of the British Islands*. Roy. 8vo, 7 vols, 421 col. pls, portr., Porter, London 1885–97 [–1898]; 2ed. 1891–97[–1898]. Parts 28–36 of both editions are identical. Two extra plates, turtle dove and marsh warbler, are sometimes present.

2245. LILLEY, A. E. V. & MIDGLEY, W. *Book of studies in plant form*. 8vo, 1 pl., Chapman & Hall, London 1895. Later editions.

2246. *Lilliputian spectacle de la nature: or, nature delineated, in conversations and letters passing between the children of a family*. 12mo, 3 vols, John Marshall, London [c.1785]. Not by N. A. Pluche.

2247. LINDLEY, John *A synopsis of the British flora, arranged according to the natural orders; containing vasculares or flowering plants. Vol. 1*. Sm. 8vo, xii+360 pp, Longman, London 1829; 2ed. 1835; 3ed. 1841; 3ed. 1859. All publ. The Vol. 1 is omitted from title after 1st ed.

2248. —— *An outline of the first principles of botany*. 12mo, viii+106 pp, 4 pls, Taylor & Walton, London 1830; 2ed. 1831; 3ed. 1835 title *A key to structural, physiological and systematic botany*; 4ed. 1841 title *Elements of botany*; 5ed.

1847; ed. 1849; ed. 1861. 3ed and later are combined with *Nixus plantarum* No. 2251. 3ed. was reissued with an appendix iv+81–89 pp in 1835 title *A key .. with an appendix containing a catalogue of medicinal plants.*

2249. —— *An introduction to the natural system of botany*. 8vo, xlviii+374 pp, Longman, London 1830; 2ed. 1836 title *A natural system of botany*.

2250. —— *An introduction to botany*. 8vo, xvi+557 pp, 6 pls, text figs, Longman, London 1832; 2ed. 1835; 3ed. 1839; 4ed. 2 vols 1848.

2251. —— *Nixus plantarum*. 8vo, v+28 pp, Ridgway, London 1833. Later eds in No. 2248.

2252. —— *Ladies' botany: or, a familiar introduction to the study of the natural system of botany*. 8vo, 2 vols, 50 col. pls, Ridgway. London 1834–37; Vol. 1 2ed. [ ]; 3ed. 1837; 4ed. 1840; Vol. 2 2ed. [ ], both vols 5ed. 1848; 6ed. [1862]; 6ed. 1865. See also No. 2255.

—— *A key to structural, physiological and systemetic botany*. 1835. *see* No. 2248.

—— *A natural system of botany*. 1836. *see* No. 2249.

2253. —— *Botany, in four parts*. 8vo, iv+222 pp, text figs, Baldwin & Cradock, London [1838]. Library of Useful Knowledge. Natural Philosophy Vol. 4.

2254. —— *School botany; an explanation of the characters of the . . . plants belonging to the flora of Europe*. 8vo, viii+218 pp, text figs, Longman, London 1839; ed. 1845 title *School botany: or, the rudiments of botanical science*; ed. 1854 title *School botany and vegetable physiology*; 1856; 1858; 1860; ed. 1862.

2255. —— *The Ladies' botany of Professor Lindley abridged by the author*. 8vo, xx+424 pp, text figs, Ridgway, London 1840; ed. 1841. See No. 2252.

—— *Elements of botany*. 1841. *see* No. 2248.

2256. —— *The vegetable kingdom*. 8vo, xlviii+908 pp, 1 pl., text figs, Bradbury and Evans for the author, London 1846; 2ed. 1847; 3ed. 1853.

2257. —— *A glossary of technical terms used in botany*. 8vo, xii+100 pp, text figs, Bradbury & Evans, London 1848. A separate issue of pt. 2 of No. 2248.

2258. —— *The symmetry of vegetation: an outline of the principles to be observed in the delineation of plants*. 8vo, 51 pp, Department of Science & Art, London 1854.

2259. —— *Descriptive botany: or, the art of describing plants correctly in scientific language*. 8vo, 31 pp, Bradbury & Evans, London 1858; 2ed. 1860. Later undated editions.
— see Paxton, *Sir* Joseph.

2260. —— & MOORE, Thomas *The treasury of botany, a popular dictionary of the vegetable kingdom*. Sm. 8vo, xx+xiv+1254 pp, 12 pls, text figs. Longman, London 1866; 1870; 2ed. 1874 [=1873]; ed. 1876 with suppl.; 1884; 1899.

2261. LINDNER, Friedrich Wilhelm & LACHMANN, F. H. A. *Pictorial natural history. Part 1*. 4to, 12+218 pp, 16 col. pls, Ewer, London [1845]. All publ. Translation of *Malerische Naturgeschichte der drei Reiche*, Brunswick 1840[–1844].

2262. LINDSAY, Archibald *Dissertatio inauguralis, de plantarum incrementi causis*. 8vo, 43 pp, Balfour & Smellie, Edinburgh 1781.

2263. LINDSAY, *Miss* B. *An introduction to the study of zoology*. 8vo, xix+356 pp, text figs, Sonnenschein, London 1895; 1899. Introductory Science Text-books.

2264. LINDSAY, William Lauder *A popular history of British lichens*. Roy. 16mo, xxxiii+352 pp, 22 col. pls, text figs, Reeve, London 1856; ed. 1858. Popular Natural History Series.

2265. —— *Mind in the lower animals in health and disease*. 8vo, 2 vols, Kegan Paul, London 1879.

2266. LINDSEY, William Henry *A season at Harwich . . . to which is added researches, historical, natural and miscellaneous*. 8vo, vi+172 pp, 12 pls, 3 maps, Simpkin Marshall, London 1851.

*Linnaeus and Jussieu; or the rise and progress of systematic botany*. 1844. *see* Carr, D. C.

2267. LINNAEUS, Carl *the father*. *Flora Anglica*. 4to, 29 pp, Höjer, Uppsala 1754. Later editions in *Amoenitates Academicae,* 1759 *etc.* Reprint London 1909. Respondent was Isaac Olaus Grufberg. Facsimile 1973.
—— *Miscellaneous tracts relating to natural history*. 1759. *see* Stillingfleet, B.

2269. —— *Termini botanici. In usum juventutis Academicae Edinensis*. 8vo, 40 pp, 5 pls, Balfour, Edinburgh 1770; ed. 1778; ed. 12mo Newcastle-upon-Tyne 1779. Editions by J. Hope of *Termini botanici*, Uppsala [1762]. Also contained in J. Berkenhout, *Clavis Anglica linnaei botanicae*, 1764, and other later botany books.

2270. —— *Institutes of botany; containing . . . descriptions of all the known genera of plants*. 4to, iv+302 pp, Griffin, London 1771[–72]. All publ. Translation of *Genera plantarum*, Leyden 1737, 6ed. Stockholm 1764.

2271. —— *Genera animalium. Ex editione duodecima Systematis naturae . . . in usus academicos*. 4to, 80 pp, University Press, Edinburgh 1771. Extracted by S. Ramsey from *Systema naturae*, 12ed. Stockholm 1766–68.

2272. —— *Genera plantarum. Ex editione duodecima Systematis naturae . . . in usus academicos*. 8vo, 88 pp, 1 tab. University Press, Edinburgh 1771. Extracted by J. Hope from *Systema naturae*, 12ed. Stockholm 1766–68. see also Nos 2275 & 2282.

2273. —— *Institutions of entomology: being a translation of Linnaeus's Ordines et genera insectorum . . . together with observations of the translator*. 8vo, viii+272 pp, Horsfield, London 1773. Translation of parts of *Systema naturae*, 12ed. Stockholm 1766–68, by Thomas Pattinson Yeats.

2274. —— *The elements of botany . . . being a translation of the Philosophia botanica, and other treatises of Linnaeus . . . to which is added an appendix wherein are described some plants lately found in Norfolk and Suffolk by Hugh Rose*. 8vo, xii+472 pp, 14 pls, Cadell, London 1775. Translation of *Philosophia botanica* Stockholm 1751, ed. Vienna 1770, by Hugh Rose.

2275. —— *A generic and specific description of British plants*. 8vo, xxviii+258 pp, 5 pls, Caslon, Kendal 1775. Translation of parts of *Genera plantarum*, Leyden 1737; 6ed. 1764, by James Jenkinson.

2276. —— *Genera plantarum; ex editione decima tertia, Systematis naturae*. 8vo, 106 pp, 1 tab., University Press, Edinburgh 1780. Extracted by John Hope from *Systema naturae*, 13ed. 1774, which is 2ed. of *Systema vegetabilium*, J. A. Murray's ed. See also Nos 2272 & 2282.

2277. —— *Select dissertations from the Amoenitates Academicae, a supplement to Mr. Stillingfleet's Tracts relating to natural history. In two volumes, Vol. 1*. 8vo, xiv+480 pp, Robinson & Robson, London 1781. All publ. 12 Linnaean theses translated by F. J. Brand from *Amoenitates Academicae*, Stockholm 1749 *etc.* See also B. Stillingfleet.

2278. —— *A system of vegetables . . . translated from the thirteenth edition . . . of the Systema vegetabilium . . . and from the*

*Supplementum plantarum*. 8vo, 2 vols, 11 pls. Vol. 1 Jackson for a Botanical Society at Lichfield, Vol. 2 Leigh & Sotheby for a Botanical Society at Lichfield. Lichfield 1782, 83; issue of Vol. 1 1783. Soulsby considered the 1782 issue of Vol. 1 to be a proof. This is not so. Vol. 2 ends with *An alphabetical catalogue of English and Scotch names of plants*, 1784. Translations of *Systema vegetabilium*, 3ed. [=13ed.] Göttingen 1774, and *Supplementum*, Brunswick 1781 by Linnaeus, C. *the son*, Translated by Erasmus Darwin.

2279. —— *Reflections on the study of nature*. 8vo, xi+40 pp, Nicol, London 1785; ed. Dublin 1786 title . . . *and a dissertation on the sexes of plants*; abridged ed. 1883 title *Linnaeus on the study of nature*. Translation of preface to *Museum Regis Adolphi Frederici*, Stockholm 1754, by *Sir* James Edward Smith.

2280. —— *A dissertation on the sexes of plants*. 8vo, xv+62 pp, by the author [i.e. by the translator *Sir* James E. Smith], London 1786; ed. Dublin 1786. Dublin ed. is in No. 2279. Translation of *Disquisitio . . . sexum plantarum*, St Petersburg 1760.

2281. —— *The families of plants with their natural characters translated from the last edition of the Genera plantarum, and the Mantissae plantarum . . . and from the Supplementum plantarum*. 8vo, 2 vols, lxxx+840 pp, Jackson printed for a Botanical Society at Lichfield. Lichfield 1787. Translations of *Genera plantarum*, ed. Frankfurt 1778. *Mantissae*, Stockholm 1767 & 1771, *Supplementum*, Brunswick 1781. Translated by Erasmus Darwin.

2282. —— *The animal kingdom, or zoological system of . . . Sir C. Linnaeus. Class I. Mammalia. (Class II. the birds)*. 4to, 2 pts, xii+644 pp, 10 pls, Murray, London 1792. All publ. Pp. 433–68 omitted. Translation of parts of *Systema naturae*, 13ed. (F. Gmelin), Leipzig 1788–93, by Robert Kerr.

2283. —— *Characteres generum plantarum*. 8vo, 2+138 pp, 1 tab., University Press, Edinburgh 1793. Extracted by Daniel Rutherford from *Systema vegetabilium*, 1784. See also Nos 2272 & 2276.

2284. —— *A general system of nature . . . translated from Gmelin's last edition of the . . . Systema naturae*. 8vo, 7 vols, 11 pls, 1 portr., Swansea 1800–04; ed. Lackington, London [1800–] 1802–06; ed. 1806. Translation of *Systema naturae*, 13ed. (F. Gmelin), Leipzig 1788–93, by William Turton.

—— *Linnaeus on the study of nature*. 1883. *see* No. 2279.

—— *see*

Bladh, A. J.
Handley, S. C. T.
Hope, J.
Lee, J.
Miller, J.
Pulteney, R.
Sherborn, C. D.
Sibly, E.
Stillingfleet, B.
Thornton, R. J.

Linnaeus, Carl *the son. see* Linnaeus, C. *the father*.

Linnean Society of London, Zoological Club. *see*

Bicheno, J. E.
Brookes, J.
Children, J. G.
Vigors, N. A.

2285. LINTON, Edward Francis *Flora of Bournemouth, including the Isle of Purbeck*. 8vo, viii+290 pp, 1 map, Bournemouth [1900].

2286. LINTON, William James *The ferns of the English lake country: with a list of varieties*. Sm. 8vo, 124 pp, 1 pl., text figs, Hamilton, London 1865; 2ed. Windermere 1878 title . . . *tables of varieties*.

*List of flowering plants . . . in the south-west of Westmorland*. [1868]. *see* Carr, E. D.

*A list of four hundred wild flowering plants . . . of Peterborough*. [1860] *see* Paley, F. A.

*List of plates of skeletons of birds that have been published*. 1874. *see* Eyton, T. C.

*A list of the natural flies that are taken by trout*. 1853. *see* Theakston, M.

*A list of the rare plants found in the neighbourhood of Tunbridge Wells*. [1801]. *see* Forster, T. F.

2287. LISTER, Arthur *Monograph of the Mycetozoa*. 8vo, 224pp, 78 pls, British Museum (Natural History), London 1894; 2ed. 1911; 3ed. 1925.

2288. —— *Guide to the British Mycetozoa*. 8vo, 42 pp, British Museum (Natural History), London 1896; 2ed. 1905; 3ed. 1909.

2289. LISTER, Martin. *Martini Lister . . . Historiae animalium Angliae tres tractatus . . . de araneis . . . de cochleis tum terrestribus tum fluviatilibus . . . de cochleis marinis*. 4to, [viii]+250 pp, 9 pls, Martyn, London 1878; appendix 23 pp, 1 pl., York 1681. There is an ed. of the appendix in Goedart, J. *De insectis*, 1685.

2290. —— *De cochleis*. Fol., t.p. + 102 pls, for private circulation, London 1685. Pls 1–82, 1–18, 9a, 13a. A forerunner of No. 2291.

2291. —— *Martini Lister Historiae conchyliorum liber primus. (Liber IV, Appendix)*. Fol., 6 pts, 8 pp, 1069 pls on 185 11, for the author, 1685[–1692–97]; another ed. 1685–92; 2ed. Oxford 1770 title *Historiae sive synopsis methodica . . .*; 3ed. Oxford 1823; index 1823. There are 1084 engravings of which the first are preliminary text engravings. Bibliography and numbering of pls is complex. 3ed. is said to be 30 copies. See also No. 2290.

2292. —— *Martini Lister Exercitatio anatomica, in qua de cochleis maxime terrestribus et limacibus agitur*. 8vo, xi+208 pp, 8 pls, Smith & Walford, London 1694.

2293. —— *Martini Lister Exercitatio anatomica altera, in qua maxime agitur de buccinis fluviatilibus & marinis*. 8vo, 267+128 pp, 6 pls, Smith & Walford, London 1695.

2294. —— *Martini Lister Conchyliorum bivalvium utriusque aquae exercitatio anatomica tertia*. 4to, 2 pts, xliii+173+51 pp, 10 pls. for the author, London 1696.

—— *see* Ray, J. *Historia insectorum*. 1710.

*The little book of nature*. 1840. *see* Wilson, T. *pseudonym* and *The little entomologist*, and *The little marine botanist*.

*The little conchologist* [1837]. *see* Clark, Samuel.

2295. *The little entomologist: intended chiefly as a guide to the knowledge of British insects*. 32mo, x–65 pp, 1 col. + 3 plain pls, Darton & Clark, London [1840]. By the author of *The little marine botanist*. Also found in a combination volume with that work and 3 others by Wilson, T. *pseudonym* with title *The little book of nature*.

The Little Library. *see* Taylor, Jeffreys; *The ocean*; Schoberl, F.; Halstead, C. A.

2296. *The little marine botanist; or, guide to the collecting and arranging of sea-weed*. 32mo, xiii+67 pp, 1 col. + 3 plain pls, Darton & Clark, London [1840]. By the author of *The little entomologist*. Also found in a combination volume with that

LIVINGSTONE, G. – British Birds: their haunts and habits. S



work and 3 others by Wilson, T. *pseudonym* with title *The little book of nature*.

*Little stories of pretty little birds*. [c.1845]. *see* Kendall, E. A.

Liverpool Marine Biological Committee Memoirs. *see*
Chadwick, H. C. *Echinus*. 1900.
Gibson, R. J. H. & Auld, *Miss* H. P. *Codium*. 1900.
Herdman, W. A. *Ascidia*. 1899.
Johnstone, J. *Cardium*. 1899.

2297. LIVERPOOL NATURALISTS FIELD CLUB *The flora of Liverpool*. 8vo, vi+178 pp, The Club, Liverpool [1867–]1872; appendix 14 pp, [1873]; 2nd appendix [1875]; ed. 1902.

2298. [LIVINGSTONE, George] *British birds their haunts and habits*. 8vo, 282 pp, for private distribution, London 1868.

2299. LLOYD, George N. *Botanical terminology; or, a dictionary*, 12mo, vii+228 pp, Duncan, Edinburgh 1826.

2300. LLOYD, W. *Essay on the nutrition of plants and the influence they exert on the atmosphere*. Sm. 8vo, Edinburgh 1838.

2301. LLOYD, William Alford *Note on the specific gravity of sea-water used for aquarium purposes*. 8vo, s. sh., London [1858].

2302. —— *A list, with descriptions, illustrations, and prices, of whatever relates to aquaria*. Sm. 8vo, 128 pp, text figs, London 1858; suppl. pp. 129–162, 1860.

2303. —— *A guide book to the marine aquarium*. 8vo, 30 pp, 1 plan, Crystal Palace Aquarium Company, London 1872; 4ed. 1873 title *Official handbook . . .*

Lloyd's Natural History Series, *editor* Sharpe, R. B. Sm. 8vo, 16 vols, Lloyd, London 1896. Relevant volumes are entered under authors. Almost identical with Allen's Naturalists' Library q.v. See also *The royal portfolio of natural history pictures*. [c.1897].

2304. [LOBB, Richard] *The contemplative philosopher: or short essays on various objects of nature throughout the year*. 12mo, 2 vols, 2 pls, Robinson, London 1800; 4ed. 1817 signed. Reprinted from *The Universal Magazine*, 1785–92.

2305. LOBLEY, James Logan *editor*. *Hampstead Hill: its structure, materials, and sculpturing . . . flora . . . insect fauna . . . birds*. 8vo, 100 pp, text figs, Roper & Drowley, London 1889.

Lockhart, Charles *see* Woodward, B. B.

*London catalogue of British mosses and hepatics*. 2ed. 1881. *see* Hobkirk, C. C. P. & Boswell, H.

*The London catalogue of British plants*. 1844. *see* Watson, H. C. & Dennes, G. E.

2306. *Longlands castle, or a visit to the grandmother containing with a variety of incidental topics, a sketch of the elements of zoology, botany, mineralogy, and other branches of natural history*. 12mo, 252 pp, M. J. Godwin, London 1816; 2ed. 1816 title *The young travellers* . . . By a lady, Juvenile Library.

2307. LORD, Thomas *Lord's entire new system of ornithology. Or oecumenical history of British birds*. Fol., vi pp, 114 col. pls with letterpress, by the author, London 1791[–1795]. Engraved title Berkhampstead 1796.

2308. LORD, William Barry *Crab, shrimp, and lobster lore*. Sm. 8vo, xvi+122 pp, 1 pl., text figs, Routledge, London 1867.

2309. [LOUDON, *Mrs* Jane née Webb] *Tales about plants*. 16mo, xii+500 pp, text figs, Tegg, London 1839. By Peter Parley *pseudonym*, attribution uncertain but not by S. G. Goodrich.

2310. —— *The young naturalist, or, the travels of Agnes Merton and her mamma*. 16mo, vii+176 pp, 1 col. pl., text figs, Routledge, London 1840; 2ed. 1850; 1851; 3ed. 1860.

2311. —— *First book of botany*. 12mo, Bell & Daldy, London 1841; ed. 1848; ed. 8vo 1870.

2312. —— *Botany for ladies; or, a popular introduction to the natural system of plants*. 8vo, xiv+493 pp, text figs, Murray, London 1842; ed. 1849; 2ed. 1851.

2313. —— *The year-book of natural history for young persons*. 16mo, xvi+264 pp, Murray, London 1842.

2314. —— *British wild flowers*. 4to, 216 pp, 48 col. pls, Orr, London 1844–45; ed. xvi+311 pp, 60 col. pls, Smith, London 1846; [1848]; ed. pts 1–11 1856; 3ed. 1859. First ed. is unfinished.

2315. —— *Glimpses of nature . . . during a visit to the Isle of Wight*. 16mo, x+232 pp, 5 pls, text figs, Grant & Griffith, London 1844; 2ed. 1848.

2316. —— *Facts from the world of nature, animated and inanimate*. Sm. 8vo, 4 pts, x+310 pp, 1 pl., text figs, Grant & Griffith, London 1848; ed. 1853.

—— *The entertaining naturalist, being popular descriptions . . . of more than five hundred animals*. 1843. *see* Boreman, T.

2317. LOUDON, John Claudius *editor. An encyclopaedia of plants*. 8vo, x+2+1159 pp, text figs, Longman, London 1829; 1836; suppl. 1841; ed. 1841; 2nd suppl. 1855; ed. 1866; ed. 1872.

2318. —— *Loudon's Hortus Britannicus. A catalogue of all the plants . . . in . . . Britain*. 8vo, xxiv+576 pp, text figs,

Longman, London 1830; 2ed. with suppl. 1832; ed. 1837; ed. [1839]; ed. 1850.

2319. —— *Arboretum et fruticetum Britannicum; or, the trees and shrubs of Britain, native and foreign*. 8vo, 8 vols, 412 col. pls, text figs, for the author, London [1835–]1838; 2ed. 1844; 2ed, 1854; abridged ed. 1842 title *An encyclopaedia of trees and shrubs; abridged ed.* 1883 title *Trees and shrubs.* 25 col. copies, rest plain.

2320. LOVELL, M. S. *The edible mollusks of Great Britain and Ireland*. 8vo, 207 pp, 12 col. pls, Reeve, London 1867; 2ed. 1884 title *The edible Mollusca* . . . Facsimile 1977. Crown Series.

2321. LOVELL, Robert [Greek] *Pambotanologia. Sive enchiridion botanicum.* 12mo, 671 pp, Davis, Oxford 1659; 2ed. 1665.

2322. —— [Greek] *Panzooruktologia. Sive panzoologico-mineralogia*. 8vo, 2 pts, 519+152 pp, Godwin, Oxford 1661.

2323. LOW, George *Fauna Orcadensis: or, the natural history of the quadrupeds, birds, reptiles, and fishes of Orkney and Shetland*. 4to, xx+230 pp, Constable, Edinburgh 1813.

2324. —— *A tour through the Islands of Orkney and Shetland, containing hints relative to their ancient, modern and natural history collected in 1774.* 8vo, lxxiv+223 pp, 1 pl., Peace, Kirkwall 1879.

2325. LOWE, Edward Joseph *The conchology of Nottingham*. 12mo, 172 pp, text figs, London 1853.

2326. —— *Ferns, British and exotic*. 4to, 8 vols, 479 col. pls, text figs, Groombridge, London 1856–60; issue 1867–71; suppl. 72 col. pls, 1871 title *A natural history of new and rare ferns none of which are included in 'Ferns, British and exotic'*; 1872; 1877.

2327. —— *A natural history of British grasses*. 8vo, vii+245 pp, 74 col. pls, London [1857–]1858; ed. 1862; ed. 1864; ed. 1868; 1871; 3ed. 1891.

2328. —— *Our native ferns*. Roy. 8vo, 2 vols, 79 col. pls, text figs., London 1865–67; ed. 1867–69; ed. 1874–76; 1880.

2329. —— *British ferns and where found*. Sm. 8vo, 168 pp, text figs, Sonnenschein, London 1891. Young Collector Series.

—— *see* Ansted, D. T.

2330. LOWE, John *The yew-trees of Great Britain and Ireland*. 8vo, xiv+270 pp, 23 pls, text figs, Macmillan, London 1897.

2331. LOWNE, Benjamin Thompson *A popular natural history of*

*Great Yarmouth and its neighbourhood*. 8vo, viii+62 pp, Great Yarmouth 1863.

2332. —— *The anatomy and physiology of the blow-fly*. Roy. 8vo, viii+121 pp, 9 col.+1 plain pls, Van Voorst, London 1870 [=1869]; 2ed. 2 vols, demy 8vo, 1890–95.

2333. —— *The philosophy of evolution*. 8vo, viii+159 pp, 8 pls, Van Voorst, London 1873.

Lowry, John Wilson & Landseer, Thomas *Illustrations of zoology*. 1851. *see Illustrations*.

2334. LOWSON, John Melvin *A text book of botany*. Sm. 8vo, viii+404 pp, text figs, Clive, London 1899; 2ed. 1900; 3ed. 1903; 4ed. 1908; 6ed. 1920; 1922; 7ed. 1932. University Tutorial Series.

2335. LUBBOCK, *Sir* John, *Baron* Avebury *Monograph of the Collembola and Thysanura*. 8vo, x+276 pp, 31 col. + 47 plain pls, Ray Society, London 1873. Ray Society Publ. No. 48 for 1871.

2336. —— *On the origin and metamorphosis of insects*. Sm. 8vo, xvi+108 pp, text figs, Macmillan, London 1874; ed. 1874; 1883; 2ed. 1890. 6 of the text figs are full page and called pls. Nature Series.

2337. —— *On British wild flowers considered in relation to insects*. Sm. 8vo, xvi+186 pp, 1 pl., text figs, Macmillan, London 1875; 2ed. 1875. Later printings. Nature Series.

2338. —— *On certain relations between plants and insects*. 8vo, text figs, Collins, London 1878.

2339. —— *Scientific lectures*. 8vo, x+187 pp, 1 col. pl., text figs, Macmillan, London 1879; 2ed. 1890; 3ed. 1906.

2340. —— *Ants, bees, and wasps*. Sm. 8vo, xix+448 pp, 5 col. pls, text figs, Kegan Paul, London 1882; 2ed. 1882; 3ed. 1882; 4ed. 1882; 5ed. 1882; 6ed. 1883; 7ed. 1885; 8ed. 1886; 9ed. 1888; 10ed. 1890; 1891; 1892; 13ed. 1893; 1894; 14ed. 1898; 16ed. 1904; 17ed. 1906; 1915; 1929. International Scientific Series Vol. XL. Based on papers in *J. Linn. Soc. Lond*, 1874–81.

2341. —— *Fifty years of science. Being the address delivered at York to the British Association. August, 1881*. 8vo, 90 pp, Macmillan, London 1882; 2ed. 1882; 4ed. 1890.

2342. —— *Chapters in popular natural history . . . for use in . . . schools*. 8vo, viii+223 pp, text figs, National Society for Promoting the Education of the Poor, London [1883]. Extracts from his other works.

2343. —— *Flowers, fruits, and leaves*. Sm. 8vo, xv+147 pp, text figs, Macmillan, London 1886; 1908. Nature Series.

2344. —— *On the senses, instincts, and intelligence of animals, with special reference to insects*. Sm. 8vo, xxix+292 pp, text figs, Kegan Paul, London 1888; 2ed. 1889; 3ed. 1891; 1899. International Scientific Series Vol. LXV.

2345. —— *A contribution to our knowledge of seedlings*. Sm. 8vo, 2 vols, text figs, Kegan Paul, London 1892; ed. 1 vol. 1896; 1907. International Scientific Series Vol. LXXIX.

2346. —— *On buds and stipules*. Sm. 8vo, xix+239 pp, 4 col. pls, text figs, Kegan Paul, London 1899. International Scientific Series Vol. LXXXVI.

2347. LUBBOCK, Richard *Observations on the fauna of Norfolk, and more particularly on the district of the Broads*. 8vo, viii+156 pp, 2 pls, 1 map, Muskett, Norwich 1845; 1848; 1860; ed. 1879; additions to pt. 1, Watford 1884. The additions by Thomas Southwell.

2348. LUCAS, Joseph *Studies in Nidderdale*. 8vo, xxvi+292 pp, text figs, Stock, London [1882].

2349. LUCAS, Richard Cockle *Facsimiles of nature from the hill and the valley, the forest, the field and the garden*. Fol., no text, col. pls, Chilworth Tower, Hants 1848. Nature printed. Very few copies.

2350. LUCAS, William John *The book of British butterflies*. Sm. 8vo, vii+246 pp, text figs, Upcott Gill, London 1893.

2351. —— *The book of British hawk-moths*. Sm. 8vo, x+157 pp, text figs, Upcott Gill, London 1895.

2352. —— *British dragonflies (Odonata)*. 8vo, xiv+356 pp, 27 col. pls, text figs, Upcott Gill, London 1900.

2353. LUCKLEY, John Lamb *Flowers of the Aln*. 12mo, 28 pp, by the author, Alnwick 1849.

2354. —— *Botany of Alnwick*. 8vo, by the author, Alnwick 1860.

2355. —— *Botanical rambles. A flora of Alnwick*. 8vo, 112 pp, Alnwick & County Gazette printed, Alnwick [1893]. Reprinted from *The Alnwick & County Gazette*.

2356. [LUKIS, Frederick Corbin] *Rambles among the Channel Islands*. Sm. 8vo, viii+368 pp, 1 pl., text figs, SPCK, London [c.1850]. By a naturalist. Attribution doubtful.

2357. LUMSDEN, James & BROWN, Alfred *A guide to the natural history of Loch Lomond and neighbourhood. Mammals and birds*. 8vo, 103 pp, 2 pls, Glasgow 1895.

2358. [LUPTON, Thomas] *A thousand notable things on various subjects . . . also, a discourse on the emigration of British birds*. 8vo, xxiv+220 pp, Walker, London 1815; ed. 1827. The discourse, pp. 175–208, is by J. Legg, q.v. for earlier editions, and is in these editions only. First edition of Lupton [1595] and later are signed, but are not natural history books.

2359. LUXFORD, George *A flora of the neighbourhood of Reigate, Surrey*. 12mo, xviii+118 pp, 1 map, Van Voorst, London 1838.

2360. LYDEKKER, Richard *Phases of animal life past and present*. 8vo, 248 pp, text figs, Longman, London 1892. Reprinted from *Knowledge*.

2361. —— *Horns and hoofs, or chapters on hoofed animals*. 8vo, xv+411 pp, text figs, Cox The Field Office, London 1894. Knowledge Series.

2362. —— *editor The Royal natural history*. Roy. 8vo, 6 vols, 72 col.+57 plain pls, text figs, Warne, London 1893–96.

2363. —— *Life and rock: a collection of zoological and geological essays*. 8vo, xi+221 pp, text figs, Universal Press, London 1894. Knowledge Series.

2364. —— *A handbook to the British Mammalia*. Sm. 8vo, xiii+339 pp, 32 col. pls, text figs, Shaw, London [1895]; ed. 1896. 1895 ed. is Allen's Naturalists' Library; 1896 ed. is Lloyd's Natural History.

2365. —— *A hand-book to the Carnivora. Pt. 1. Cats, civets and mungooses.* Sm. 8vo, viii+312 pp, 32 col. pls, text figs, Shaw, London [1895]; ed. 1896. 1895 ed. is Allen's Naturalists' Library; 1896 ed. is Lloyd's Natural History.

2366. —— *A geographical history of mammals.* Sm. 8vo, xii+400 pp, 1 col. map, text figs, University Press, Cambridge 1896.

2367. —— *The deer of all lands*. 4to. xx+329 pp, 24 col. pls, text figs, Rowland Ward, London 1896.

2368. —— *Wild oxen, sheep, and goats of all lands*. 4to, xiv+318 pp, 28 col. pls, text figs, Rowland Ward, London 1898.
—— *see* Flower, *Sir* William H.

2369. LYELL, *Sir* Charles *The geological evidences of the antiquity of man, with remarks on theories of the origin of species by variation*. 8vo, xii+520 pp, 2 pls, text figs, Murray, London 1863; 2ed. 1863; 3ed. 1863; 4ed. 1873.

2370. LYELL, *Mrs* Katherine Murray *A geographical handbook of the known ferns*. 8vo, x+225 pp, Murray, London 1870.

2371. LYONS, Israel *Israelis Lyons jun. Fasciculus plantarum circa Cantabrigium nascentium*. 8vo, xvi+56 pp, Millar, London 1763.

# M

M. A. *see* Venning, *Miss* Mary Ann.

M., A. D. *A description of more than three hundred animals*. 1812. *see* Boreman, T.

2372. M., M. K. *Leaves from nature's book*. 16mo, 2 ser., Macintosh, London [1866–]1872.

2373. —— *The birds we see: and the story of their lives*. 16mo, text figs, Hamilton, London 1868; ed. 1877.

2374. —— *What the wild flowers teach us*. 16mo, col. pls, Nelson, London 1875.

2375. —— *With the birds; the homes and habits of our feathered friends*. 16mo, 12 col. pls, Nelson, London 1875; ed. 12mo 1887.

2376. —— *In the woods*. Sm. 8vo, viii+186 pp, 1 pl., text figs, Nelson, London 1879.

M., T. B. *Zoological sketches*. 1844. *see* Murray, T. B.

2377. MCALDOWIE, Alexander Morison *The birds of Staffordshire, with illustrations of local bird haunts*. 8vo, vi+146 pp, 7 pls, privately printed, Stoke-on-Trent 1893.

2378. MACALISTER, Alexander *An introduction of animal morphology and systematic zoology. Pt. 1 Invertebrata*. 8vo, xvi+461 pp, text figs, Longman, London 1876. All publ.

2379. —— *An introduction to the systematic zoology and morphology of vertebrate animals*. 8vo, viii+36 pp, text figs, Longman, Dublin 1878. Dublin University Press Series.

2380. —— & M'NAB, William Ramsay *editors Guide to the County of Dublin: its geology, industries, flora and fauna*. 8vo, vi+219+100 pp, 2 maps, Hodges, Dublin 1878. British Association Handbook.

2381. MCALPINE, Archibald N. *How to know grasses by the leaves*. 8vo, vi+92 pp, 18 pls, Douglas, Edinburgh [1890].

—— *see* M'Alpine, D.

2382. M'ALPINE, Daniel *Zoological atlas . . . Vertebrata. (Invertebrata).* Obl. 4to, 2 vols, 40 col. pls with letterpress, Johnston, Edinburgh 1881. Usually bound in one volume.

2383. —— *The botanical atlas*. Fol., 2 vols, 52 col. pls with letterpress, 1 tab., Johnston, Edinburgh 1883.

2384. —— *Life histories of plants*. 8vo, viii+296 pp, text figs, Sonnenschein, London [1886].

2385. —— & M'ALPINE, Archibald N. *Biological atlas*. 4to, ix+49 pp, 24 col. pls, Johnston, Edinburgh 1880; 2ed. 1881.

2386. MCANDREW, Robert *On the geographical distribution of testaceous molluscs in the North Atlantic and neighbouring seas*. 4to, 51 pp, Liverpool 1854. Reprinted from *Proc. Liverpool lit. phil. Soc*. Vol. 8.

2387. —— *List of the British marine invertebrate fauna.* 4to, 68 pp, British Association Dredging Committee, London 1860.

2388. MACARTNEY, James *Prospectus of . . . lectures on comparative anatomy, and the laws of organic existence.* 12mo, 21 pp, n.p., 1806.

2389. MACAULEY, Kenneth *The history of St Kilda*. 8vo, iv+278 pp, 1 map, Becket & De Hondt printed, London 1764; 2ed. 12mo, Dublin 1765 title *A voyage to, and history of, St Kilda.* Although signed by Macauley, perhaps by John Macpherson.

2390. MCCOOK, Henry Christopher *Tenants of an old farm; leaves from the note-book of a naturalist*. 8vo, xv+460 pp, text figs, Hodder & Stoughton, London 1888. First edition is New York 1885.

2391. M'COSH, James & DICKIE, George *Typical forms and special ends in creation*. Sm. 8vo, viii+539 pp, text figs, Constable, Edinburgh 1856; 2ed. 1857.

2392. M'CRIE, Thomas *editor The Bass Rock*. 8vo, viii+139+436 pp, 1 col. + 6 plain pls, text figs, Grieg, Edinburgh 1848; ed. [1855].

2393. M'DIARMID, John *Sketches from nature*. Sm. 8vo, 388 pp, Oliver & Boyd, Edinburgh 1830.

2394. MACDONALD, Duncan George Forbes *Cattle, sheep and deer (containing also remarks on the game laws and grouse moors*). 8vo, vii+745 pp, portr., Steel & Jones, London 1872; 2–5eds 1872.

2395. MACDONALD, George & ALLAN, James *The botanist's word-book*. 8vo, viii+36 pp, Reeve, London 1853.

MacEarlean, John C. *see* Hogan, E. I.

2396. MACGILLIVRAY, Paul Howard *A catalogue of the flowering plants and ferns growing in the neighbourhood of Aberdeen.* 8vo, viii+44 pp, Aberdeen 1853.

2397. MACGILLIVRAY, William *Descriptions of the rapacious birds of Great Britain*. 8vo, vii+482 pp, 2 pls, text figs, Maclachlan & Stewart, Edinburgh 1836.

2398. —— *A history of British birds, indigenous and migratory*. 8vo, 5 vols, 29 pls, text figs, Vols 1–3 Scott, Vols 4–5 Orr, London 1837–52.

2399. —— *A history of British quadrupeds*. Sm. 8vo, 309 pp, 34 col. pls, 1 portr., Lizars, Edinburgh 1838; ed. [1845–46]; ed. 1848. Jardine's Naturalists Library, Vol. XXII, 2ed. Vol. XVII.

2400. —— *A manual of botany*. Sm. 8vo, xii+280 pp, 16 pls, text figs, Scott, London 1840; 2ed. 1853.

2401. —— *A manual of British ornithology*. Sm. 8vo, 2 pts, text figs, Scott, London 1840–42; 2ed. 1846 title *British ornithology; a manual of British Birds*.

2402. —— *A history of the molluscous animals of the Counties of Aberdeen, Kincardine, and Banff*. 12mo, xxiv+372 pp, Cunningham & Mortimer, London 1843; 2ed. 1844 title *A history of the molluscous animals of Scotland, as found in the north-eastern district*.

2403. —— *The natural history of Deeside and Braemar*. 8vo, xx+507 pp, 5 pls, 2 maps, text figs, privately printed [for Queen Victoria], London 1855.

—— *The conchologist's text-book*. 6ed. 1865. *see* Brown, T. *Captain*.

—— *see* Withering, W.

2404. MCINTOSH, William Carmichael *Observations and experiments on the Carcinas moenas*. 8vo, 64 pp, 4 pls, Williams & Norgate, London 1861; ed. Dundee 1926.

2405. —— *A monograph of the British annelids*. Roy. 4to, 4 vols, 48 col. + 94 plain pls, Ray Society, London 1873–1923. 6 plain pls in Vol. 3 (1915) reissued (1920) in colour. Ray Society Publ. Nos 49, 50, 76, 87, 91, 97, 100, 106, 107 for 1872, 73, 89, 1907, 10, 14, 16, 20 & 21.

2406. —— *The marine invertebrates and fishes of St Andrews*. 4to, vi+186 pp, 5 col. + 4 plain pls, text figs, Black, Edinburgh 1875; additions 130 pp, 3 col. pls, London 1927. Mostly reprinted from *Ann. Mag. nat. Hist*., 1874.

2407. —— & MASTERMAN, Arthur Thomas *The life-histories of*

*the British marine food fishes*. 8vo, xv+516 pp, 20 col. + 1 plain pls, text figs, Clay, London 1897.

2408. MACIVOR, William Graham *Hepaticae Britannicae*. 8vo, 18 11, specimens, New Brentford 1847.

2409. MACKAY, James Townsend *Flora Hibernica: comprising the flowering plants, ferns, Characeae, Musci, Hepaticae, Lichenes and Algae of Ireland*. 8vo, 3pts, xxxiv+354+279 pp, Curry, Dublin 1836.

2410. MACKENZIE, Charles, *The natural history of all the most remarkable quadrupeds, birds, fishes, reptiles and insects*. 12mo, viii+208 pp. 52 pls, Reynolds, London [1839]; 1859; 1860; ed. [ ].

2411. MCLACHLAN, Robert *Instructions for the collection and preservation of neuropterous insects*. 8vo, 23 pp, text figs, Van Voorst, London 1873. Reprinted from *Ent. mon. Mag*, Vol. 9.

2412. —— *A monographic revision and synopsis of the Trichoptera of the European fauna*. 8vo, iv+522+ciii pp, 59 pls, Van Voorst, London 1874–80; suppl. iv+76, 7 pls, 1884; reprint 1968.

2413. —— & EATON, Alfred Edwin *A catalogue of British Neuroptera*. 8vo, viii+42 pp, Roworth for the Entomological Society of London, London 1870.

2414. MACLAREN, James *A critical examination of some of the principal arguments for and against darwinism*. 8vo, viii+479 pp, Bumpus, London 1876.

2415. MACLEAY, William Sharp *Horae entomologicae: or, essays on the annulose animals*. Vol. 1. 8vo, 2 pts, xxx+523 pp, 3 pls, Bagster, London 1816, 21. All publ.

2416. —— *An essay on the dying struggle of the dichotomous system*. 8vo, 35 pp, London 1830.

2417. MACLISE, Joseph *Comparative osteology; being morphological studies to demonstrate the archetype skeleton of vertebrated animals*. Fol., [180] pp, 54 pls, Taylor & Walton, London 1847.

2418. MACLOC, J. *New, complete and universal natural history; of all the most remarkable, birds, fishes, reptiles, and insects in the known world*. 12mo, xii–13–336 pp, 1 pl., text figs, Dean & Munday, London 1813; Minerva Press for Newman, London 1813; ed. 1815; 3ed. 1820 title *A natural history of all the most remarkable quadrupeds, birds, fishes, serpents, reptiles, and insects in the known world*; 4ed. 1827.

2419. MACMILLAN, Hugh *Footnotes from the page of nature*.

8vo, x+294 pp, text figs, Macmillan, Cambridge 1861; 2ed. 1874 title *First forms of vegetation*.

2420. MACMILLAN, William *A list of the butterflies and moths found in or near Castle Cary, Somerset, 1871–95*. 8vo, 28 pp, Castle Cary [1896–97].

2421. [MCNAB, Catherine Mary] *Juvenile conversations on natural history illustrative of the wisdom and goodness of God. The moss and lichen*. 16mo, 38 pp, Paton & Ritchie, Edinburgh 1850.

2422. MACNAB, William Ramsey *Botany, outlines of classification of plants*. 16mo, 193 pp, text figs, Longman, London 1878; 4ed. 1886: 5ed. 1889. London Science Class Books.

2423. —— *Botany, outlines of morphology and physiology*. 16mo, 161 pp, text figs, Longman, London 1878; 2ed. 1879; 3ed. 1880; 4ed. 1883; 6ed. 1888. London Science Class Books.

—— *see* Macalister, A.

2424. MCNICOLL, David Hudson *Handbook for Southport . . . with copious notes on the natural history of the district*. 8vo, v+175 pp, 3 col. pls, Churchill, London 1859; 2ed. 1881; 3ed. 1883.

2425. —— *Dictionary of natural history terms*. Sm. 8vo, vii+584 pp, Reeve, London 1863.

2426. MACPHERSON, Hugh Alexander *The visitation of Pallas's sand-grouse to Scotland in 1888*. 8vo, viii+38 pp, London 1889.

2427. —— *An introduction to the study of British birds*. Sm. 8vo, 120 pp, text figs, Sonnenschein, London 1891; ed. 1907. Young Collector Series.

2428. —— *A vertebrate fauna of Lakeland, including Cumberland and Westmorland, with Lancashire north of the sands*. 8vo, xiv+552 pp, 2 col. + 7 plain pls, 1 map, text figs, Edinburgh 1892; reprint 1972.

2429. —— & DUCKWORTH, William *The birds of Cumberland critically studied, including some notes on the birds of Westmorland*. 8vo, xx+206 pp, 1 col. pl., 1 col. map, Thurnham, Carlisle 1886.

Macpherson, John *see* Macaulay, K.

Macquin, Ange Denis *editor A description of more than three hundred animals*. 1812. *see* Boreman, T.

2430. MACREIGHT, Daniel Chambers *Manual of British botany*. 8vo, xvii+296 pp, Churchill, London 1837.

2431. MCWILLIAM, Robert *An essay on the origin of the dry rot*. 4to, xx+420 pp, 3 pls, Taylor, London 1818.

Mahon, *Lady* Jane *see* Ward, *Hon*. Mary.

2432. MAILLET, Benoît de *Telliamid: or, discourses between an Indian philosopher, and a French missionary, on the diminution of the sea, the formation of the earth, the origin of men and animals*. 8vo, lii+284 pp, London 1750. Translated from *Telliamid*, The Hague 1748.

2433. MAIN, James *Illustrations of vegetable physiology practically applied*. 8vo, xiii+328 pp, text figs, Orr, London 1833; 2ed. 1836.

2434. —— *Popular botany*. 8vo, xii+227 pp, 27 col. pls, Orr, London 1835. Reprinted from *British cyclopaedia of natural history*, 1835–38.

2435. MAJOR, Henry *Botany*. 12mo, 192 pp, Stewart, London [1875]. Extra Subject Series.

2436. MALAN, Cesar Jean Salomon *writing as* Salomon Caesar Malan *A systematic catalogue of the eggs of British birds*. 8vo, viii+170 pp, Van Voorst, London 1848.

2437. MALPIGHI, Marcello *Marcelli Malpighi ... Dissertatio epistolica de bombyce*. 4to, 100 pp, 12 pls, Martyn & Allestry, London 1669; ed. 1673 in *Dissertationes epistolicae duae una de formatione pulli in ovo altera de bombyce.*

2438. —— *Anatome plantarum. Cui subjungitur appendix iteratis et auctas ejusdem authoris de quo incubatio observationes continens*. Fol., 2 pts, 100 pls, Martyn, London 1675–79; reprint 1901.

2439. —— *Marcelli Malpighi ... Opera omnia*. Fol., 2 vols, 123 pls, Scott & Wells, London 1689. There are issues with imprints of Wells, of Sawbridge and of Littlebury; and an issue of Vol. 1 dated 1687.

2440. —— *Marcelli Malpighi ... Opera posthuma*. Fol., 110+187 pp, 19 pls, 1 portr., Churchill, London 1697.

*Mamma's entertaining description of the trees*. [c.1835]. *see* Taylor, Joseph.

Manchester Museum *see* Marshall, A. M.; Sherborn, C. D.; Weiss, F. E.

2441. MANCHESTER SOCIETY OF NATURAL HISTORY *Aves Britannicae. A systematic catalogue of British birds in the Manchester Society of Natural History*. 8vo, 26 pp, Manchester 1836.

Mandeville, Bernard de *Zoologia medicinalis Hibernica*. 1744 *see* Keogh, J.

2442. MANGIN, Arthur *The mysteries of the ocean*. 8vo, xii+470 pp, 15 pls, text figs, Nelson, London 1868; ed. 1870. Translation of *Les mystères de l'ocean*, Tours 1864, by William Henry Davenport Adams.

2443. MANGLES, James *The floral calendar, monthly and daily*. 12mo, xx+156 pp, 12 col. pls, text figs, for private distribution, London 1839.

Mansel Pleydell, John Clavell *see* Pleydell, J. C. M.

2444. MANSON, Richard Taylor *Zig-zag ramblings of a naturalist*. 8vo, 159 pp, Darlington 1884; 2ed. 1898.

2445. MANTELL, Gideon Algernon *Outlines of the natural history of the environs of Lewes*. 4to, 24 pp, 3 pls, Baxter, Lewes 1824. Also published as an appendix to Horsfield, T. W. *History of Lewes*, 1824.

2446. —— *Thoughts on animalcules*. Sm. 4to, xvi+144 pp, 12 col. pls, Murray, London 1846; ed. 1850 title *The invisible world revealed by the microscope: or, thoughts on animalcules*.

2447. MAPLET, John *A greene forest; or, a naturall historie*. 12mo, 111 11, Denham, London 1567; reprint 1930.

[Marcet, *Mrs* Jane] *Conversations on botany*. 1817. *see* Fitton, Elizabeth & Fitton, Sarah.

2448. [——] *Conversations on vegetable physiology*. 12mo, 2 vols, 4 pls, Longman, London 1829; ed. 1830; 2ed. signed 1832; 1834; 3ed. 1839.

Maquin, Ange Denis *see* Boreman, Thomas.

2449. MARCH, Henry Colley *Darwinism and the evolution of man, with Ancestral man*. 8vo, 47+34 pp, 2 pls, text figs, Heywood, London 2ed. 1883.

2450. MARCHANT, James Robert Vernam & WATKINS, Watkin *Wild birds protection acts, 1880–96*. 8vo, xiv+174 pp, Porter, London 1897.

2451. MAREY, Étienne Jules *Animal mechanism: a treatise on terrestrial and aerial locomotion*. 8vo, xvi+283 pp, text figs, King, London 1874; 2ed. 1874; 4ed. 1876. Translation of *La machine animal*, Paris 1873. International Scientific Series, Vol. XI.

2452. —— *Movement*. Sm. 8vo, xv+323 pp, text figs, Heinemann, London 1895. Translation of *Le mouvement*. Paris 1894, by Eric Law Pritchard.

2453. MARGESSON, *Lady* Isabel Augusta *editor A handbook to the study of natural history for the use of beginners*. Sm. 8vo, xx+232 pp, Philip, London 1894.

Marilaun, Anton Kerner von *see* Kerner von Marilaun, A.

Marine Biological Association of the United Kingdom *see* Cunningham, J. T.

2454. *Marine botany and sea-side objects*. Sm. 8vo, iv+128 pp, text figs, Ward & Lock, London 1861. Later undated editions.

Marnock, Robert *see* Deakin, R.

2455. MARSDEN, Herbert W. *A list of British birds with . . . the graduated list for labelling eggs*. Sm. 8vo, vii+35 pp, Marsden, Gloucester 1881. The graduated list also sold alone.

2456. MARSH, Sylvester *Section cutting. A practical guide to the preparation and mounting of sections for the microscope*. Sm. 8vo, viii+87 pp, text figs, Churchill, London 1878; 2ed. 1882.

2457. MARSHALL, Arthur Milnes *The frog: an introduction to anatomy and histology*. 8vo, iv+86 pp, text figs, Cornish, Manchester 1882; 2ed. 1885; 3ed. 1888; 4ed. 1891; 5ed. 1894; 6ed. London 1896; 7ed. 1900; 9ed. 1906; 10ed. 1909; 12ed. 1928. Later issues. The Owens Course of Elementary Biology, Part 1.

2458. —— *Outline classification of the animal kingdom*. 12mo, 15 pp, Cornish, Manchester 1891; 2ed. 1892; 3ed. 1897; 4ed. 1911; 5ed. 1924. Manchester Museum Handbook No. 3. Eds 3–5 are by Sydney John Hickson.

2459. —— *Biological lectures and addresses*. 8vo, viii+363 pp, Nutt, London 1894.

2460. —— *Lectures on the darwinian theory*. 8vo, xx+236 pp, Nutt, London 1894.

2461. —— & HURST, Charles Herbert *A junior course of practical zoology*. 8vo, xxiv+440 pp, text figs, Smith Elder, London 1887; 2ed. 1888; 3ed. 1892; 4ed. 1895; 5ed. 1899; 6ed. 1905; 7ed. 1912; 8ed. 1918; 9ed. 1920; 10ed. 1924; 1928; 11ed. 1930.

2462. —— & MARSHALL, William P. *Report on the Pennatulida collected in the Oban dredging excursion of the Birmingham Natural History and Microscopical Society . . . 1881*. 8vo, 77 pp, 4 pls, Birmingham 1882. Reprinted from the *Transactions* of the Society.

Marshall, Edward Shearburn *see* Hanbury, F. J.

2463. MARSHALL, Thomas Ansell *Ichneumonidum Britannicorum catalogus*. 8vo, 22 pp, London 1870.

2464. —— *A catalogue of the British Hymenoptera. Chrysididae etc., (Oxyura).* 8vo, 2 pts, Napier for the Entomological Society of London, 1872, 73.

2465. MARSHALL, William *The new water-weed Anacharis alsinastrum; some account of it*. 8vo, 16 pp, London 1852. Reprinted from *The Cambridge Independant Press.*

2466. MARSHAM, Thomas *Entomologia Britannica, sistens insecta Coleoptera Britanniae indigena . . . Tomus I.* 8vo, xxxi+547 pp, 30 col. pls, White, London 1802; another issue [1802] title *Coleoptera Britannica.* All publ. The plates are often absent.

2467. MARSTON, Alfred *A guide to the ferns and many of the rarer plants growing around Ludlow*. 12mo, 45 pp, Ludlow [1870]; 2ed. 1882.

2468. MARTIN, Benjamin *Grammatica philosophica: the philosophical grammar, being a view of the present state of experimented philosophy*. 8vo, vii+362 pp, 16 pls, 2 tabs, Noon, London 1735; 2ed. 1738; 3ed. 1748; 4ed. 1753; 6ed. 1762; 7ed. 1769; 18ed. 1778. Eds 8–17 non-existent.

2469. —— *The description and use of a new invented pocket reflecting microscope with a micrometer*. 8vo, 29 pp, 2 pls, Chichester [1738].

2470. —— *Micrographia nova, or a new treatise on the microscope and microscopic objects*. 8vo, 62 pp, 2 pls, Reading 1742.

2471. —— *The natural history of England*. 8vo, 2 vols, 7 pls, 41 maps, 1 plan, text figs, Owen for the author, London 1759–63. All publ. Part of *The general magazine of arts and sciences, II, the natural history of the world.*

2472. —— *Optical essays*. 8vo, ii+50 pp, 1 pl., for the author, London [?1770].

2473. —— *The description and use of the graphical perspective and microscope . . . to which is added, a short account of an opake solar microscope*. 8vo, 12 pp, 1 pl., for the author, London 1771.

2474. —— *Typographia naturalis: or, the art of printing or taking impressions from natural subjects, as leaves, shells, fossils, etc.* 8vo, 8 pp, for the author, London 1774.

2475. —— *The description and use of an opake solar microscope, in which all opake bodies, whether of animal, vegetable, fossil, or marine production, are shown*. 8vo, 8 pp, 1 pl, for the author, London 1774.

2476. —— *The young gentleman and lady's philosophy, in a con-*

*tinued survey of the works of nature and art*. 8vo, 2 vols, 70 pls, 5 maps, London 1759–63; 2ed. 3 vols 1772–82; 3ed. 1781–82. 1st ed. is part of *The general magazine of arts and sciences.*

2477. MARTIN, Edward Alfred *Glimpses into natures secrets; or, strolls on beach and down*. Sm. 8vo, x+125 pp, text figs, Eliot Stock, London 1890.

2478. —— *Amidst nature's realms*. Sm. 8vo, 157 pp, text figs, Simpkin, London 1892.

2479. —— *Nature chat*. Sm. 8vo, 141 pp, Taylor, London 1897. Rambler's Library.

2480. MARTIN, George Anne *The undercliff of the Isle of Wight; its climate, history and natural productions*. 12mo, 366 pp, 1 pl., Churchill, London 1849.

2481. MARTIN, John H. *Microscopic objects, figured and described*. 8vo, 114 pp, 97 pls, Van Voorst, London 1870.

2482. —— *A manual of microscopic mounting, with notes on the collection and examination of objects*. 8vo, vi+200 pp, 12 pls, Churchill, London 1872; 2ed. 1878.

2483. —— & WESTON, James *A handbook to the fernery and aquarium*. Sm. 8vo, 48+43 pp, text figs, Fisher Unwin, London 1880.

2484. MARTIN, Martin *A late voyage to St. Kilda . . . with a history of the island, natural, moral, and topographical*. 8vo, 158 pp, 1 pl., 1 map, Brown & Goodwin, London 1698; 3ed. 1749 title *A voyage to St. Kilda*; 4ed. 1753; ed. Glasgow 1818. 1818 ed. is in *Miscellanea Scotica* Vol. 2.

2485. —— *A description of the Western Islands of Scotland*. 8vo, 392 pp, 1 pl., 1 map, Bell printed, London 1703; 2ed. 1716; ed. 1884. Copies of first ed. occur with date misprinted 1673.

2486. MARTIN, Matthew *The aurelian's vade mecum: containing an English alphabetical and Linnaean systematical catalogue of plants affording nourishment to butterflies, hawk-moths, and moths in the state of the caterpillar*. 12mo, xii+[36] pp, 1 pl., White, London 1785.

2487. —— *Observations on marine vermes, insects: Fasc. 1 (2)*. 4to, x+26 pp, 2 pls, Exeter 1786–87. All publ.

2488. [MARTIN, William Charles Linnaeus] *An introduction to the study of birds. (Mammals, Reptiles)*. 8vo, 3 vols, 3 pls, text figs, RTS, London [1835]. Later eds signed.

2489. —— *A general introduction to the natural history of mammiferous animals, with a particular view of the physical history of man and . . . monkeys*. 8vo, 545 pp, 13 pls, text figs, Wright printed, London [1840–]1841. All signatures say Vol. 1 but not t.p. Alternative title *Natural history of quadrupeds and other mammiferous animals including the principal varieties of the human race*. Vol. 1.

2490. [——] *Our song birds*. Sm. 8vo, vi+192 pp, RTS, London [1846]. Monthly volume No. 5.

2491. [——] *Comparisons in structure of animals. The hand and arm*. Sm. 8vo, iv+192 pp, RTS, London [1848]. Monthly Volume No. 28.

2492. [——] *British fish and fisheries*. Sm. 8vo, vi+192 pp, RTS, London [1850], Monthly Volume No. 49.

2493. MARTINDALE, J. A. *The study of lichens*. 8vo, 53 pp, 2 pls, Ambleside 1889.

2494. MARTINEAU, Harriet *A complete guide to the English Lakes . . . to which is added an appendix containing an account of the flowering plants, ferns, and mosses*. 8vo, iv+233 pp, 21 pls, 1 col. map, Windermere [1858]; 3ed. 1861; 4ed. 1871; 5ed. 1876. First edition [1855] is without the appendix and the 2nd also occurs without it.

2495. MARTINET, Johannes Florentius *Catechism of nature*. 12mo, London 1790; ed. Dublin [1790] title *A display of nature*. A large number of later editions. Translation of *Katechismus der Natuur*, 3ed. Amsterdam 1778–79, by John Hall.

2496. MARTYN, John *Methodus plantarum circum Cantabrigiam nascentium*. 8vo, vii+132 pp, Reilly, London 1727; ed. [1729] 24 pp., all publ.

2497. —— *J. Martyni Historia plantarum rariorum*. Fol., 5 decads, 52 pp, 50 col. pls, Reily, London 1728; English ed. 1728. L.P.

2498. —— *The first lecture of a course of botany; being an introduction to the rest*. 8vo, viii+23 pp, 14 pls, Strahan, London 1729.

—— *see* Tournefort, J. P. de.

2499. MARTYN, Thomas *draughtsman The universal conchologist*. Fol., 4 vols, 160 col. pls, by the author, London 1784–87; 2ed. 1784–92. The bibliography is complex. There is an issue dated 1789, title *About 120 plates of nondescript shells*, probably a remainder of later date.

2500. —— *The English entomologist, exhibiting all the coleopterous insects found in England*. Roy. 4to, 33+44 pp, 2 plain + 42 col. pls, Shakespeare Press for the author, London 1792 [=1793]. Text in English and French.

2501. —— *Aranei, or the natural history of spiders, including the principal parts of the well-known work on English spiders by E. Albin, as also the whole of the celebrated publication on Swedish spiders by C. Clerk* [i.e. Clerck]. Imp. 4to, 2 pts, ii+vii+70+31 pp, 29 col. + 2 plain pls, at his Academy, London 1793.

2502. —— *Psyche. Figures of nondescript lepidopterous insects or rare moths and butterflies from different parts of the world*. 4to, 12 pp, 32 col. pls, for the author, London [1797]. Very few copies.

2503. MARTYN, Thomas *F.R.S. Plantae Cantabrigienses: or, a catalogue of the plants which grow wild in the County of Cambridge . . . to which are added lists of the more rare plants growing in many parts of England and Wales*. 8vo, xiii+114 pp, for the author, London 1763.

2504. —— *Heads of a course of lectures in botany*. 8vo, vii+19 pp. Thurlbourn & Woodyer, for private circulation, Cambridge 1766.

2505. —— *Elements of natural history*. 8vo, 70 pp, Cambridge 1775. All publ.

2506. —— *Heads of a course of lectures in natural history*. 8vo, Cambridge 1782.

2507. —— *Thirty-eight plates with explanations; intended to illustrate Linnaeus's System of vegetables: and particularly adapted to the Letters on the elements of botany*. 8vo, vi+72 pp, 38 col. pls, White, London 1788; ed. 1794; ed. 1799; ed. 1817. See Rousseau, J. J. for Martyn's translation of the letters, 1785.

2508. —— *Flora rustica; exhibiting accurate figures of such plants as are either useful or injurious to husbandry*. 8vo, 4 vols, 144 col. pls with letterpress, Nodder, London 1792–94; ed. Harding [ ].

2509. —— *The language of botany: being a dictionary of the terms made use of in that science*. 8vo, xxviii+[320] pp, White, London 1793; 2ed. 1796; 3ed. 1807.

2510. MARTYN, William Frederick *A new dictionary of natural history: or, compleat universal display of animated nature*. Fol., 2 vols, 100 col. pls, Harrison, London 1785. Lowndes states that this was written by W. F. Mavor. Based on the collections in the Leverian Museum.

2511. MASEFIELD, John Richard Beech *Wild bird protection and nesting boxes*. 12mo, 129 pp, 9 pls, Leeds 1897.

2512. MASIUS, Hermann *Studies from nature*. 8vo, viii+196 pp, 8 pls, text figs, Chapman & Hall, London 1855. Translation of Vol. 1 of *Naturstudien-Skizzen*, Leipzig 1852–57 by Charles Boner.

Massart, Jean *see* Demoor, J.

2513. MASSEE, George Edward *A monograph of the British Gasteromycetes*. 8vo, 103 pp, 4 pls, [Oxford] 1889. Reprinted from *Ann. Bot*. Vol. 4.

2514. —— *British Fungi. Phycomycetes and Ustilagineae*. Sm. 8vo, xv+232 pp, 8 pls, Reeve, London 1891. Crown Series.

2515. —— *The evolution of plant life. Lower forms*. 8vo, viii+242 pp, text figs, Methuen, London 1891. University Extension Series.

2516. —— *A monograph of the Myxogastres*. Roy. 8vo, 367 pp, 12 col. pls, Methuen, London 1892.

2517. —— *The plant world; its past present and future*. Sm. 8vo, x+212 pp, text figs, Whittaker, London 1891; 2ed. 1898. Library of Popular Science.

2518. —— *British fungus-flora. A classified text book of mycology*. Sm. 8vo, 4 vols, text figs, Bell, London 1892–95.

2519. [MASSON, David] *The British Museum*. 8vo, viii+432 pp, text figs, Chambers, Edinburgh 1850. Chamber's Instructive and Entertaining Library.

Masterman, Arthur Thomas *see* Mackintosh, W. C.

2520. MASTERS, Maxwell Tylden *Vegetable teratology, an account of the principal deviations from the usual construction of plants*. 8vo, xx+534 pp, text figs, Ray Society, London 1869. Ray Society Publ. No. 45 for 1868.

2521. —— *Botany for beginners*. 8vo, xiii+185 pp, text figs, Bradbury, London [1872].

2522. MATEAUX, Clara L. *Odd folks at home*. Roy. 8vo, text figs, Cassell, London [1878].

2523. MATHEW, Murray Alexander *The birds of Pembrokeshire and its islands*. 4to or 8vo, lii+131 pp, 4 pls, 2 maps, Porter, London 1894.

—— *see* D'Urban, W. S. M.

2524. MATHEWS, William *The flora of the Clent and Licky hills, and neighbouring parts of the County of Worcester*. Sm. 8vo, 54 pp, Mark & Moody, Stourbridge, 1881.

2525. MATON, William George *Observations relative chiefly to the natural history . . . of the western counties of England*. 8vo, 2 vols, 16 pls, 1 map, Easton, Salisbury 1797.

2526. —— *The natural history of a part of the County of Wilts, comprehended within the distance of ten miles round the city of Salisbury*. 8vo, 74 pp, Nicols printed, London 1843.

2527. MATTHEW, Patrick *On naval timber and arboriculture*. 8vo, xvi+391 pp, text figs, Longman, London 1831.

2528. MATTHEWS, Andrew *Trichopterygia illustrata et descripta*. Imp. 8vo, xiii+188 pp, 31 pls, London 1872; suppl. 112 pp, 7 pls, 1900.

2529. —— *An essay on the genus Hydroscapha*. 8vo, London 1876.

2530. —— *A monograph of the coleopterous families Corylophidae and Sphaeriidae*. 8vo, 220 pp, 9 pls, Janson, London 1899.

2531. —— & FOWLER, William Weekes *Catalogue of British Coleoptera*. 4to, 47 pp, London 1883.

2532. MAUNDER, Samuel *The treasury of natural history; or a popular dictionary of animated nature*. 12mo, xiv+812 pp, 1 pl., text figs, Longman, London 1848; 2ed. 1849; 4ed. 1854; 6ed. 1862; 1870; 1873; ed. 1874; 1878.

2533. MAVOR, William Fordyce *The lady's and gentleman's botanical pocket book, adapted to Withering's arrangement of British plants*. 12mo, x+186 pp, 1 col. + 1 plain pls, Vernor & Hood, London 1800.

2534. —— *Natural history for the use of schools*. 12mo, x+392 pp, 66 pls, Phillips, London 1800; 2ed. 1801; ed. Dublin 1802 title *A complete system of natural history:* 5ed. 1806 title *The elements of natural history*; 6ed. 1811; 8ed. 1815; 9ed. 1816; ed. 1828; French ed. Newbery, London 1801.

2535. —— *The circle of the arts and sciences, for the use of schools and young persons*. 12mo, 475 pp, text figs, Phillips, London 1808.

2536. —— *Catechism of animated nature, or an easy introduction to the animal kingdom*. 12mo, 72 pp, text figs, Lackington, London 1810; 2–4 eds 1810.

2537. —— *Catechism of botany*. 12mo, 72 pp, text figs, Lackington, London 1810.

—— *see* Martyn, W. F.

2538. MAWE, John *The voyager's companion, or shell collector's pilot . . . also for preserving the skins of animals*. Sm. 8vo, xi+56 pp, 2 col. pls, for the author, London 3ed. 1821; 4ed. 8vo 1825. The first 2 editions are anonymous with title

*Directions to Captains of ships, officers and travellers, particularly those engaged in the south-sea fishery.*

2539. —— *The Linnaean system of conchology*. 8vo, xv+207 pp, 37 col. pls, Longman for the author, London 1823.

2540. MAXWELL, *Sir* Herbert Eustace, *Bart Memories of the months*. Sm. 8vo, 7 ser., pls, 2 portrs, Arnold, London 1897–1922. Later editions.

2541. MAY, William *The little book of British birds*. 32mo, 191 pp, 7 pls, Tilt, London [c.1840]. Hand-books for Children.

2542. —— *The little book of British quadrupeds*. 32mo, 191 pp, 7 pls, Tilt, London [c.1840]. Hand-books for children.

2543. MAYALL, John *Cantor lectures on the microscope . . . 1885.* 8vo, 97 pp, text figs, London 1886. Reprinted from *J. Soc. Arts*.

2544. —— *Cantor lectures on the modern microscope . . . 1888*. 8vo, 21 pp, text figs, London 1888. Reprinted from *J. Soc. Arts*.

2545. MAYNE, Robert Gray *An expository lexicon of the terms . . . in medical and general science.* 8vo, x+1504 pp, Churchill, London 1860. Later editions are only medical.

2546. [MAYO, Elizabeth] *Lessons on shells, as given to children between the ages of eight and ten, in a Pestalozzian school, at Cheam, Surrey*. 8vo, x+228 pp, 10 pls, Seeley & Burnside, London 1832; 2ed. 1838; 3ed. 1846.

2547. MEADE, Richard Henry *List of species of Araneida at present known to inhabit Great Britain*. 8vo, 6 pp, Butterfield printed for private circulation, Northampton 1853. Printed on one side.

2548. —— *Descriptive list of the British Anthomyidae (Diptera)*. 8vo, 2 pts, 79 pp, London 1897. A suppl. to list in *Ent. mon. Mag*., 1881–83.

*The meadow queen*. [1836]. *see* Waring, *Miss* S.

2549. MELVILL, James Cosmo *and others The flora of Harrow . . . with notices of the birds of the neighbourhood . . . and of the butterflies and moths*. 8vo, vi+127 pp, 1 map, Longman, London 1864; 2ed. Harrow 1876.

2550. *The menagerie; a reward for good children*. 12mo, 69 pp, 1 pl., text figs, Houlston, Wellington 1821; 2ed. 1822. Later editions.

2551. *The menagerie, or a peep at the quadruped race; being an exhibition of one hundred and ten animals*. 12mo, 34 pp, 40 col. pls, Newbery, London 1800.

*The menageries*. 1829–40. *see* Rennie, J.

2552. *The menageries, the natural history of monkeys. Vol. 1*. 12mo, text figs, Knight, London 1838. All publ. Library of Entertaining Knowledge.

2553. MENAULT, Ernest *The intelligence of animals*. 12mo, xiv+370 pp, text figs, Cassell, London [1869]; ed. Sm. 8vo [ ] title *Wonders of animal instinct*. Translation of *L'intelligence des animaux*, Paris 1868.

2554. [MENDES DA COSTA, Emanuel] *Conchology, or natural history of shells*. 4to, 26 pp, 12 col. pls, for the author, London [1771]. Nos. 1–6. All publ.

2555. —— *Elements of conchology: or, an introduction to the knowledge* . 8vo, viii+vi+318 pp, 7 col. pls, 1 tab., White, London 1776.

2556. —— *Historia naturalis testaceorum Britanniae, the British conchology*. 4to, xii+254+vii pp, 17 col. pls, for the author, London 1778.

Meredith, *Mrs* C. *see* Twamley, *Miss* L. A.

2557. MERRETT, Christopher *Pinax rerum naturalium Britannicarum*. 8vo, [xxviii]+221 pp, Pulleyn, London 1666; ed. 1667; 2ed. 1667; ?ed. 1704. The first 1667 ed. is not called 2ed; the second differs only in a cancel title page. Ed. 1704 is probably a ghost.

2558. MERRIFIELD, *Mrs* Mary Philadelphia *A sketch of the natural history of Brighton and its vicinity*. 8vo, xi+227 pp, 1 pl., 1 map, Pearce, Brighton, 1860; 1864.

2559. MERRIN, Joseph *The lepidopterist's calendar: giving the time of appearance of the British Lepidoptera*. Sm. 8vo, v–116 pp, Newman, London 1860; 2ed. Gloucester 1875. – Robert Marsden (?)

2560. MERTON, Francis S. *Handbook to the aquarium*. 8vo, 16 pp, Whiteley, London 1856.

2561. MESSER, Frederick A. *A new and easy method of studying British wild flowers by natural analysis*. 8vo, 146 pp, text figs, Bogue, London 1860.

2562. MEYEN, Franz Julius Ferdinand *Outlines of the geography of plants*. 8vo, x+422 pp, 1 chart, text figs, Ray Society, London 1846. Ray Society Publ. No. 7 for 1846. Translation of *Grundriss der Pflanzen Geographie*, Berlin 1836, by Margaret Johnston.

2563. MEŸER, Henry Leonard *Illustrations of British birds*. Imp. 4to, 4 vols, 330 col. pls, Longman, London [1835–41]; issue 1837–44; 2ed. 1838–44; ed. 8vo 7 vols 1842–50; issue 8vo 1853–57 title *Coloured illustrations of British birds and their eggs*. Bibliography is complex and the number of plates varies from copy to copy; 313 was the usual number issued, the rest being added. Author's name is spelt Meyer in 2ed. and later.

2564. MEYRICK, Edward *A handbook of British Lepidoptera*. 8vo, vi+843 pp, text figs, Macmillan, London 1895; ed. [1928] title *A revised* . . .; reprint of 1928 ed. 1969.

2565. —— *A label list of British Lepidoptera, compiled after A handbook of British Lepidoptera*. 8vo, 24 pp, Watkins & Doncaster, London 1896. Printed on one side.

2566. MIALL, Louis Compton *Summary notes on vegetable anatomy and physiology and the classification of British plants*. 8vo, Bradford 1867.

2567. —— *The geology, natural history, and prehistoric antiquities of Craven in Yorkshire*. 4to, 41 pp, 1 col. pl., 1 col. map, Dodgson, Leeds 1878. Reprinted from Whitaker, T. D. *History of Craven*, 3ed. 1878.

2568. —— *Handbook for Leeds and Airedale*. 8vo, vii+140 pp, 2 maps, MacCorquodale, Leeds 1890. British Association Handbook.

2569. —— *Object lessons from nature: a first book of science*. 8vo, 302 pp, Cassell, London 1890; 2ed. 1891.

2570. —— *The natural history of aquatic insects*. Sm. 8vo, ix+395 pp, text figs, Macmillan, London 1895; 1903; 1911; 1924.

2571. —— *Round the year*. Sm. 8vo, viii+295 pp, text figs, Macmillan, London 1896; 1898; ed. 1906.

2572. —— & CARRINGTON, Benjamin *The flora of the West Riding*. 8vo, xii+97 pp, 2 col. maps, Pamplin, London 1862.

2573. —— & DENNY, Alfred *The structure and life-history of the cockroach (Periplaneta orientalis)*. 8vo, 224 pp, text figs, Reeve, London 1886. Studies in Comparative Anatomy, III.

2574. —— & HAMMOND, Arthur Rashdall *The structure and life-history of the harlequin fly (Chironomus)*. 8vo, vi+196 pp, 1 pl., text figs, Clarendon, Oxford 1900.

2575. MICHAEL, Albert Davidson *British Oribatidae*. 8vo, 2 vols, 49 col. & 13 plain pls, Ray Society, London 1884–88. Ray Society Publ. Nos 61 & 65 for 1883 & 1887. Plates I–LIV, XLVII*, A–G.

2576. MICHELET, Athanais, *Mrs* Jules *Nature, or the poetry of earth and sea.* Roy, 8vo, xxxi+431 pp, 1 pl., text figs, Nelson, London 1872; 1884. Translated from French by William Henry Davenport Adams.

2577. MICHELET, Jules *The bird*. Roy. 8vo, x–14–340 pp, text figs, Nelson, London 1868 [=1867]; 1869; 1870; 1872; 1874. Translation of *L'oiseau*, Paris 1856; 8ed. 1867; by A. E.

2578. —— *The insect*. Roy. 8vo, xii+368 pp, 1 pl., text figs, Nelson, London 1875 [=1874]. Translation of *L'insecte,* Paris 1858; 6ed. 1867, by William Henry Davenport Adams.

2579. *The microscope and its wonders*. 16mo, 94 pp, Chambers, London [1866]. Chambers's Useful Handbooks.

*Microscopic objects . . . with instructions for preparing and viewing them*. 1847. *see* Goring, C. R. & Pritchard, A.

2580. *Migratory birds; or such as visit Britain at different seasons of the year*. 8vo, iv+64 pp, Craddock, London 1847.

2581. MILES, Alfred Henry *editor Natural history in anecdote, illustrating the nature, habits, manners and customs of animals*. 8vo, xii+384 pp, pls, Hutchinson, London 1895.

2582. —— *editor The concise knowledge natural history*. 8vo, xvi+771 pp, text figs, Hutchinson, London 1897.

2583. MILLAIS, John Guille *Game birds and shooting sketches, illustrating the habits, modes of capture, stages of plumage, and the hybrids and varieties*. Imp. 4to, 172 pp, 16 col. + 18 plain pls, 1 portr., text figs, Sotheran, London 1897.

2584. —— *British deer and their horns*. Imp. 4to, viii+224 pp, 1 col. + 10 plain pls, text figs, Sotheran, London 1897.

2585. MILLAR, George Henry *A new complete and universal body, or system of natural history*. Fol., iv+618 pp, 86 pls, Hogg, London [1785].

2586. MILLAR, James *A guide to botany; or, a familiar illustration of the principles of the Linnaean classification*. 8vo, Edinburgh 1818; 2ed. 1821.

2587. MILLARD, Simeon Warner *Outline of British entomology in prose and verse*. 12mo, xxiii+231 pp, 3 pls, Culverwell, Bristol 1821.

2588. MILLER, John *Collection of plants and insects . . . according to the system of Dr. Linnaeus*. Fol., 7 11, 10 col. pls, for the author, London 1759–60. Nos i–iv ? all publ. Title from *Proposals for this work in fifty parts*. Author's name Müller, Johann Sebastian.

2589. —— *Illustratio systematis sexualis Linnaei. An illustration of the sexual system of Linnaeus*. Fol., 108 col. + 108 plain pls, 4 tabs, London [1770–]1777; suppl. 8 col. + 8 plain pls 1780; issue 1794; ed. roy. 8vo, 2 vols, 1779–89 title in English alone; ed. 8vo, 1 vol. 1794. The second vol. of the 8vo ed. is not present in the fol. ed.; its title *An illustration of the Termini botanici of Linnaeus*; published at Lambeth.

2590. MILLER, J. S. *A natural history of the Crinoidea, or lily-shaped animals: with observations on the genera Asteria, Euryale, Comatula & Marsupites*. Imp. 4to, viii+150 pp, 50 tinted pls, Frost for the author, Bristol 1821. See also Thompson, J. V.

2591. MILLER, John Frederick *Icones animalium et plantarum. Various subjects of natural history*. Fol. 10 11, 60 pls, London [1776–92]; ed. 1796 title *Cimelia physica. Figures of rare and curious quadrupeds*. Text of 1796 ed. by Shaw, George.

2592. MILLER, Phillip *A short introduction to the knowledge of the science of botany*. 8vo, 49 pp, 5 pls, Rivington, London 1760. There are later eds in his *Gardener's Kalendar* from its 12ed. 1760 onwards.

2593. MILLER, Samuel Henry & SKERTCHLY, Sydney Barber Josiah *The fenland: past and present*. Roy. 8vo, xxxii+646 pp, 1 col. + 20 plain pls, 1 map, text figs, Leach, Wisbech 1878.

2594. MILLER, Thomas *Beauties of the country; or, descriptions of rural customs, objects, scenery, and the season*. Sm. 8vo, xv+425 pp, text figs, Van Voorst, London 1837.

2595. —— *Common wayside flowers*. 4to, 185 pp, 2 pls, col. text figs, Routledge, London 1841. Later editions.

2596. —— *The country year book. Spring and summer. (Autumn and winter)*. Sm. 8vo, 2 vols, 4 pts, 4 col. pls, text figs, Chapman & Hall, London 1847. Later editions, some in one vol.

2597. MILLER, William *A dictionary of English names of plants . . . English-Latin and Latin-English*. 8vo, 2 pts, viii+264 pp, Murray, London 1884.

2598. MILLER, William John Clark, *Essays and nature studies, with lectures*. Roy. 8vo, xv+220 pp, 1 portr., Elliot Stock, London 1899. 200 copies.

2599. MILLS, Alfred *Natural history of 48 birds*. 64mo, 96 pp, 48 pls, Darton, London 1810; 1812; ed. 1816.

2600. —— *Natural history of 48 quadrupeds*. 64mo, 96 pp, 48 pls, Darton, London 1810; 1812; ed. 1815.

2601. MILLS, Frederick William *An introduction to the study of Diatomaceae*. 8vo, xi+243 pp, Illiffe, London 1893.

2602. MILLS, Wesley *The nature and development of animal intelligence*. 8vo, xii+307 pp, Unwin, London 1898.

2603. MILNE, Colin *A botanical dictionary: or, elements of sys-*

*tematic and philosophical botany*. 8vo, [ii+450 pp,] 2 tabs, Lowndes, London 1770; suppl. [68] pp, 9 pls, 1778; ed. with suppl. 1778; 3ed. 1805.

2604. —— & GORDON, Alexander *Indigenous botany; or habitations of English plants . . . chiefly in Kent, Middlesex, and the adjacent counties: Vol. 1*. 8vo, viii+476 pp, Lowndes, London [1792–]1793. All. publ.

Milne-Edwards, Henri *see* Edwards, H. M.

2605. MILTON, John Laws *The stream of life on our globe*. Sm. 8vo, Hardwicke, London 1864; 2ed. 1872.

*The minstrelsey of the woods*. 1832. *see* Waring, *Miss* C.

*Mr. Mygale's hobby*. 1889. *see* Steele, Francesca Maria.

Mitchell, David William *see* Zoological Society of London.

2606. MITCHELL, Frederick Shaw *The birds of Lancashire*. 8vo, 12+xviii+224 pp, 2 col. + 9 plain pls, 1 map, text figs, Van Voorst, London 1885; 2ed. 1892.

2607. —— & HOWARD, R. J. *Birds of the Stoneyhurst district*. 8vo, 14 pp, 1 map on wrapper, Market Weighton 1888.

2608. MITCHELL, John *Dendrologia; or, a treatise of forest trees, with Evelyn's Sylva revised*. 8vo, 280 pp, text figs, Keighley 1827.

2609. MITCHELL, *Sir* Peter Chalmers *Outlines of biology*. Sm. 8vo, xv+297 pp, text figs, Methuen, London 1894; 2ed. 1902; 3ed. 1911.

2610. MITCHELL, John Mitchell *The herring: its natural history and national importance*. 8vo, xii+372 pp, 6 pls, Edmonston & Douglas, Edinburgh 1864.

2611. MIVART, St George Jackson *On the genesis of species*. Sm. 8vo, xv+342 pp, text figs, Macmillan, London 1871; 2ed. 1871.

2612. —— *Lessons in elementary anatomy*. Sm. 8vo, xxvi+535 pp, text figs, Macmillan, London 1873; ed. 1879.

2613. —— *Man and apes: an exposition of structural resemblances bearing upon questions of affinity and origin*. Sm. 8vo, vi+200 pp, 40 pls, Hardwicke, London 1873.

2614. —— *The common frog*. Sm. 8vo, viii+158 pp, 1 pl., text figs, Macmillan, London 1874. Nature Series.

2615. —— *Lessons from nature, as manifested in mind and matter*. 8vo, xvi+462 pp, Murray, London 1876.

2616. —— *The cat. An introduction to the study of backboned animals, especially mammals*. 8vo, xxiii+557 pp, 1 pl., text figs, Murray, London 1881.

2617. —— *Nature and thought; an introduction to a natural philosophy*. 8vo, vi+261 pp, Kegan Paul, London 1882; 2ed. 1885.

2618. —— *Birds: the elements of ornithology*. Sm. 8vo, [viii]+329 pp, text figs, Porter, London [1892].

2619. —— *Types of animal life*. 8vo, viii+374 pp, text figs, Osgood McIlvaine, London 1893; 1894.

2620. —— *An introduction to the elements of science*. 8vo, xii+392 pp, Osgood McIlvaine, London 1894 [=1893].

2621. —— *Essays and criticisms.* 8vo, 2 vols, Osgood McIlvaine, London 1895.

2622. MOFFATT, C. B. *Life and letters of A. G. More . . . with selections from his zoological and botanical writings*. 8vo, xii+642 pp, 1 portr., 1 col. map, Hodges & Figgis, Dublin 1898.

2623. [MOFFATT, Jeannie C.] *The ferns of Moffat*. Fol., 92 pp, Moffat 1863. By O.

2624. MOFFETT, Thomas *Insectorum sive minimorum animalium theatrum*. Fol., xx+326 pp, 2 pls, text figs, Cotes printed, London 1634. There are 3 variant title pages. For English ed. *see* Topsell, E.

2625. —— *Healths improvement*. 4to, 8+296 pp, Thomson, London 1655; ed. 12mo 1746.

2626. MOHL, Hugo von *Principles of anatomy and physiology of the vegetable cell*. 8vo, viii+159 pp, 1 pl., text figs, Van Voorst, London 1852. Translation of *Grundzüge der Anatomie und Physiologie der vegetablischen Zells*, Braunschweig 1851, by Arthur Henfrey.

2627. [MONRO, Alexander] *first of the name An essay on comparative anatomy*. 8vo, 22+138 pp, Nourse, London 1744; 1775; ed. Edinburgh 1782; ed. 12mo Edinburgh 1783 title *A treastise on comparative anatomy*. First ed. is unauthorized. First authorized is in *Works,* Edinburgh 1781. 1782 ed. is in *The anatomy of the human bones*. For 2 further eds see Fyfe, A.

2628. MONRO, Alexander *second of the name The structure and physiology of fishes explained, and compared with those of man and other animals*. Fol., 128 pp, 50 pls on 44 leaves, Elliot, Edinburgh 1785. Plates 1–44, 3*, 9*, 10*, 14*, 15*, 50*.

2629. *The monsters of the deep, and curiosities of animal life*. Sm. 8vo, text figs, Nelson, London 1874.

2630. MONTAGU, George *Ornithological dictionary; or alphabetical synopsis of British birds*. 8vo, 2 vols, 1 col. pl., White, London 1802; suppl. 460 pp, 24 pls, 1813; 2ed. 1 vol. 1831 title *Ornithological dictionary of British birds*; 1833; ed. 4to 1866 title *A dictionary of British birds*; ed. 8vo [1881]; 1882; 1889.

2631. —— *Testacea Britannica, or natural history of British shells*. 4to, 2 pts, xxxvii+606 pp, 16 col. pls, White, London 1803; suppl. v+183 pp, 13 col. + 1 plain pls, 1808; ?ed. 1823.

2632. MONTROSE NATURAL HISTORY AND ANTIQUARIAN SOCIETY *List of British birds in the Society's collection together with the birds known to frequent Montrose district*. 8vo, 11 pp, Walker printed, Montrose 1881.

2633. MOORE, David *Concise notes on the indigenous grasses of Ireland*. Fol., 47 specimens, Dublin 1843; 2ed. 1850.

2634. —— *Contributions to the British and Irish floras of Musci and Hepaticae. On some plants supposed to be additions to the Irish flora.* 8vo, 10 pp, for the author, [Dublin 1863].

2635. —— & MORE, Alexander Goodman *Contributions towards a Cybele Hibernica*. 8vo, lv+399 pp, 1 map, Dublin 1866; 2ed. 1898. 2ed. is by Colgan, Nathaniel & Scully, Reginald William. For a suppl. see More, A. G.

2636. MOORE, Peter George *British birds systematically arranged in five tables*. Fol., 15 pp, 5 tabs, Van Voorst, London 1879.

2637. MOORE, Thomas *A handbook of British ferns*. 8vo, iv+156+viii pp, text figs, Groombridge, London 1848; 2ed. 1853; 3ed. 1857.

2638. —— *A popular history of the British ferns and allied plants*. Roy. 16mo, xii+354 pp, 20 col. pls, Reeve, London 1851; 2ed. 1855; 3ed. 1859; ed. 1862; ed. [ ]. Popular Natural History Series.

2639. —— *The ferns of Great Britain and Ireland*. Imp. fol., [159] pp, 51 col. pls, Bradbury & Evans, London 1855[–56]; ed. 1857; ed. 8vo 2 vols. [1850–]1859[–60] title *Nature-printed British ferns*.

2640. —— *British ferns and their allies; an abridgement of The popular history of British ferns*. 16mo, iv+187 pp, 12 col. pls, text figs, Routledge, London 1859; ed. 1861; ed. 1862; 3ed. 1872. Later eds.

2641. —— *The field botanist's companion*. 8vo, xxvii+424 pp, 24 col. pls, Van Voorst, London 1862; ed. 1867 title *British wild flowers familiarly described in the four seasons*.

—— *The elements of botany*. 10ed. 1865. *see The elements of the science of botany*. 1833.

2642. —— *Baildon's nature-printed ferns prepared according to his new patented process. The descriptions by Thomas Moore*. Fol., [10] pp, 4 col. pls, Reeve, London 1869. Very few copies.

—— *see Lindley, J.*

2643. MOQUIN-TANDON, Christian Horace Bénédict Alfred *Elements of medical zoology*. 8vo, Baillière, London 1861. Translation of *Elémens de zoologie médicale*, Paris 1860, by Robert Thomas Hulme.

2644. —— *The world of the sea*. Roy. 8vo, viii+448 pp, col. pls, London [1869]; ed. 1882. Translation of *Le monde de la mer*, Paris 1865, by Henry Martyn Hart. Originally published under the pseudonym Fredol, Alfred.

2645. MORE, Alexander Goodman *Outlines of the natural history of the Isle Wight*. 18mo, 107 pp, 1 map, Spottiswoode, London 1860. Reprinted from Venables, Edmund *The Isle of Wight*, 1860.

2646. —— *A comparative list of British plants*. 8vo, 38 pp, London 1863. Reprinted from *The Phytologist*.

2647. —— *A catalogue of the flowering plants and ferns growing wild in the Isle of Wight*. 8vo, 11 pp, Isle of Wight Philosophical and Scientific Society, Ryde [1868].

2648. —— *A supplement to the Flora Vectensis* [of W. A. Bromfield] 8vo, 30 pp, Taylor, London 1871. Revised reprint from *J. Bot.*

2649. —— *Recent additions to the flora of Ireland; being a first supplement to the Cybele Hibernica* [of D. Moore & A. G. More]. 8vo, 40 pp, Dublin 1872. Reprinted from *Proc. R. Irish Acad*. Vol. 1.

2650. —— *Enquiries respecting some Irish birds*. 8vo, 4 pp, Dublin 1885.

2651. —— *A list of Irish birds*. 8vo, 32 pp, National Museum of Ireland, Dublin 1885; 2ed. 1889; additions 2 pp, [1892].

2652. —— *Botany: characteristic and rare plants* [of Cork]. 8vo, 4 pp, [Cork 1893].

—— *see* Moore, D.

—— *see* Moffat, C. B.

2653. MORGAN, Conwy Lloyd *The springs of conduct*. Sm. 8vo, viii+317 pp, Kegan Paul, London 1885.

2654. —— *Animal biology*. Sm. 8vo, xxii+370 pp, text figs, Rivington, London 1887; 2ed. 1889; 4ed. 1911.

2655. —— *Animal life and intelligence*. 8vo, xvi+512 pp, 1 pl., text figs, Arnold, London 1890–91; 1891.
2656. —— *Animal sketches*. 8vo, 312 pp, text figs, Arnold, London [1891].
2657. —— *Habit and instinct*. 8vo, 351 pp, 1 pl., Arnold, London 1896.
2658. —— *Animal behaviour*. 8vo, viii+344 pp, text figs, Arnold, London 1900.
2659. MORGAN, John *The flora of Llanbrynmair*. 8vo, Oswestry [1893].
2660. MORGAN, Thomas Owen *Flora Cereticae superioris. A catalogue of the plants indigenous in the neighbourhood of Aberystwith*. 8vo, 18 pp, Aberystwith 1849.
2661. MORISON, Robert *Hortus Regius Blesensis auctus, cum notulis durationis & charactismis Plantarum ... Praeludiorum botanicorum pars prior ... pars altera*. 8vo, [xxxii]+499 pp, Allestry, London 1669.
2662. —— *Plantarum umbelliferarum distributio nova*. Fol., 91 pp, 12 pls, 8 tabs, Sheldonian, Oxford 1672. Laters eds. in No. 2663.
2663. —— *Plantarum historiae universalis Oxoniensis; pars secunda (tertia)*. Fol., 2 vols, portr., 292 pls, Sheldonian, Oxford 1680, 99; ed. 3 vols, 1715. ed. 1738; New ed. of No. 2662 contained in pt. 3, sect. 9.
2664. MORLEY, Charles & FRIEDERICHS, Hulda *Half-holidays at the Zoo*. 4to, photographs, Westminster Gazette Office, London 1895.
2665. MORLEY, Claude *The Coleoptera of Suffolk*. 8vo, xiv+113 pp, 1 map, Plymouth 1899.
2666. —— *The Hymenoptera of Suffolk. Pt. 1*. 8vo, viii+22 pp, Plymouth 1899. All publ.
2667. MORRIS, Beverley Robinson *British game-birds and wild-fowl*. Roy. 4to, iv+254 pp, 60 col. pls, Groombridge, London 1855; 2ed. 1873; 3ed. 1891; 4ed. 8vo 2 vols 1895; 5ed. 1897; issues [ ].
2668. MORRIS, Francis Orpen *A guide to the arrangement of British birds*. 8vo, 20 pp, Groombridge, London [1834]. Printed on one side.
2669. —— *An essay on scientific nomenclature*. 8vo, 10 pp, Groombridge, London 1850.
2670. —— *A history of British birds*. Roy. 8vo, 6 vols, 358 col. pls, Groombridge, London [1850–]1851–57; 1862; 1863–64;

1865; 1866; 1868; 2ed. 1870; 2ed. 1870–71; 3ed. 1891; 4ed. 1895–96; 5ed. 1893; 5ed. revised 1903; cabinet ed. sm. 8vo, 8 vols, [1863–67].

2671. —— *Book of natural history; containing a description of animals and birds*. Imp. 16mo, ii+318 pp, 160 col. pls, Groombridge, London 1852.

2672. —— *A history of British butterflies*. Sm. 4to, 106 pp, 48 col. pls, Groombridge, London [1852–]1853; 1857; 1860; 1865; 3ed. 1870; 4ed. 1876; 6ed. 1890; 7ed. 1893; 8ed. 1895; 9ed. 1904.

2673. —— *A natural history of the nests and eggs of British birds*. Roy. 8vo, 3 vols, 226 col. pls, Groombridge, London 1853–56; 1856–61; 1863–64; 1866–67; 2ed. 1870–71; 1875; 3ed. 1892; 4ed. 1896; 1898. No pls numbered 11, 55, 56, 60, 69, 94, 119.

2674. —— *'Fact is stranger than fiction'. Anecdotes in natural history*. Sm. 8vo, xii+311 pp, Longman, London 1860.

2675. —— *Records of animal sagacity and character*. 12mo, xxvii+302 pp, Longman, London 1861.

2676. —— *A natural history of British moths*. Roy. 8vo, 4 vols, 132 col. pls, Knox, London [1859–]1861–70; 1871; 1872; 1874; 3ed. 1891; 1894; 5ed. 1896 title *A history . . .*; 6ed. 1903.

2677. —— *A catalogue of British insects in all the orders*. 8vo, 125 pp, Van Voorst, London 1865[–1867]. Most copies end at p. 56, the end of the Lepidoptera with a few Diptera, and in this form was issued to subscribers to the parts issue of No. 2676. In most copies Van Voorst's name is struck out and Longman substituted in ink.

2678. —— *The difficulties of darwinism. Read before the British Association*. 8vo, 68 pp, Longman, London 1869.

2679. —— *A double dilemma in darwinism*. 8vo, 15 pp, Poole, London [1870].

2680. —— *Anecdotes in natural history*. 8vo, text figs, Partridge, London [1872]; 2ed. [1889]. A different work from No. 2674.

2681. —— *All the articles of the Darwin faith*. 8vo, 50 pp, Poole, London 1875; 1877; [1882]; [1890].

2682. —— *Simple lessons for home use. – Birds*. 12mo, iv+32 pp, Stanford, London 1877.

2683. —— *A guard against the Guardian*. 8vo, London [1877].

2684. —— *The Darwin craze. A paper on the plumage of birds and butterflies*. 8vo, 14 pp, Poole, London 1880.

2685. —— *Letters to the Times about birds*. 8vo, 268 pp, Poole, London [1880].

2686. —— *The cui bono of hereditism*. 8vo, ii+20 pp, Kensit, London [1890].

2687. —— *The demands of darwinism on credulity*. 8vo, 68 pp, Partridge, London [1890].

2688. MORRIS, Joseph William *Handbook to Bath, prepared for the occasion of the visit of the British Association*. 8vo, v+262 pp, 3 maps, Bath 1888. British Association Handbook.

2689. MORRIS, Richard *The botanist's manual*. 12mo, 189 pp, Sherwood, London 1824.

2690. MORSE, Edward Sylvester *The first book of zoology*. 8vo, xvi+190 pp, text figs, King, London 1876; ed. 1880. First ed. is New York 1875.

2691. MORTIMER, Cromwell *editor Some papers lately read before the Royal Society concerning the fresh-water polypus*. 4to, xvii pp, London 1743. Reissued in *Phil. Trans*., Vol. 42, pp. 281–88. The papers are by Abraham Trembley.

2692. [MORTON, Charles] *An essay towards a probable solution to this question. Whence come the stork and the turtle, the crane and the swallow, when they know and observe the appointed time of their coming. Or where those birds do probably make their recess and abode, which are absent from our climate, at some certain times and seasons of the year*. 8vo, 50 pp, Crouch, London 1703; ed. 1744 in *Harleian Miscellany*, 2. By a person of learning and piety. Also ascribed to Francis Roberts. Editions of 1698 or even 1686 may have been published.

2693. MORTON, John *The natural history of Northampton-shire*. Fol., iv+55+46+x pp, 14 pls, 1 map, Knaplock & Wilkin, London 1712.

2694. MORWOOD, Vernon S. *Facts and phases of animal life*. Sm. 8vo, xvi+286 pp, 1 pl., text figs, Hogg, London 1882.

2695. —— *Wonderful animals: working, domestic and wild*. Sm. 8vo, xix+288 pp, 1 pl., text figs, Hogg, London 1883.

2696. MOSLEY, *Sir* Oswald *Bart* & BROWN, Edwin *The natural history of Tutbury*. Imp. 8vo, xiii+408 pp, 6 col. + 3 plain pls, Van Voorst, London 1863.

2697. MOSLEY, Seth Lester *Illustrations of European butterflies*. 8vo, 253 col. pls, Mosley Huddersfield 1879–94. 60 parts, ?all publ. ?6 copies.

2698. —— *A history of British birds, their nests and eggs*. 8vo, 3 vols,

161 col. pls of birds & 116 col. pls of eggs, with letterpress, Mosley, Huddersfield [1881–] 1892. Vol. 3 pts i & ii all publ. Plates of birds 1–157, 160–163; of eggs 1–115, 109*. Superior or ordinary issues. Young Naturalist Series.

2699. —— *Illustrations of varieties of British Lepidoptera*. 8vo, 73 col. pls with letterpress, Mosley, Huddersfield 1878–82 [–1885]. Superior or ordinary issues.

2700. —— *Illustrations of varieties of British Lepidoptera. Second series*. 8vo, 30 col. pls with letterpress, Mosley, Huddersfield [1896–1903]. Pts 1–6, all publ. This is a partial 2ed. of No. 2699.

2701. —— *Annotated catalogue of British birds, nests, and eggs*. 8vo, 48 pp, text figs, Mosley, Huddersfield 1896. Issued with *Nature study*, Vol. V.

2702. —— *An illustrated catalogue of the varieties of British Lepidoptera*. 8vo, 28 pp, 16 pls, text figs, Mosley, Huddersfield 1896–97. Issued with *Nature study*, Vols V & VI.

Moss Exchange Club. *see* Wadell, C. H.

2703. MOSSOP, John *Ornithologia Britannica, A synopsis of British birds in verse and prose*. Sm. 8vo, xviii+259 pp, Jackson & Walford, London 1841.

2704. MOTT, Frederick Thompson *Charnwood forest*. 8vo, 144 pp, 2 col.+4 plain pls, 1 col. map, text figs, Kent, London 3ed. 1868. This is the first ed. by Mott.

2705. —— *et al. The flora of Leicestershire, including the cryptogams*. 8vo, xxvi+372 pp, col. maps, Leicester Literary & Philosophical Society, London 1886.

2706. MUDD, William *A manual of British lichens*. Roy. 8vo, viii+309+xxii pp. 5 col. pls, Penny for the author, Darlington 1861. See also No. 2707.

2707. —— *Herbarium lichenum Britannicorum*. 4to, 3 fasciculi, 301 specimens with printed names, Great Ayton 1861. A few copies as companions to the *Manual* No. 2706.

2708. —— *A monograph of British Cladoniae*. 4to, viii+36 pp, 35 11 of specimens, Cambridge 1865. Very few copies.

2709. [MUDIE, Robert] *The British naturalist; or, sketches of the more interesting productions of Britain and the surrounding seas*. Sm. 8vo, 2 vols, 1 pl., text figs, Whittaker, London 1830; 2ed. 1835.

2710. —— *The botanic annual; or familiar illustrations . . . of plants*. 8vo, xv+446 pp, 1 pl., Cochrane, London 1832.

2711. —— *First lines of zoology, by question and answer*. 12mo, 7 pts, Whittaker, London 1832.

2712. —— *A popular guide to the observation of nature*. 12mo, xxii+372 pp, text figs, Whittaker, London 1832. Constable's Miscellany, Vol. LXVII.

2713. —— *The feathered tribes of the British Islands*. 12mo, 2 vols, 18 col. pls, text figs, Whittaker, London 1834; 2ed. 1835; 3ed. 1841; 4ed. 1853; 1854; 1861; 1878; ed. 1888.

2714. —— *The natural history of birds*. Sm. 8vo, viii+xii+408 pp, 1 col. pl., text figs, Orr, London 1834.

2715. —— *The sea*. 12mo, Ward, London 1835.

2716. —— *Spring; or, the causes, appearances, and effects of the seasonable renovations of nature, in all climates*. 12mo, xii+280 pp, 1 col. pls, Ward, London 1837; [1860]. May also be found bound with No. 2717.

2717. —— *Summer; or, the causes, appearances, and effects of the grand nuptials of nature in all its departments*. 12mo, xi+287 pp, 1 col. pl., Ward, London 1837; [1860]. May also be found bound with No. 2716.

2718. —— *Autumn; or, the causes, appearances, and effects of the seasonal decay and decomposition of nature*. 12mo, xi+288 pp, 1 col. pl., Ward, London 1837; [1860]. May also be found bound with No. 2719.

2719. —— *Winter; or, the causes, appearances, and effects of the great seasonal repose of nature*. 12mo, xi+276 pp, 1 col. pl., Ward, London 1837; [1860]. May also be found bound with No. 2718.

2720. —— *Gleanings of nature*. Roy. 8vo, ix+108 pp, 13 col.+39 plain pls, Orr, London 1838.

2721. Müller, Hermann Heinrich Ludwig *The fertilization of flowers.* Roy. 8vo, x+669 pp, text figs, Macmillan, London 1883. Translation of *Die Befruchtung der Blumen durch Insekten*, Leipzig 1873, by D'Arcy Wentworth Thompson.

2722. Müller, Johann Friedrich Theodor *known as* Fritz *Facts and arguments for Darwin*. Sm. 8vo, viii+144 pp, text figs, Murray, London 1869. Translation of *Für Darwin*, Leipzig 1864, by William Sweetland Dallas.

Müller, Johann Sebastian *see* Miller, John.

2723. Müller, Johannes *On certain variations in the vocal organs of the Passeres that have hitherto escaped notice*. 4to, 74 pp, 8 pls, University Press, Oxford 1878. Trans of papers in *Phys.*

*Abhand. K.-Preussische Akad. Wiss. Berlin*, 1845–47, by Francis Jeffrey Bell.

Muffett, Thomas *see* Moffett, T.

2724. MUIRHEAD, George *The birds of Berwickshire*. Roy. or demy 8vo, 2 vols, 4 col. pls, 2 col.+1 plain maps, text figs, Douglas, Edinburgh 1889–95. L.P. 100 copies.

2725. MULLEN, Allan *An anatomical account of the elephant accidentally burnt in Dublin . . . together with a relation of new anatomical observations on the eyes of animals*. 8vo, 72 pp, 2 pls, Smith, London 1682.

2726. MURCHE, Vincent T. *Elementary botany for beginners*. 8vo, 144 pp, Blackie, London [1855]. Elementary Text Books.

2727. MURRAY, Alexander *The northern flora; or, a description of the wild plants belonging to the north and east of Scotland. Part 1*. 8vo, xvii+150+xvi pp, 2 pls, Edinburgh 1836. All publ.

2728. MURRAY, Andrew *Catalogue of the Coleoptera of Scotland*. Sm. 8vo, viii+145 pp, Blackwood, Edinburgh 1853. Printed on one side.

2729. —— *On the disguises of nature; being an inquiry into the laws which regulate external form and colour in plants and animals*. 8vo, Edinburgh 1859. Reprinted from *Edinburgh new phil. Jl.*

2730. —— *The geographical distribution of mammals*. 4to, xvi+420 pp, 2 col. pls, 103 col. maps on 61 leaves, text figs, Day, London 1866.

2731. —— *Economic entomology. Aptera*. Sm. 8vo, xxiii+433 pp, text figs, Chapman & Hall, London [1877]. South Kensington Museum, Science Handbooks.

2732. —— *Catalogue of the Doubleday collection of Lepidoptera. Part I, British Lepidoptera. (Part II, European collection)*. Sm. 8vo, 32+36 pp, Chapman & Hall, London [1877].

2733. MURRAY, George Robert Milne *An introduction to the study of seaweeds*. Sm. 8vo, xvi+271 pp, 8 col. pls, text figs, Macmillan, London 1895.

—— *see* Bennett, A. W.

2734. MURRAY, Hugh *An encyclopaedia of geography: comprising a complete description of the earth . . . exhibiting . . . the natural history of each country*. 8vo, xii+1567 pp, text figs, Longman, London 1834; 2ed. 1844. Assisted by William Wallace, Robert Jameson, W. J. Hooker, William Swainson.

2735. MURRAY, John *Experimental researches on the light and luminous matter of the glow-worm, the luminosity of the sea, the phenomena of the chameleon, the ascent of the spider into the atmosphere, and the torpidity of the tortoise etc*. 12mo, 177 pp, text figs, M'Phun, Glasgow 1826.

2736. —— *Researches in natural history*. 12mo, Whittaker, London 1830; 2ed. 1830.

2737. [——] *The physiology of plants; or the phenomena and laws of vegetation*. 12mo, xi+298 pp, 2 pls, text figs, Murray, London 1833. Ascription uncertain.

2738. [——] *The economy of vegetation, or phenomena of plants*. 8vo, 175 pp, 2 pls, text figs, Relfe, London 1838. By a Fellow of the Linnean Society.

2739. MURRAY, Richard Paget *The flora of Somerset*. 8vo, lxi+iv+437+418a–d pp, Taunton [1892–]1896; suppl. iv+242 pp, Taunton 1914. Suppl. by E. S. Marshall. *Proc. Somerset archaeol. nat. Hist. Soc.*, Vols 38–42, strictly therefore not a book.

2740. [MURRAY, Thomas Boyles] *Zoological sketches, consisting of descriptions of one hundred and twenty animals*. 8vo, ix+249 pp, 120 col. pls, SPCK, London 1844. Plates derive from a set of 150, 4to, coloured sheets by J. W. Whimper issued by SPCK c.1840. Text by T.B.M.

*The museum of natural history*. [1859–62]. *see* Richardson, *Sir* John.

*My microscope and some objects from my cabinet*. 1887. *see* Needham, J. T.

*Mr. Mygale's hobby*. 1889. *see* Steele, F. M.

# N

N., E. *All the British butterflies arranged in their respective families*. [1868]. *see* Newman, E.

N., T. *An account of some new microscopical discoveries*. 1745. *see* Needham, J. T.

2741. NAEGELI, Carl Wilhelm von & SCHWENDERER, S. *The*

*microscope in theory and practice.* 8vo, xii+382 pp, text figs, Sonnenschein, London 1887; 1892; 2ed. 1899. *Translation of Das Mikroscop*, Leipzig 1867, by *Sir* Frank Crisp.

2742. NALL, John Greaves *Great Yarmouth and Lowestoft . . . with chapters on archaeology, natural history*. Sm. 8vo, ii+728 pp, 1 map, Longman, London 1866; issue 2 vols 1866; ed. 1867.

2743. NAPIER, Charles Ottley Groom *earlier* Groom *Variations in the colour and size of the eggs of birds*. 4to, 7 pp, London [1868]. Reprinted from *Naturalist's note book*, 1868.

2744. —— *The food, use and beauty of British birds, an essay accompanied by a catalogue of all the British birds.* Sm. 8vo, 83 pp, 1 pl., London 1865; ed. 1870.

2745. —— *The book of nature and the book of man*. 8vo, vi+479 pp, pls, text figs, Hotton, London 1870.

2746. —— *Natural history rambles. Lakes and rivers*. 12mo, 247 pp, text figs, SPCK, London 1879.

*Narratives of nature, and history book for young naturalists*, [?1845]. *see* Strickland, *Miss* Agnes.

2747. NASH, Joseph *A practical treatise on British song birds*. 12mo, vi+102 pp, 8 col. pls, Sherwood Jones, London 1824; ed. 1876 title *British song birds*.

2748. *The national natural history*. 8vo, vi+384, 90 col. pls, text figs, Warne, London [1875].

2749. *Natural history*. 24mo, 154 pp, Chidley, London 2ed. 1838.

2750. *Natural history. A popular introduction to the study of quadrupeds, or of the class Mammalia*. 12mo, xii+492 pp, RTS, London 1833.

2751. *Natural history. Birds.* 4to, Goode, London [c.1850].

2752. *Natural history. Birds and beasts*. 32mo, London [1854].

2753. *A natural history book for children*. 4to, pls, Warne, London 1891.

2754. *Natural history. Fishes*. 4to, London [?1854].

Natural history for beginners. *see* Crown Series of Natural History.

2755. *Natural history for children. Quadrupeds, birds, insects, fishes and reptiles, trees and plants*. 12mo, 5 vols, 5 pls, text figs, Baldwin, London 1819.

*Natural history for the young*. [1869]. *see* Laurie, *Mrs* J. W.

*Natural history for young observers. see* Crown Series of Natural History.

*Natural history in anecdote.* [1895]. *see* Miles, A. H.

2756. *Natural history notes and anecdotes. Series 1.* 16mo, RTS, London 1887. RTS Library.

2757. *Natural history of animals. Containing an account of remarkable beasts.* 12mo, 180 pp, text figs, Goodwin printed, Dublin 1818; ed. Wetton London [?1820].

*A natural history of animals containing the anatomical descriptions of several creatures dissected by the Royal Academy of Sciences at Paris.* 1688. *see* Perrault, C.

2758. *The natural history of animals, containing an account of remarkable fishes.* 12mo, text figs, Grace, Dublin 1818.

2759. *A natural history of beasts and birds.* 32mo, Otley, [?1815].

2760. *A natural history of beasts and birds.* 12mo, 47 col. pls, Richardson, Derby [c.1825].

*The natural history of beasts (birds) compiled from the best authorities.* 1793. *see* Jones, S.

2761. *The natural history of birds and beasts; showing the character and properties of each animal.* 32mo, 32 pp, 16 pls, Wallis, London 1800; 1801; 1803.

2762. *Natural history of birds; containing a familiar survey of the feathered creation.* 32mo, 40 pp, 1 col. pl., col. text figs, Chappell, London [c.1825].

*The natural history of birds; containing a variety of facts . . . intended for the amusement and instruction of children.* 1786–91. *see* Galton, S. J.

*Natural history of birds, fish, insects and reptiles.* 1793. *see* Buffon, G. L. Le Clerc *Conte* de, Barr's Buffon.

2763. *The natural history of birds, from the works of the best authors, ancient and modern.* 8vo, 2 vols, 162 col. pls, Brightly & Childs, Bungay 1815; ed. Offor, London 1815; ed. Bungay 1820.

*A natural history of birds, quadrupeds, fishes, reptiles, serpents, butterflies and insects.* [?1815]. *see A natural history of British quadrupeds, foreign quadrupeds* etc.

*A natural history of British birds.* [ ]. *see A natural history of British quadrupeds, foreign quadrupeds* etc.

*Natural history of British birds.* [ ]. *see A new book of natural history.*

*A natural history of British quadrupeds.* [ ]. *see A natural history of British quadrupeds, foreign quadrupeds* etc.

2764. *A natural history of British quadrupeds, foreign quadrupeds, British birds, water birds, foreign birds, fishes, reptiles, serpents, and insects.* 12mo, 7 parts of 36 pp each, fpce, text figs,

Apollo Press – W. Davison, Alnwick 1809; ed. [?1815]. The date 1909 is spurious, perhaps 1815, or misprint for 1819. The parts were also issued separately [?1814] by Davison, but not Apollo Press. Alternative cancel titles 1. *A cabinet of natural history, containing pretty pictures*. 2. *A natural history of birds, quadrupeds, fishes, reptiles, serpents, butterflies and insects.* Wood engravings by Thomas Bewick.

*A natural history of fishes.* [ ]. *see A natural history of British quadrupeds, foreign quadrupeds* etc.

*Natural history of fishes, and of reptiles, insects . . . vegetables*. 1793. *see* Jones, S.

*A natural history of foreign birds.* [ ]. *see A natural history of British quadrupeds, foreign quadrupeds* etc.

*A natural history of foreign quadrupeds.* [ ]. *see A natural history of British quadrupeds, foreign quadrupeds* etc.

*The natural history of four footed beasts*. 1802. *see A pretty book of pictures.*

*The natural history of insects*. 1829, 35. *see* Rennie, J. & Westwood, J. O.

2765. *The natural history of insects compiled from Swammerdam, Brookes, Goldsmith*. 8vo, 210 pp, 20 pls, Morison Perth 1792. For other editions of this compilation see Buffon, G. L. Le Clerc *Conte de*, to which it is a companion volume.

*A natural history of Ireland, in three parts. By several hands*. 1726. *see* Boate, G.

2766. *The natural history of quadrupeds, and cetaceous animals, from the works of the best authors, ancient and modern*. 8vo, 2 vols, 120 col. pls, Brightly, Bungay 1811.

2767. *Natural history of quadrupeds, for children*. 12mo, 116 pp, 26 pls, Darton, London 1813; 2ed. 1824; ed. 1829. 1st ed. by the author of *The decoy*; others by A lady.

2768. *The natural history of quadrupeds, including all the linnaean class Mammalia. To which is prefixed a general view of nature.* 12mo, 2 vols, 88 col. pls, Johnson, London 1801. Up to p. 157 of Vol. 1 probably by Samuel Galton.

*The natural history of quadrupeds. To which is prefixed, the history of the earth . . . and man.* 1804. *see* Buffon, G. L. Le Clerc *Conte de.*

2769. *The natural history of remarkable beasts.* 16mo, 164 pp, text figs, Jones, Dublin 1820; 12ed. [?1825]; ed. [1830].

*A natural history of remarkable birds and fishes . . . abridged from Dr. Goldsmith's Natural history.* [c.1825]. *see* Goldsmith, O.

2770. *The natural history of remarkable birds, with their habits and instincts.* 12mo, viii+171 pp, text figs, Graisberry & Campbell, Dublin 1819; Espy printed 1821; Jones, Dublin 1821; ed. Wetton, London [?1821]; ed. [?1825]. Also occurs in SPCK bindings. 1836 (per WW182:832)

2771. *The natural history of remarkable insects, with their habits and instincts.* 12mo, viii+169 pp, text figs, de Veaux, Dublin 1819; ed. 1822; ed. [?1825]; ed. Wetton, London [?1825].

2772. *The natural history of remarkable trees, shrubs, and plants.* 12mo, Jones, Dublin 1819; ed. [?1825]; ed. Wetton, London [?1825].

2773. *The natural history of reptiles and serpents . . . an account of worms, corals and of sponges.* 12mo, 178 pp, text figs, Bentham, Dublin 1821; ed. 1824; ed. Wetton, London [?1825].

*A natural history of reptiles, serpents and insects.* [ ]. *see A natural history of British quadrupeds, foreign quadrupeds*, etc. 1809.

2774. *A natural history of sea anemones.* 12mo, 2 ll, [?London ?1795].

*The natural history of Selborne.* 1789. *see* White, Gilbert.

*A natural history of singing birds.* 1754. *see* Albin, E.

Natural History of the British Islands Series, (Van Voorst). *see*

Bate, C. S. & Westwood, J. O. *Sessile-eyed Crustacea.* 1863–68.
Bell, T. *Quadrupeds.* 1837.
—— *Reptiles.* 1839.
—— *Stalk-eyed Crustacea.* 1853.
Donkin, A. S. *Diatomaceae.* 1841.
Forbes, E. *Starfishes.* 1841.
—— & Hanley, S. C. T. *Mollusca.* 1853.
Gosse, P. H. *Sea-anemones.* 1860.
Hewitson, W. C. *Eggs of birds.* 2ed. 1846.
Hincks, T. *Hydroid zoophytes.* 1868.
—— *Polyzoa.* 1880.
Johnston, G. *Zoophytes.* 2ed. 1847.
Jones, T. R. *Organization of the animal kingdom.* 1841.
Newman, E. *Ferns.* 1844.
Selby, P. J. *Forest trees.* 1842.
White, G. *Natural history of Selborne.* ed. 1877.
Yarrell, W. *Fishes.* 1836.
—— *Birds.* 1837–43.

*Natural history of the most remarkable beasts (birds)*. 1824. *see* Harrison, J.

2775. *A natural history of the three kingdoms*. Obl. 8vo, col. pls with letterpress, Cassell, London [1868].

*The natural history of the year for children*. 1852. *see* Woodward, B. B.

*A natural history of water birds*. [ ]. *see A natural history of British quadrupeds, foreign quadrupeds* etc.

2776. *The natural history of wonderful fishes*. 12mo, Jones, Dublin [?1823].

2777. *Natural history; or, a descriptive account of the animal world. Beasts*. 12mo, 12 pp, Scoggins, Ipswich [?1860].

2778. *Natural history; or, a short introduction to animated nature*. 8vo, Darton, London [1854].

Natural History Society of Glasgow. *see* Glasgow Natural History Society.

Natural History Society of Manchester. *see* Manchester Natural History Society.

*The naturalist's and traveller's companion*. 1772. *see* Lettsom, J. C.

Naturalist's Library. *see* Jardine, *Sir* William *editor*.

2779. *The naturalist's pocket magazine*. 12mo, 8 vols, 432 col. pls, Harrison Cluse, London 1799–1802 [–1803]. Vol. 8 of 4 parts only, all publ.

*Nature delineated*. 1739. *see* Pluche, N. A. de la.

*Nature display'd*. 1733. *see* Pluche, N. A. de la.

2780. *Nature displayed*. 12mo, Dublin, 1828; ed. [?1830]. Partial trtanslation of Martinet, J. F. *Katechismus der Natuur*, Amsterdam 1778–79. q.v.

Nature Series, (Macmillan). *see*

Allen, C. G. B. *Colours of flowers*. 1882.
Ball, W. P. *The effects of use and disuse*. 1890.
Bernard, H. B. *The Apodidae*. 1892.
De Varigny, H. *Experimental evolution*. 1892.
Lankester, E. R. *Degeneration*. 1880.
Lubbock, *Sir* John *Bart Origin and metamorphoses of insects*. 1874.
—— *British wild flowers in relation to insects*. 1875.
—— *Flowers, fruits and leaves*. 1886.
Mivart, St G. *The common frog*. 1874.
Romanes, G. J. *Organic evolution*. 1882.

2781. NAVE, Johann. *A handy-book to the collection and preparation of Algae, diatoms, desmids, Fungi, lichens, mosses, and other lower Cryptogamia*. Sm. 8vo, xii+203 pp, 26 pls, text figs, Hardwicke, London 1867; 3ed. [1882]; ed. 1904. Translation of *Anleitung zum Einsammeln, Prëpariren und Untersuchungen der Pflanzen*, Dresden [1864], by William Webb Spicer.

2782. NECKHAM, Alexander *De naturis rerum libri duo*. 8vo, lxxviii+521 pp, Longman, London 1863. Rolls Series No. 24.

2783. [NEEDHAM, John Turbeville] *An account of some new microscopical discoveries founded on an examination of the calamary, also observations on the farina foecundens of plants*. 8vo, viii+126 pp, 6 pls, F. Needham, London 1745; issue 1745 signed, title *Some new* . . . First issue signed T.N.

2784. —— *Observations upon generation, composition, and decomposition of animal and vegetable substances*. 4to, 53 pp, 1 pl., London 1749. Reprinted from *Phil. Trans*., Vol. 45.

2785. NEIL, James *Rays from the realm of nature; or, parables of plant life*. 8vo, 1 col. pl., text figs, Cassell, London [1879].

2786. NEILL, Patrick *Tour through some of the islands of Orkney and Shetland, with a view chiefly to objects of natural history*. 8vo, ix+239 pp, Constable, Edinburgh 1806. Reprinted from *The Scots Magazine*, 1804–05.

2787. [NELSON, John] *Pinaceae: being a handbook of the firs and pines*. 8vo, xix+223 pp, Hatchard for the author, London 1866. By Johannes Senilis *pseudonym*.

2788. NELSON, Thomas Hudson *The ornithology of the district around Redcar*. 8vo, 34 pp, Bishop Auckland 1877.

2789. *A new book of natural history of various British birds*. 12mo, 24 pp, 1 pl., text figs, Marsden, Chelmsford [c.1800]. Alternative title *Natural history of British birds*.

*New cyclopaedia of botany*. [1854]. *see* Brooke, R.

*A new general history of birds*. 1745. *see Ornithologia nova.*

2790. *A new moral system of natural history: or, the beauties of nature displayed . . . quadrupeds and birds*. 16mo, xii+224 pp, 1 pl., text figs, Riley, London 1789; 1790. Pocket Library Vol. 6. See also Riley, G.

2791. *The new museum of natural history*. 8vo, 100 pp, text figs, Oliver & Boyd, Edinburgh 1810; 1811; 1814.

2792. *A new natural history, intended as a present for good boys and girls*. 32mo, 16 pp, 1 pl., text figs, Walker, Otley [c.1820].

2793. *New statistical account of Scotland*. 8vo, 15 vols, Edinburgh 1834–45. By the ministers of the parishes.

2794. *A new system of natural history of quadrupeds, birds, fishes, and insects*. 8vo, 3 vols, 149 pls, Hill, Edinburgh 1791–92.

2795. NEWBERY, E. A. *An exchange list of the British Hemiptera-Heteroptera*. Sm. 8vo, Keys, Plymouth 1896. Printed on one or both sides. Based on Saunders, E. *Hempitera-Heteroptera*, 1892.

2796. NEWBEGIN, Marion Isabel *Colour in nature; a study in biology*. Sm. 8vo, xii+344 pp, text figs, Murray, London 1898.

Newcastle Museum. *see* Fox, G. T.

2797. NEWELL, Robert *The zoology of the English poets, corrected by the writings of modern naturalists*. Sm. 8vo, xvi+160 pp, text figs, Longman, London 1845; ed. Lingham, London [?1845] title *The modern naturalist*.

2798. NEWMAN, Edward *Sphinx vespiformis: an essay*. 8vo, 54 pp, 3 diagrams, Westley, London 1832; ed. [1842–]1843 title *The system of nature*.

2799. —— *The grammar of entomology*. 8vo, xvi+304 pp, 2 col.+2 plain pls, Westley, London 1835; ed. 1841 title *A familiar introduction to our history of insects.*

2800. —— *A history of British ferns*. 8vo, xxxiv+104 pp, text figs, Van Voorst, London 1840. See also No. 2802.

2801. —— *Notes on Irish natural history, more especially ferns*. 8vo, iv+64 pp, Van Voorst, London 1840.

—— *A familiar introduction to the history of insects.* 1841. see No. 2799.

—— *The system of nature*. [1842–]1843. see No. 2798.

2802. —— *A history of British ferns, and allied plants.* Roy, or demy 8vo, xxxii+424 pp, text figs, Van Voorst, London 1844; 2ed. 1851; 3ed. 1854; 4ed. [1865]; 5ed. 1882; condensed ed. 1880. Natural History of the British Islands Series. Considered by Newman to be distinct from No. 2800, though based on it.

2803. —— *A catalogue of British ferns*. 8vo, 24 pp, London [1845].

2804. [——] *A catalogue of British vertebrated animals, the names derived from Bell's Quadrupeds and Reptiles and Yarrell's British birds and Fishes*. 8vo, ii+77 pp, Van Voorst, London 1845.

2805. —— *editor The letters of Rusticus on the natural history of Godalming*. 8vo, viii+163 pp, text figs, Van Voorst, London 1849; ed. [ ]. Extracted from *Mag. nat. Hist.*, *Ent. Mag* and *Entomologist*.

2806. —— *A complete list of British ferns*. Fol., London 1854.

2807. —— *An essay on the employment of physiological characters in the classification of animals*. 8vo, 32 pp, London [?1856].

2808. [——] *The insect hunters; or, entomology in verse*. Sm. 8vo, viii+86 pp, Newman, London [1857]; 2ed. [1861] signed.

2809. —— *Birds-nesting: being a complete description of the nests and eggs of birds which breed in Great Britain and Ireland*. 8vo, 52 pp, Van Voorst, London 1862; ed. 1888 title *Birds-nesting and bird-skinning*. Reprinted from *The Zoologist*.

2810. —— *An illustrated natural history of British moths*. Roy. 8vo, viii+486 pp, text figs, Tweedie, London 1869; ed. 1874; 1884; [ ]. Originally published as 6 extra numbers to *Young England*.

2811. [——] *All the British butterflies arranged in their respective families*. Fol., s. sh., Tweedie, [London 1868.]. By E.N.

2812. —— *An illustrated natural history of British butterflies*. Roy. 8vo, xvi+176 pp, text figs, Tweedie, London 1871 [=1870]; ed. 1872; [ ].

—— *An illustrated natural history of British butterflies and moths*. Roy. 8vo, 2 vols, text figs, Glaisher, London [ ]. A continuation volume of Nos 2810 & 2812.

2813. NEWPORT, George *Insecta*. 8vo, 128 pp, text figs, Marchant, London 1839. Reprinted from Todd, R. B. *editor Cyclopaedia of anatomy and physiology*. 1835–59, pts 17 & 18.

2814. NEWTON, Alfred *Suggestions for forming collections of birds' eggs*. 8vo, text figs, Newman, London 1860. Enlarged reprint of a circular of The Smithsonian Institution of Washington.

2815. —— *Abstract of Mr. J. Wolley's researches in Iceland respecting the gare-fowl or great auk*. 8vo, 26 pp, Macmillan, London 1861. Reprinted from *Ibis*, Vol. 3.

2816. —— *On the zoology of ancient Europe*. 8vo, 31 pp, Macmillan, London 1862.

2817. —— *Ootheca Wolleyana; an illustrated catalogue of the collection of birds' eggs, formed by John Wolley*. 8vo, 2 vols, 21 col. + 16 tinted pls, 1 map, text figs, Porter, London 1864[–1907].

2818. —— *Zoology*. Sm. 8vo, 128 pp, 1 map, text figs, SPCK, London 1874; ed. 1891. Manuals of Elementary Science.

2819. —— *A dictionary of birds*. 8vo, xii+124+viii+1088 pp, 1 col. map, text figs, Black, London 1893–96; ed. 1896. Reprinted with additions from *Encyclopaedia Britannica*, 9ed. 1875–89.

2820. NEWTON, Edwin Tully *The typical parts in the skeleton of a cat, duck, and codfish.* 8vo, 21 pp, 1 pl., Chapman & Hall, London 1875.

2821. —— *Classification of animals: being a synopsis of the animal kingdom, with special reference to fossil forms*. 8vo, 15 pp, Philip, London 1887.

2822. NEWTON, James *Enchiridion universale plantarum; or, an universal and compleat history of plants*. 8vo, 40 pp, 15 pls, [?London ?1689]. All publ. No title leaf.

2823. NEWTON, James *A complet herbal, containing the prints and English names of several thousand trees, plants, shrubs, flowers*. 8vo, 19 pp, 176 pls, Cave, London 1752; 2ed. 1798; 6ed. 1802; ed. 1805; ed. 4to, Blackburn 1808 title *A new herbal.*

2824. NICHOLSON, Henry Alleyne *Advanced text book of zoology for the use of schools*. 8vo, xii+340 pp, text figs, Blackwood, Edinburgh 1870; 2ed. 1873; 5ed. 1894 title *A text book of zoology*. Blackwood's Class Books, Natural Science Series.

2825. —— *A manual of zoology for the use of students; with a general introduction on the principles of zoology*. 8vo, xvi+322 pp, text figs, Hardwicke, London 1870; ed. 2 pts xx+622 pp, Blackwood, Edinburgh 1870; 2ed. 1871; 3ed. 1873; 4ed. 1875; 5ed. 1878; 6ed. 1880; 7ed. 1887. The London ed. is called Vol. 1, and is the first part only.

2826. —— *An introductory text-book of zoology*. 8vo, Blackwood, Edinburgh 1871; ed. 1873; 2ed. 1875; 3ed. 1878; 6ed. 1887.

2827. —— *Introduction to the study of biology*. 8vo, x+163 pp, Blackwood, Edinburgh 1872.

2828. —— *Outlines of natural history for beginners*. 16mo, Blackwood, Edinburgh 1873.

2829. —— *The nature of life: being the introductory lecture to the course of natural history delivered in the United College, St Andrews*. 8vo, Edinburgh 1875.

2830. —— *Synopsis of the classification of the animal kingdom*. 8vo, vi+131 pp, text figs, Blackwood, Edinburgh 1882.

2831. —— *Natural history, its rise and progress in Britain*. 8vo, iv+312 pp, 4 pls, text figs, Chambers, London 1886.

2832. NICOLS, Arthur *The puzzle of life, and how it has been put together*. Sm. 8vo, text figs, Longman 1877; 2ed. 1877.

2833. —— *Zoological notes on the structure, affinities, habits, and mental faculties of wild and domestic animals*. 8vo, vii+370 pp, Upcott Gill, London 1883.
2834. —— *Natural history sketches among the Carnivora, wild and domestic*. Sm. 8vo, 247 pp, pls, Upcott Gill, London 1885.
2835. —— *Snakes, marsupials and birds: a book of anecdotes*. 8vo, vii+370 pp, text figs, Upcott Gill, London [1889].
2836. NIEUWENTIJDT, Bernard *The religious philosopher: or, the right use of contemplating the works of the creator*. 8vo, 3 vols, 1,100 pp, 29 pls, Senex, London 1718–19; 2ed. 1719–21; 2ed. 1 vol. 4to 1724; 3ed. 1730; 4ed. 1730. Translation of *Het regt Gebruik der werelt Beschouwingen*, Amsterdam 2ed. 1717, by John Chamberlayne.
2837. NISBET, John *Our forests and woodlands*. Sm. 8vo, pls, Dent, London 1900; ed. 1909. Haddon Hall Library.
2838. NITZSCH, Christian Ludwig *Nitzsch's Pterylography*. Imp. 4to, xii+118 pp, 10 pls, text figs, Ray Society, London 1867. Ray Society Publ. No. 41 for 1866. Translation of *System der Pterylographie*, Halle 1840, by William Sweetland Dallas.
2839. *The noble lyfe & natures of man of bestes, serpentys, fowles & fisshes yt be moste knowen*. Fol., 82 11, text figs, John of Doesborch, Antwerp [?1521]; facsimile 1954. Translation of *Der Dieren Palleys,* Antwerp 1520 which is an abridged version of the Tractatus de animalibus in *Hortus sanitatis* (editio princeps Mainz 1491.), by Lawrence Andrewes.
2840. NOBLE, Heatley *List of European birds*. Sm. 8vo, 66 pp, London 1893.

Nodder, Frederick P. *see* Shaw, G.

Nodder, Richard P. *see* Shaw, G.

Noll, Fritz *see* Strasburger, E.

2841. NORMAN, Alfred Mearle *Museum Normanianum, or a catalogue of the Invertebrata of Europe.* 8vo, 7 numbers, for the author, Houghton-le-Spring 1886–1900; 2ed. Nos 1–3, Durham 1901, 05.

—— *see* Bowerbank, J. S.

2842. —— & BRADY, George Stewardson *A monograph of the British Entomostraca, belonging to the families Bosminidae, Macrothricidae, and Lynceidae*. 8vo, 57 pp, 6 pls, Williams & Norgate, London 1867. Reprinted from *Nat. Hist. Trans. Northumb, Durham*, Vol. 1.
2843. NORMAN, George *List of Diatomaceae occurring in the neighbourhood of Hull*. 8vo, 23 pp, [Hull 1859]; 2ed. [1865].

2844. NORTH, Isaac William *A week on the Isles of Scilly*. 8vo, ix+199 pp, 4 pls, Penzance 1850.

Norwell, J. Stewart *see* Smith, W. R.

2845. NOTCUTT, W. Lowndes *A handbook of the microscope and microscopic objects*. 12mo, 152 pp, 9 col.+3 plain pls, Lumley, London 1859.

2846. —— *Handbook of British plants*. 8vo, 213 pp, Journal of Horticulture Office, London 1865; ed. 1870.

2847. *Notes of lessons on animals and common things*. 8vo, 118 pp, Stewart, London 1882. Educational Series.

*Notes explanatory of the View of nature in ascending regions*. 1853. *see* Emslie, J.

2848. *Notes on birds, original and selected*. 16mo, Whittaker, London 1868.

2849. *Notes on wild flowers*. 8vo, 520 pp, Chapman & Hall, London 1864.

2850. NOTT, John Fortune *Wild animals photographed and described*. Roy. 8vo, xi+568 pp, 38 pls, Sampson Low, London 1886.

Nowell, John *see* Baines, H.

2851. [NOYCE, Elisha] *The marvels of nature, or outlines of creation*. 8vo, xii+338 pp, 1 pl., text figs, Ward & Lock, London 1858; [1862]. By the author of *The boy's book of industrial information*.

# O

O. *The ferns of Moffat*. 1863. *see* Moffatt, J. C.

O., J. C. *Guide to Newquay*. 1872. *see* Oliver, J. C.

2852. O'BRIEN, Charlotte & PARKINSON, C. *Wild flowers of the Undercliff, Isle of Wight*. Sm. 8vo, vi+143 pp, 8 col. pls, Reeve, London 1881.

*Observations on the brumal retreat of the swallow*. 1808. *see* Forster, T. I. M.

*The observing eye; or letters to children on the three lowest divisions of animal life*. [1851–52]. *see* Wright, *Mrs* Anne.

2853. *The ocean; a description of wonders and important products of the sea*. 16mo, 180 pp, 4 pls, text figs, Harris, London 1833; 2ed. 1835. Little Library, Vol. 12.

*Ocean flowers and their teachings*. 1846. *see* Howard, M. M.

Ochsenheimer, Ferdinand *see* Children, J. G.

*Odd stories about animals*. 1871. *see* Burrows, E.

2854. OGILVIE, George *The master builder's plan; or, the principles of organic architecture as indicated in the typical forms of animals*. 8vo, xvi+196 pp, 3 pls, text figs, Longman, London 1858.

2855. —— *The genetic cycle in organic nature; or, the succession of forms in the propagation of plants and animals*. 8vo, xv+296 pp, 6 pls. Brown, Aberdeen 1861.

Ogilvie-Grant, William Robert *see* Grant, W. R. O.

Oldham, Charles *see* Coward, T. A.

2856. OLIVER, Daniel *Lessons in elementary botany: the part of systematic botany . . . by the late Professor Henslow*. Sm. 8vo, viii+317 pp, text figs, Macmillan, Cambridge 1864; 2ed. 1869. Many later editions.

2857. —— *Illustrations of the principal natural orders of the vegetable kingdom*. Obl. 4to, 152 pp, 109 col. pls, Chapman, London 1874; ed. 1893.

2858. [OLIVER, John Cardell] *Guide to Newquay*. 8vo, 42 pp, Lake, Truro 1872; 2ed. 1877; 3ed. Newquay 1879 title . . . *together with notes on its geology . . . flora, fauna*; ed. 1884.

2859. OLIVER, Joseph W. *Synopsis of the natural orders of British flowering plants*. 16mo, 18 pp, Simpkin Marshall, London [1886].

2860. —— *Elementary botany*. 8vo, 208 pp, Blackie, London 1891.

2861. —— *The student's introductory handbook of systematic botany*. 8vo, xvi+366 pp, text figs, Blackie, London 1894; 2ed. 1896; 3ed. 1898; 4ed. 1903. Later editions. Science Text-books Series.

*On Surrey hills*. 2ed. 1892. *see* Jordan, D.

2862. ORD, William Miller *Notes on comparative anatomy: a syllabus of a course of lectures delivered at St. Thomas's Hospital*. Sm. 8vo, Churchill, London 1871.

2863. ORDOYNO, Thomas *Flora Nottinghamiensis; or, a systematic arrangement of the plants growing naturally in the County of Nottingham*. Sm. 8vo, 7+v+344 pp, Ridge for the author, Newark 1807.

2864. O'REILLY, Bernard *A catalogue of the subjects of natural history in the Museum of the . . . Dublin Society*. 8vo, 130 pp, Dublin 1813.

2865. *Original anecdotes of British quadrupeds*. 12mo, viii+130 pp, pls, for the author, London 1842. By the author of *Poetic sketches*.

2866. ORMEROD, Edward Latham *British social wasps*. Sm. 8vo, xi+270 pp, 4 col. + 10 plain pls, Longman, London 1868.

*Ornithologia Britannica*. 1771. *see* Tunstall, M.

2867. *Ornithologia nova: or, a new general history of birds*. 16mo, 2 vols, text figs, Warren printed, Birmingham 1743; ed. 1744; ed. 1745 without latin title.

2868. ORR, William Somerville *editor Orr's Circle of the sciences*. 8vo, 9 vols, pls, text figs, Orr, London 1854–56; ed. 1854–60. In the first ed. Vols I, II & VI are organic nature; in 2ed. Vols I, II & III. Vols II & III have additional title pages *A system of natural history . . . in two volumes*. For parts published separately see Ansted, D. T. *et al*., Dallas, W. S., Owen, R. & Smith, E.

2869. OSWALD, Felix Leopold *Zoological sketches; a contribution to the outdoor study of natural history*. 8vo, 266 pp, 1 pl., text figs, Allen, London 1883 [=1882]. English edition of USA book, Philadelphia 1883.

2870. OTTÉ, Joachim *Twelve British ferns*. 8vo, 12 specimens with letterpress, Hardwicke, London 1859. Also available as a portfolio without printing.

2871. OUEDEMANS, Antoine Corneille *The great sea-serpent*. Roy. 8vo, xv+591 pp, text figs, Luzac for the author, London 1892.

2872. *Our animal friends*. 8vo, 224 pp, pls, Chambers, London 1888.

*Our song birds*. [1846]. *see* Martin, W. C. L.

2873. *Outlines of botany, first sketched for the use of his nieces*. 12mo, London 1827.

2874. *Outlines of natural history. Quadrupeds embossed for the use of the blind*. 4to, Glasgow 1842.

Outlines of the natural history of Europe Series. *see*
Forbes, E. 1859.
Henfrey, A. 1852.

2875. OWEN, Charles *An essay towards a natural history of serpents*. 4to, xxiii+240 pp, 7 pls, for the author, London 1742.

2876. OWEN, James *The stepping stone to natural history. Verte-*

*brate or backboned animals*. Sm. 8vo, vii+191 pp, 12 pls, Longman, London 1856; ed. 1857.

2877. —— *Natural history for beginners*. 18mo, 2 pts, Longman, London 1859. Gleig's School Series.

2878. OWEN, Jean Allan, *Mrs* Owen Visger *Forest, field and fell*. Sm. 8vo, 221 pp, Lawrence & Bullen, London 1893.

—— *see* A son of the marshes *pseudonym*.

2879. —— & BOULGER, George Simonds *The country month by month*. Sm. 8vo, 4 vols, Bliss, London 1894–95.

2880. OWEN, *Sir* Richard *Odontography: or, a treatise on the comparative anatomy of the teeth*. 4to, or Roy. 8vo, xix+lxxiv+665+37 pp, atlas vol. 168 pls, Baillière, London 1840–45.

2881. —— *Lectures on comparative anatomy and physiology of the invertebrate animals*. 8vo, 392 pp, text figs, Longman, London 1843; 2ed. 1855.

2882. [——] *Synopsis of the arrangement of the preparations in the Museum*. 8vo, 84 pp, Royal College of Surgeons, London 1845.

2883. —— *Lectures on the comparative anatomy and physiology of the vertebrate animals. Part 1. Fishes*. 8vo, x+308 pp, text figs, Longman, London 1846. All publ.

2884. —— *On the archetype and homologies of the vertebrtate skeleton*. 8vo, viii+203 pp, 6 pls, 3 tabs, text figs, Van Voorst for the author, London 1848. Reprinted from *Rep. Brit. Assoc. Adv. Sci*. for 1846, 47.

2885. —— *On parthenogenesis; or, the successive production of procreating individuals from a single ovum*. 8vo, 76 pp, 1 pl., Van Voorst, London 1849.

2886. —— *On the nature of limbs*. 8vo, 119 pp, 3 pls, text figs, Van Voorst, London 1849.

2887. —— *Descriptive catalogue of the osteological series*. 4to, 2 vols, xlv+914 pp, Royal College of Surgeons, London 1853.

2888. —— *The principal forms of the skeleton and the teeth as a basis for a system of natural history and comparative anatomy*. Sm. 8vo, text figs, Houlston & Stoneman, London 1856. Reprinted from Orr's *Circle of the sciences*, Vol. 1, pp. 161–304.

2889. —— *On the classification and geographical distribution of the Mammalia*. 8vo, iv+103 pp, text figs, Parker, London 1859. Rede Lecture, Cambridge.

2890. —— *On the extent and aims of a national Museum of Natural*

*History*. 8vo, 126 pp, 2 plans, Saunders, London 1862; 2ed. 1862.

2891. —— *Instances of the power of God as manifested in his animal creation: a lecture*. 8vo, 35 pp, 2 pls, text figs, Young Men's Christian Association, London 1863; ed. 1864.

2892. —— *On the anatomy of vertebrates*. 8vo, 3 vols, text figs, Longman, London 1866–68.

2893. —— *Derivative hypothesis of life and species*. 8vo, 40 pp, Longman, London 1868. Reprinted from No. 2892.

2894. —— *Essays on the conario-hypophysial tract, and on the aspects of the body in vertebrates and invertebrates*. 8vo, 48 pp, Taylor & Francis, London 1883.

2895. OXFORD AND CAMBRIDGE UNIVERSITIES ENTOMOLOGICAL SOCIETIES *An accentuated list of the British Lepidoptera*. 8vo, xliv+118 pp, by the Societies, London 1858.

# P

P., J. A. *A guide to the Mount's Bay and the Land's End*. 1816. *see* Paris, J. A.

2896. P., J. P. *The microscope and its revelations; a lecture*. 8vo, 50 pp, Solva [Glam.] 1859.

P., R. *The natural history of Oxfordshire*. 1677. *see* Plot, R.

P., W. & I., A. *A botanical tour in the Highlands of Perthshire*. 1857. *see* Pamplin, W. & Irvine, A.

2897. PACKER, J. J. *British mosses*. Fol., Thirsk 1855.

2898. PADDOCK, George Harrie *Notes on some of the birds found in the neighbourhood of Newport, Salop*. 8vo, 42 pp, Wellington 1890.

2899. —— *Catalogue of Shropshire birds*. 8vo, 104 pp, Wellington 1897.

2900. PAGE, David *The past and present life of the globe*. 8vo, 256 pp, text figs, Blackwood, Edinburgh 1861.

2901. PAGE, Leigh *Stars of the earth; or, wild flowers of the months*. 8vo, 256 pp, 1 pl., text figs, Johnstone Hunter, Edinburgh 1868 [=1867].

2902. PAGET, Charles John & PAGET, *Sir* James *Bart Sketch of the natural history of Yarmouth and its neighbourhood*. 8vo, xxxii+68 pp, Skill, Yarmouth 1834; ed. London 1834.

Paget, *Sir* James *Bart see* Paget, C. J.

2903. PAINTER, William Hunt *A contribution to the flora of Derbyshire.* 8vo, 156 pp, 1 map, Bell, London 1889; suppl. [112] pp, Leeds 1902. Suppl. reprinted from *The Naturalist*, 1899 & 1902, with original pagination.

2904. [PALEY, Frederick Apthorp] *The wild flowers of Dover*. 4to, xii+76 pp. Dover [?1855].

2905. —— *A few words on wheat-ears. A lecture*. 8vo, London 1859; ed. Cambridge 1869.

2906. —— *A list of four hundred wild flowering plants ... of Peterborough.* 16mo, xxviii+26pp, Gardner, Peterborough 1860.

2907. PALEY, William *Natural theology; or, evidences of the existence and attributes of the Deity.* 8vo, Faulder, London 1802; 2ed. 1802; 3ed. 1803; 4ed. 1803; 5ed. 1803; 6ed. 1803; 7ed. 1804; 11ed. 1807; 12ed. 1809; 13ed. 1811; 14ed. 1813; ed. 12mo Alnwick 1818; 16ed. 1819; 19ed. 1819; 20ed. 1820; ed. 1822; ed. 1822 title *Natural theology and tracts*; ed. 2 vols 1826; ed. 2 vols, Oxford 1828; ed. 1830; ed. 5 vols 1835–39; ed. 1845; ed. 1848; ed. Edinburgh 1849; ed. [1875]; ed. [1879]; ed. 1880 title ... *revised to harmonize with modern science*; abridged ed. 1820. In the 5 vol. ed. the first two are by Lord Brougham. Also in *Collected Works*.

—— *see* Hack, *Mrs* Maria.

—— *see* Kidd, J.

—— *see* Paxton, J.

2908. PALMER, Shirley *Popular lectures on vertebrated animals of the British Islands. Part first, on the British Mammifera*. 8vo, x+96 pp, 1 tab, Longman, London 1832. All publ. There is probably an ed. Wrightson, Birmingham 1831.

2909. PAMPLIN, WIlliam *A catalogue of the rarer species on indigenous plants of Battersea and Clapham*. 12mo, 17 pp, Clapham 1827.

2910. [—— & IRVINE, Alexander] *A botanical tour in the Highlands of Perthshire.* 8vo, viii+76 pp, London 1857. By W. P. & A. I. Reprinted from *The Phytologist.*

2911. PANTON, Jane Ellen *Country sketches in black and white*. 8vo, x+313 pp. Bogue, London [1882].

2912. PARDON, George Frederick *editor Stories about animals.*

*(Birds)*. Sm. 8vo, 2 vols, pls, text figs, Blackwood, London [1858].

2913. PARFITT, Edward *The fauna of Devon*. 8vo, Plymouth 1866–91. 21 offprints with original pagination but new general title.

2914. [PARIS, John Ayrton] *A guide to the Mount's Bay and the Land's End; being a sketch of the topography, natural history*. 12mo, ix+156 pp, Vigurs, Penzance 1816; 2ed. 1824; ed. 1828. First ed. by J. A. P., 2ed. by Physician.

2915. PARK, Carton Moore *A book of birds*. 4to, 26 pls with letterpress, Blackie, London 1900.

2916. PARK, John James *The topography and natural history of Hampstead, in the County of Middlesex*. 8vo, xxi+359+xxxix pp, 1 pl., 1 map, White Cochrane, London 1814; ed. 1818.

2917. PARKER, Thomas Jeffery *A course of instruction in zootomy (Vertebrata)*. Sm. 8vo, xxiii+397 pp, text figs, Macmillan, London 1884. Manuals for Students.

2918. —— *Lessons in elementary biology*. 8vo, xxii+408 pp, text figs, Macmillan, London 1891; 2ed. 1893; 3ed. 1897; 1898; 1900; 1903. Manuals for Students.

2919. —— & HASWELL, William Aitcheson *A text-book of zoology*. Roy. 8vo, 2 vols, text figs, Macmillan, London 1897; 2ed. 1910; 3ed. 1921; 4ed. 1928; 5ed. 1930; 6ed. 1940; 7ed. Vol. 1, 1962.

2920. —— & —— *A manual of zoology*. 8vo, xxv+563 pp, text figs some col., Macmillan, London 1900; ed. 1918; ed. 1937.

2921. —— & PARKER, William Newton *An elementary course in practical zoology*. 8vo, xii+608 pp, text figs, Macmillan, London 1900; 2ed. 1908; 3ed. 1911; 1920.

2922. PARKER, William Kitchen *A monograph on the structure and development of the shoulder-girdle and sternum in the Vertebrata*. Imp. 4to, xii+239 pp, 30 col. pls, text figs, Ray Society, London 1868. Ray Society Publ. No. 42 for 1867.

2923. —— *On mammalian descent*. 8vo, xii+229 pp, text figs, Griffin, London 1885. Hunterian Lecture.

2924. —— & BETTANY, George Thomas *The morphology of the skull*. 8vo, xv+368 pp, text figs, Macmillan, London 1877.

Parker, William Newton *see* Parker, T. J.

Parkinson, C. *see* O'Brien, C.

2925. PARKINSON, John *Theatrum botanicum; the theatre of plants: or, an herball of a large extent*. Fol., 1755 pp, text figs, Cotes printed, London 1640.

Parley, Peter *pseudonym Tales about animals.* 1832. *see* Goodrich, S. G.

Parley, Peter *pseudonym Tales about plants.* 1839. *see* Loudon, *Mrs* Jane

*The parlour menagerie.* 1875. *see* Hogg, J.

2926. PARNELL, Richard *The natural and economical history of the fishes of the river district of the Firth of Forth.* 8vo, 360 pp, 38 pls, Blackwood, Edinburgh [?1837]. Reprinted from *Mem. Wernerian Soc.* Vol. 7.

2927. —— *The grasses of Scotland. (of Britain).* Roy. 8vo, 2 pts, xxvii+xxi+310 pp, 142 pls, Blackwood, Edinburgh 1842, 45.

2928. PARRY, Richard *The Llandudno visitor's handbook.* 8vo, vi+107 pp, Williams, Llandudno 1855; 2ed. 1861 title *Llandudno: its history and natural history.*

2929. PARSONS, James *Microscopic theatre of seeds. Vol. 1.* 4to, 348 pp, 8 pls, Needham, London [1744–] 1745. All publ.

2930. —— *Philosophical observations on the analogy between the propagation of animals and that of vegetables.* 8vo, xvi+276 pp, 1 pl., Davis, London 1752.

2931. PARTINGTON, Charles Frederick *editor The British cyclopaedia of natural history.* 8vo, 3 vols, 50 col. pls, Orr & Smith, London [1834] 1835–37.; issue Vols 1 & 2 1842. Forms part of *The British Cyclopaedia of arts and sciences*, 10 vols, 1835–38.

2932. —— *Introduction to the science of botany.* 12mo, vii+144 pp, 1 col. pl., Harvey, London 1835. Portfolios of either 9 small or 14 large col. pls were available with the work.

2933. PASCOE, Francis Polkinghorne *Zoological classification: a handy book of reference.* 8vo, iv+204 pp, Van Voorst, London 1877; 2ed. 1880.

2934. —— *Hints for collecting and preserving insects.* 8vo, 3 pp, London 1882.

2935. —— *The student's list of British Coleoptera.* 8vo, viii+120 pp, Taylor & Francis, London 1882.

2936. —— *Notes on natural selection and the origin of species.* 12mo, 20 pp, Taylor & Francis, London 1884.

2937. —— *List of British vertebrate animals.* 12mo, xi+85 pp, Taylor & Francis, London 1885.

2938. —— *Analytical lists of the classes, orders etc of the animal kingdom.* Sm. 8vo, vi+73 pp, Taylor & Francis, London 1886.

2939. —— *The darwinian theory of the origin of species.* 8vo, 130 pp, Gurney & Jackson, London 1890.

2940. —— *A summary of the darwinian theory of the origin of species.* 8vo, 16 pp, for private distribution, London 1891.

2941. PATISON, Jane M. *Gleanings among the British ferns.* 8vo, 69 pp, 16 specimens, London 1858.

2942. PATRICK, William *A popular description of the indigenous plants of Lanarkshire, with an introduction to botany.* 12mo, xxiv+397 pp, Edinburgh 1831; 2ed. 1832.

2943. PATTERSON, Robert *Letters on the natural history of insects mentioned in Shakespeare's plays. With incidental notes on the entomology of Ireland.* 8vo, xv+270 pp, text figs, Orr, London 1838; ed. 1841 title *The natural history of insects mentioned in Shakespeare's plays*; 1842.

2944. —— *On the study of natural history as a branch of general education in schools and colleges.* 8vo, 28 pp, Belfast 1840; ed. London 1847.

2945. —— *First steps in zoology.* 8vo, text figs, London 1848; 2ed. 1850; 3ed. Belfast 1860; 4ed. 1863.

2946. —— *Tabular view of the classification of animals. Invertebrata. (Vertebrata)*. Fol., 2 sheets, London 1848; 2ed. 1849; 3ed. 1851.

2947. —— *Introduction to zoology, for the use of schools.* 8vo, 2 pts, text figs, Simms, London & Belfast 1845, 48; 2ed. pt. 1 1849; 3ed. pt. 1 1849 [=1850]; 2ed. pts 2 1849 [=1850]. Later editions. Copies of both parts were printed without the figures, for perusal of his friends, before publication.

2948. —— *Animals; how they are classified.* 8vo, 50 pp, Day, London 1857.

2949. PATTERSON, *Sir* Robert Lloyd *The swimming birds of Belfast Lough.* 8vo, [88] pp, privately printed, Belfast 1876. Reprinted, with old pagination, from *Rep. Belfast nat. Hist. phil. Soc.*, 1874–76.

2950. —— *On the birds, fishes, and Cetacea commonly frequenting Belfast Lough.* 8vo, vi+267 pp, 1 col. chart, Bogue, London 1880; 2ed. 1881.

2951. PAXTON, James *Illustrations of Paley's Natural theology.* 8vo, 36 pls with letterpress, Vincent, Oxford 1826. Was also published with Duncan, J. S. *Botanical theology*, 1826 with title *Natural theology . . . collected from the appearances of nature.*

2952. PAXTON, *Sir* Joseph & LINDLEY, John *A pocket*

*botanical dictionary*. Sm. 8vo, xii+354 pp, Bradbury & Evans, London 1840; 2ed. with suppl. 72 pp, 1849; ed. 1853; ed. 1869.

Payne-Gallwey, *Sir* Ralph William Frankland *Bart see* Gallwey, *Sir* Ralph W. F. P. *Bart*

2953. [PEACOCK, Lucy] *The visit for a week; or hints on the improvement of time, containing original tales, anecdotes of natural and moral history*. 12mo, 328 pp, Hookham & Carpenter, London 1794; 2ed. 1795; 1801; 4ed. 1804; 6ed. 1806; 1814; 8ed. 1815; 9ed. 1818; 10ed. 1823; 11ed. 1838; 12ed. 1841.

2954. PEARCE, F. L. *The illustrated universal natural history: First division – mammalians*. 4to, xviii+222 pp, 18 col. pls, Protestant Book Association [c.1870]. ?All publ.

2955. PEARSON, William Henry *Hepaticae of the British Isles*. 8vo, 2 vols, vi+vii+520 pp, atlas vol. 228 pls, London [1890–] 1902.

2956. PEASE, *Sir* William Alfred *Bart The badger*. 8vo, 119 pp, 1 pl., text figs, Lawrence & Bullen, London 1898.

2957. PEEK, *Sir* Henry William *Bart A collection of birds at Rousden*. 4to, [29] pp, 6 pls, London [1888].

2958. PENA, Pierre & L'OBEL, Matthias *Stirpium adversaria nova, prefacilis vestigatio luculentaque accessio ad Priscorum presertim Dioscoridis & recentiorum materiam medicam*. Fol., 455 pp, text figs, Purfoot, London 1570. Colophon 1571.

2959. [PENNANT, Thomas] *British zoology*. Fol., 162 pp, 132 col. pls, Cymmrodorion Society, London 1761–66; 2ed. 8vo, 3 vols, signed, London & Chester 1768–69; additional volume Chester 1770; 4ed. 4to or 8vo, 4 vols, Warrington & London 1776–77; ed. Roy. 8vo, 4 vols 1812; ed. 8vo, 4 vols, Dublin 1818–21. The 3 vols of 2ed. with the additional vol. is treated as 3ed.

2960. —— *Synopsis of quadrupeds*. 8vo, xxv+382 pp, 32 pls, text figs, Monk printed, Chester 1771; ed. 4to, 2 vols, London 1781 title *History of quadrupeds*; 3ed. 4to, 2 vols 1793. Plates I–XXXI, XIIIa.

2961. —— *A tour in Scotland, 1769*. 8vo, viii+316 pp, 18 pls, for the author, Chester 1771; 2ed. London 1772; suppl. 18 pp, Chester 1772; 3ed. 4to, Warrington 1774; suppl. 8vo, London 1774; 4ed. 2 vols, Dublin 1775; 4ed. 4to, London 1776; 5ed. 4to, London 1790.

2962. —— *Genera of birds.* 8vo, xxxi+66 pp, Balfour & Smellie printed, Edinburgh 1773; ed. 4to, 16 pls, London 1781.
—— *Sketch of Caledonian zoology.* 1777. *see* Lightfoot, J.
—— *History of quadrupeds.* 1781. *see* No. 2960.
2963. —— *Arctic zoology.* Roy. or demy 4to, 2 vols, 21 pls, 2 maps, text figs, Hughs printed, London 1784–85; suppl. viii+162 pp, 1787; 2ed. 3 vols, 1793. L.P. sometimes coloured.
2964. —— *Histoire naturelle des oiseaux par le Conte de Buffon and Les planches enluminees, systematically disposed.* 4to, vii-+112 pp, White, London 1786.
2965. PENNELL, Harry Cholmondeley *The angler-naturalist.* 8vo, viii+425 pp, text figs Van Voorst, London 1863; ed. 1875; ed. [1884] title *A popular history of British fresh-water fish . . . or, the angler-naturalist.*
2966. —— *Can fish feel pain?* 8vo, 14 pp, Warne, London 1870.
—— *A popular history of British fresh-water fish.* [1844]. *see* No. 2965.
2967. —— *The sporting fish of Great Britain with notes on ichthyology.* Roy. or demy 8vo, 185 pp, 16 col. pls, text figs, Sampson Low, London 1886. L.P. 150 copies.
2968. PENNINGTON, Arthur S. *British zoophytes.* Sm. 8vo, xvi+363 pp, 24 pls, text figs, Reeve, London 1885. Crown Series.
2969. PERKINS, *Mrs* Elizabeth Steele *The elements of botany.* 8vo, xxiv+268 pp, 18 col. pls, Hurst, London 1837.
2970. PERRAULT, Claude *The natural history of animals containing the anatomical descriptions of several creatures dissected by the Royal Academy of Sciences at Paris.* Fol., 267+40 pp, 35 pls, J. Streater for T. Basset, London 1688; issue 1701; 1702. Translation of *Memoires pour servir a l'histoire naturelle des animeaux*, Paris 1671. Translated by Robert Pitfield and often listed under his name.

Perrott, *Mrs* Charlotte Louisa Emily *see* Havell, R.

2971. PERRY, George *Arcana; or, the museum of natural history.* 4to, 84 col. pls with letterpress, Miller, London [1810–] 1811.
2972. —— *Conchology, or the natural history of shells.* Fol., 61 col. pls with letterpress, Miller, London 1811; ed. [c.1825].
2973. PERRY, William Groves *Plantae Varvicenses selectae; or, botanist's guide through the County of Warwick.* 8vo, 120 pp, Warwick 1820.

A person of learning and piety. *see* Morton, C.

2974. PERSON, Claude *Elements of anatomy and animal oeconomy.* 8vo, xv+396 pp, 3 pls, Wilkie, London 1775; 2ed. 1781. Translation of *Nouveaux elemens d'anatomie*, Paris 1749 with additions by Simmons, Samuel Foart.

2975. PETIVER, James *Brief directions for . . . making . . . collections of all natural curiosities.* Fol., s.sh., [London ?1709].

2976. —— *Herbarii Britannici . . . Raii Catalogus cum iconibus . . . A catalogue of Mr. Ray's English herbal.* Fol., 8 engr. sheets, 50 pls, [London 1713]; 2 engr. sheets, pls 51–72 [London 1715]; ed. 1732.

2977. —— *Botanicum Anglicum, or, the English herbal.* Fol., 7 sheets, [London ?1713]. Printed on one side.

2978. —— *Graminum, muscorum, fungorum, submarinorum . . . Britannicorum concordia.* Fol., 12 pp, [London 1716].

2979. —— *Papilionum Britanniae icones . . . containing the figures, names, places, seasons etc. of above eighty English butterflies.* Fol., 2 pp, 6 pls, for the author, London 1716.

2980. —— *J. Petiveri Opera, historiam naturalem spectantia.* Fol. & 8vo, 3 vols, pls, Millan, London 1741; ed. fol. 2 vols, 1767. Vols 1 & 2 are fol., Vol. 3 8vo. These collected works contain bound up papers originally published in serials or separately from 1693. Bibliography is complex.

2981. PETTIGREW, James Bell *Animal locomotion: or, walking, swimming and flying.* Sm. 8vo, xii+264 pp, 3 pls, text figs, King, London 1873; 2ed. 1874; 3ed. 1883; 4ed. 1891. International Scientific Series Vol. VII.

*Peziza cuticulosa.* [1792]. *see* Forster, B. M.

2982. [PHELAN, *Mrs* Charlotte Elizabeth *later* Tonna] *Kindness to animals.* 12mo, 108 pp, 1 pl., text figs, RTS, London [1844]. Later editions. By Charlotte Elizabeth.

2983. [——] *Chapters on flowers.* 8vo, 303 pp, Seeley & Burnside, London 1836. Many later editions. By Charlotte Elizabeth.

2984. PHELPS, William *Calendarium botanicum; or, a botanical calendar.* Roy. or demy. 8vo, xiii+186 pp, 5 pls, Lackington, London 1810.

Philalethes, Eugenius *pseudonym see* Vaughan, Thomas

2985. PHILLIPS, Edwards Cambridge *The birds of Breconshire.* 8vo, London 1882; ed. xxi+158 pp, 2 pls, Brecon 1899. First edition is a reprint of a series of papers in The *Zoologist*, 1881–82.

2986. PHILLIPS, John *Life on earth: its origin and succession.* 8vo, x+224 pp, 1 col. pl., text figs, Macmillan, Cambridge 1860.

2987. PHILLIPS, William *Filices, Lycopodiaceae . . . of Shropshire.* 8vo, privately printed, Shrewsbury 1877.

2988. —— *A manual of British Discomycetes.* Sm. 8vo, xii+462 pp, 12 pls, Kegan Paul, London 1887; 2ed. 1893. International Scientific Series, Vol. LXI.

2989. —— *The Hymenomycetes of Shropshire.* 8vo, 40 pp, 2 pls, Shrewsbury [1880]. Reprinted from *Trans. Shropshire. archaeolog. nat Hist. Soc.*, Vol. 7.

2990. —— editor *Guide to the botany, ornithology and geology of Shrewsbury.* 8vo, 65 pp, Bunny & Davies, Shrewsbury 1878.

Philochelidon *pseudonym see* Forster, T. I. M.

*A philosophical survey of animal creation.* 1768. *see* Bruckner, J.

2991. [PHILIP, Robert Kemp *compiler*] *The reason why.* Sm. 8vo, xxvi+320 pp, text figs, Houlston & Wright, London [1856]. Later editions.

2992. [——] *The reason why . . . general science.* Sm. 8vo, xxvi+320 pp, text figs, Houlston & Wright, London 1857. Later editions.

2993. [——] *The reason why . . . natural history.* Sm. 8vo, xxiv+356 pp, text figs, Houlston & Wright, London 1860.

2994. PHILPOTS, John Richards *Oysters and all about them.* 8vo, 2 vols, 1370 pp, text figs, Richardson, London 1890–91.

2995. PHIPSON, Emma *Animal-lore in Shakespeare's time.* 8vo, xvi+476 pp, 1 pl., Kegan Paul, London 1883.

2996. PHIPSON, Thomas Lamb *Phosphorescence or, the emission of light by minerals, plants, and animals.* Sm. 8vo, xv+210 pp, 1 col. pl., text figs, Reeve, London 1862; 1870. Rewritten from *J. Med. Pharmacol.*, 1858.

2997. —— *The utilization of minute life.* Sm. 8vo, xix+282 pp, 9 pls, text figs, Groombridge, London 1864.

*Phocaena, or the anatomy of a porpess.* 1680. *see* Tyson, E.

*Physiology of animal and vegetable life.* 1854. *see* Bushnan, J. S.

*The physiology of plants; or the phenomena and laws of vegetation.* 1833. *see* Murray, J.

*Phytologia Britannica.* 1650. *see* How, W.

Pickard Cambridge, Octavius *see* Cambridge, O. P.

2998. *Pictorial beauties of nature; or, sketches in various departments of natural history.* 4to, 92 pp, col. pls, Ward Lock, London [1873].

2999. *The pictorial museum of animated nature.* 4to, 2 vols, 2 col. pls, text figs, Knight, London [1843–45].

3000. *Pictorial natural history.* 8vo, 5 pts, 35 col. pls with letterpress, Harvey & Darton, London [c.1845]. Pictorial Museum of Entertainment.

*Picture book for children, containing upwards of sixty of the most remarkable birds and beasts.* 1834. *see* Harrison, J.

3001. *A picture book of animals.* 12mo, 12 pp, text figs, Walker Otley [c.1825].

3002. *A picture book of animals.* 16mo, London [1876].

3003. *A picture book of birds.* 12mo, 12 pp, text figs, Walker, Otley [c.1825].

3004. *A picture book of natural history for young people, Birds. (Quadrupeds).* 4to, 2 vols, pls, Low, London 1858.

3005. *Picture-book of natural history in three languages.* Obl. 4to, pls, Myers, London [1869]. German, French, English.

3006. *A picture-book of wild animals. With illustrations in colour and plain.* Fol., 64 pp, col. pls, Routledge, London [1885].

3007. *Picture-books for children. Animals.* 12mo, pls, SPCK, [London 1859].

3008. *Picture lessons in natural history.* Fol., 24 col. pls with letterpress, Bacon, London [1878–79].

3009. *Pictures and stories about animals.* 32mo, text figs, Grout, Worcester [c.1840].

3010. *Pictures and stories of natural history.* 8vo, 3 pts, London 1871 [=1870].

3011. *Pictures and stories of natural history. Series 1. (Series 2).* 4to, 2 ser., London 1875.

3012. *Pictures of animal and vegetable life in all lands.* Fol., 48 col. pls with letterpress, Stanford, London 2ed. [c.1880].

3013. PIDSLEY, William E. Helman *Birds of Devonshire.* 8vo, 30+xx+xxx+194 pp, 1 col. pl., 1 map, Gibbings, London 1891.

3014. —— *Collecting and preserving birds (Waterton's method).* 8vo, 8 pp, Naturalist's Publishing Co., Birmingham [1891].

3015. PIERS, Laetitia E. C. *Leaves from the book of nature: or, stepping stones in creation.* 8vo, 17 pp, 1 tab., Ridgway, London 1892.

3016. PIGOTT, B. A. F. *Flowers and ferns of Cromer and its neighbourhood.* 8vo, 99 pp, London [1885].

3017. PIGGOTT, *Sir* Thomas Digby *London birds and London insects.* 8vo, vi+88 pp, text figs, Harrison, London [1884]; ed. 1892; ed. 1902.

3018. PIKE, Oliver Gregory *In birdland with field-glasses and camera.* Sm. 8vo, xvi+210 pp, pls, Unwin, London 1900; ed. 1902.

3019. PILKINGTON, James *A view of the present state of Derbyshire.* 8vo, 2 vols, 3 pls, 1 map, Drewry, Derby 1789.

Pilkington, Mary *see* Goldsmith, O.

*Pinaceae: being a handbook of the firs and pines.* 1866. *see* Nelson, J.

3020. PINNOCK, William *A catechism of ichthyology.* 12mo, 72 pp, 1 pl., Pinnock, London 1820. Many later editions, mostly undated.

3021. —— *A catechism of ornithology.* 12mo, 69 pp, 1 pl., Pinnock, London 1820. Many later editions, mostly undated.

3022. —— *A catechism of natural history.* 12mo, 68 pp, 1 pl., Pinnock, London 1821. Many later editions, mostly undated.

3023. —— *A catechism of botany.* 12mo, 70 pp, 1 pl., Pinnock, London 1822. Many later editions, mostly undated.

3024. —— *A catechism of entomology.* 12mo, 66 pp, 1 pl., Pinnock, London 1822. Many later editions, mostly undated.

3025. —— *A catechism of conchology.* 12mo, 72 pp, 1 pl., Pinnock, London 1824. Many later editions, mostly undated.

3026. PIRIE, Mary *Flowers, grasses and shrubs. A popular book on botany.* Sm. 8vo, text figs, Blackwood, London [1860]; ed. 1879.

3027. PITCAIRNE, Archibald *A. Pitcairnii Dissertatio de legibus historiae naturalis.* 12mo, Carruthers, Edinburgh 1696.
—— *see* Sibbald, *Sir* Robert.

Pitfield, Robert *see* Perrault, C.

Pitton de Tournefort, Joseph *see* Tournefort, J. P. de

*A plain and easy account of British ferns.* 1854. *see* Bosanquet, E.

3028. *Plant-life. Popular papers on the phenomena of botany.* 8vo, xii+216 pp, text figs, Japp, London 1881.

3029. *Plants: containing in one volume, the seed, the leaf, the flowers, the fruit, the grass.* 16mo, vi+32+32 pp., 65+96+32+32 pp, 1 pl., text figs, RTS, London [1841]. The parts also sold separately.

3030. *Plants of scripture, and the geography of plants.* 12mo, RTS, London [1850]. Monthly Series, double volume. See also *The geography of plants.*

3031. *Plates illustrative of natural history.* 4to, 151 col. pls with letterpress, SPCK, London 1845. Plates 1–150, 19bis.

*Plates to the natural history of birds.* 1793. *see* Galton, S. J.

3032. PLEYDELL, John Clavell Mansell *Flora of Dorsetshire.* 8vo, ii+320 pp, 1 col. map, for private circulation, London 1874; 2ed. Dorchester 1895.

3033. —— *Ornithology and conchology of the County of Dorset.* 8vo, 119 pp, privately printed, London [1874]. Reprinted from *The Atheneum* and from Hutchins, John *History and antiquities of the County of Dorset*, 3ed. 1861–74.

3034. —— *The birds of Dorsetshire.* 8vo, xvi+179 pp, 1 pl., text figs, Porter for the author, London [1888].

3035. —— *The Mollusca of Dorsetshire.* 8vo, xxxii+110 pp, 1 map, Dorset County Chronicle, Dorchester 1898 [=1899]

3036. PLINIUS SECUNDUS, Caius [*Natural history*].

1. *A summarie of the antiquities and wonders of the worlde, abstracted out of the sixtene first bookes.* 8vo, H. Denham for T. Hacket, London [?1565]; ed. 4to 1585 title *The secrets and wonders of the world* ed. 1587. Translation by I.A. of *Sommaire des singularitez de Pline*, Paris 1559, translated by P. de Changy.

2. *The historie of the world.* Fol., 2 vols, Islip, London 1601; ed. 1634; 1635; ed. 8vo 2 vols, 1847–48. Translated by Philemon Holland from ed. Lyons 1587. The 1847–48 ed., by the Wernerian Club, is unfinished, books I–VII only.

3. *The first and thirty-third books of Pliny's Natural history; a specimen of a proposed translation.* 8vo, 72 pp, Baldwin & Cradock, London 1828. Translated by J. Bostock.

4. *The natural history of Pliny.* 8vo, 6 vols, Bohn, London 1855–57. Translated by J. Bostock and H. T. Riley, a completion of No. 3. Bohn's Classical Library.

3037. PLOT, Robert *Quae's to be propounded to the most ingenious in each County in my travels through England.* Fol., s.sh., [Oxford ?1673]; ed. 1678/9.

3038. [——] *The natural history of Oxfordshire; being an essay toward a natural history of England.* Fol., 358 pp, 16 pls, 1 map, Sheldonian, Oxford 1677; 2ed. 1705 signed; facsimile of 2ed. 1972. By R.P.

3039. —— *The natural history of Stafford-shire.* Fol, 450 pp, 37 pls, 1 map, Sheldonian, Oxford 1686. Some copies have an extra plate 'Armes omitted'.

3040. PLOWRIGHT, Charles Bagge *A monograph of the British Uredineae and Ustilagineae.* 8vo, vii+347 pp, 8 pls, text figs, Kegan Paul, London 1889.

3041. PLUCHE, Noël Antoine [*Spectacle de la nature.*]

1. *Spectacle de la nature; or nature display'd.* 8vo, 7 vols, 209 pls, Pemberton, London 1733–48. Translation of *Spectacle de la nature*, 8 vols, Paris 1732–50, by Samuel Humphreys;

the last volume was not translated. Editions of each volume as follows:–

Vol. 1. 1733; 2ed. 1735; 3ed. 1736; 4ed. 1739; 5ed. 1740; 6ed. 1744; 8ed. 1754; 10ed. 1766; ed. 1776. First 2 eds not called Vol. 1 on t.p.

Vol. 2, 1736; 2ed. 1737; 3ed. 1739; 3ed. 1740; 4ed. 1739; 5ed. 1740; 6ed. 1743; 8ed. 1757; 10ed. 1766; ed. 1776.

Vol. 3, 1736; 2ed. 1737; 3ed. 1739; 3ed. 1740; 4ed. 1739; 5ed. 1740; 6ed. 1743; 8ed. 1757; 10ed. 1766; ed. 1776.

Vol. 4, 1739; 2ed. 1739; 3ed. 1740; 4ed. 1743; 4ed. 1747; 4ed. corrected 1748; 8ed. 1757; 10ed. 1766; ed. 1776.

Vols 5–7, 1748; 3ed. 1753; 4ed. 1763 (except Vol. 6 where this is called 3ed); 5ed. 1776.

2. *Spectacle de la nature: or nature delineated.* 12mo, 4 vols, 84 pls, Hodges, London 1739; 3ed. 1743–44. Translated by John Kelly and others.

3. *Lilliputian Spectacle de la nature: or, nature delineated, in conversations and letters passing between the children of a family.* 12mo, 3 vols, Marshall, London [c.1785]. A much altered adaptation.

Pluc'net Leonard *see* Plukenet, L.

3042. PLUES, Margaret *Rambles in search of ferns.* 12mo, Houlston, London 1861.

3043. —— *Rambles in search of mosses.* 12mo, Houlston, London 1861.

3044. —— *Rambles in search of wild flowers, and how to distinguish them.* Sm. 8vo, 16 col. pls, Cottage Gardener Office, London 1863; 2ed. 1864; 3ed. 1879; 4ed. 1892.

3045. —— *Rambles in search of flowerless plants.* Sm. 8vo, viii+316 pp, 20 col. pls, text figs, Cottage Gardener Office, London 1864; 2ed. 1865; 3ed. 1868.

3046. —— *British ferns.* Sm. 8vo, x+281 pp, 16 col. pls, text figs, Reeve, London 1866. Crown Series.

3047. —— *A selection of the eatable funguses of Great Britain.* 8vo, vii pp, 24 col. pls with letterpress, Cottage Gardener Office, London [1866].

3048. —— *British grasses.* Sm. 8vo, vi+307 pp, 16 col. pls, text figs, Reeve, London 1867. Crown Series.

3049. PLUKENET, Leonard *Leonardi Plukenetii Phytographia, sive stirpium illustriorum & minus cognitarum icones . . . Pars prior (-tertia)*. 4to, 3 pts, 250 pls, 1 portr., for the author, London 1691–92. Plates i-col.

3050. —— *Almagestum botanicum, sive phytographiae Pluc'netianae onomasticon methodo synthetica divestum . . . adjiciuntur & aliquot novarum plantarum icones. (Leonardi Pluc'netii Phytographiae pars quarta).* 4to, 402 pp, 78 pls, portr., for the author, London 1696. Plates ccli–cccxxviii.

3051. —— *Almagesti botanici mantissa plantarum novissime detectarum ultra millenarium numerum complectens.* 4to, 192 pp, 21 pls, for the author, London 1700. Plates cccxxix–cccl.

3052. —— *Leonardi Plukenetii Amaltheum botanicum (i.e.) stirpium indicarum alterum copiae cornu.* 4to, 214 pp, 104 pls, for the author, London 1705. Plates cccli–cccliv.

—— [Collected editions of the above.] *Leonardi Plukenetii . . . Opera omnia botanica.* 4to, 6 vols, 454 pls, 1 portr., Innys, London 1720; 2ed. 4 vols 1769 title *Plukenetii Opera.* Edition of 1720 is made up of copies of the original works with new title leaves. For index to whole see Giseke, P. D.

Pocket Library *see A new moral system of natural history.*

*A poem on the pleasures and advantages of botanical pursuits* [?1825]. *see* Hoare, Sarah.

3053. POLE, Thomas *The anatomical instructor; or, an illustration of the modern and most approved methods of preparing and preserving the different parts of the human body, and of quadrupeds, by injection, corrosion, maceration, distension, articulation, modelling.* 8vo, lxxx+304 pp, 10 pls, Couchman & Fry for the author, London 1790; ed. 12mo 1813; ed. 12mo 1824.

3054. POLEHAMPTON, Edward Thomas William & GOOD, John Mason *The gallery of nature and art; or, a tour through creation and science.* Roy. or demy 8vo, 6 vols, 100 pls, text figs, Wilks, London [1813–] 1814; 2ed. 1818; 1819; ed. 1821.

3055. POLLARD, Evelyn H. *The birds of my parish* [Haynford, Norfolk]. 8vo, xiii+294 pp, 6 pls, Lane, London 1900.

3056. *Popular botany. Showing the elementary principles of the science.* Fol., s.sh., Reynolds, London 1854.

3057. *Popular cyclopaedia of natural science. Vegetable physiology.* 8vo, viii+537 pp, text figs, SPCK, 1841.

3058. *A popular history of reptiles: or an introduction to the study of the class Reptilia.* 12mo, xii+328 pp, 36 pls, text figs, RTS, London [?1842].

3059. *A popular introduction to the study of birds.* 12mo, text figs, RTS, London [?1840].

3060. *A popular introduction to the study of quadrupeds.* 12mo, xii+492 pp, text figs, RTS, London 1823.

Popular Natural History Series (Reeve to 1858, later Routledge).
*see*

Catlow, *Miss* A. *Field botany.* 1848.
Catlow, *Miss* M. *British entomology.* 1848.
—— *Geography of plants.* 1855.
Gosse, P. H. *British ornithology.* 1849.
Laishley, R. *British birds' eggs.* 1858.
Landsborough, D. *British seaweeds.* 1849.
—— *Zoophytes.* 1852.
Lindsay, W. L. *British lichens.* 1856.
Moore, T. *British ferns.* 1851.
Roberts, *Miss* M. *The woodlands.* 1850.
—— *Mollusca.* 1851.
Sowerby, G. B. [II] *British conchology.* 1854.
—— *The aquarium.* 1857.
Stark, R. M. *British mosses.* 1854.
White, A. *Mammalia.* 1850.
—— *Birds.* 1855.
—— *British Crustacea.* 1857.

3061. *Popular zoology, comprising memoirs and anecdotes of the quadrupeds, birds and reptiles in the Zoological Society's Menagerie.* 12mo, text figs, Hatchard, London 1832.

3062. POTTER, Thomas Rossell *The history and antiquities of Charnwood Forest.* 4to, xii+192 pp, 19 pls, 1 map, text figs, Hamilton Adams, London 1842.

3063. POUCHET, Felix Archimede *The universe; or, the infinitely great and the infinitely little.* Roy. 8vo, xx+791 pp, 4 col. pls, text figs, Blackie, London 1870; ed. 1871; ed. 1874; 3ed. 1876. Later editions. Translation of *L'Univers*, Paris 1865, 2ed. 1868.

3064. POULTON, *Sir* Edward Bagnall *The colours of animals, their meaning and use, especially considered in the case of insects.* Sm. 8vo, xiii+360 pp, 1 col. pl., 1 tab., text figs, Kegan Paul, London 1890; 2ed. 1890. International Scientific Series, Vol. LXVIII.

3065. —— *Charles Darwin and the theory of natural selection.* 8vo, viii+224 pp, 1 portr., Cassell, London 1896; 1901. Century Science Series.

3066. POWER, Henry *Experimental philosophy in three books: containing new experiments, microscopical, mercurial, mag-*

*netical.* 4to, 193 pp, 2 pls, text figs, Martin & Allestry, London 1664. Each book has a title page dated 1663.

3067. POWER, J. *Calendar of flora at Market Bosworth.* 4to, Hinckley 1807.

Power, Thomas *see* Cuvierian Society of Cork.

3068. [POWNALL, Henry] *Some particulars relating to the history of Epsom . . . to which is added . . . a botanical survey of the neighbourhood.* 8vo, xiv+208 pp, 1 col.+5 plain pls, Dorling, Epsom 1825. By an inhabitant.

Powys, Thomas Littleton *see* Lilford *Baron*

3069. POYNTING, Frank *Eggs of British birds, with an account of their breeding habits. Limicolae.* 4to, xvi+[254] pp, 54 col. pls, Porter, London 1895–96. All publ. The parts are separately paginated.

3070. PRAEGER, Robert Lloyd *Open-air studies in botany; sketches of British wild flowers in their homes.* 8vo, xiii+266 pp, 8 pls, text figs, Griffin, London 1897; 2ed. 1910.
—— *see* Stewart, S. A. & Corry, T. H.

3071. PRANTL, Karl Anton Eugen *An elementary text book of botany.* 8vo, vii+322 pp, text figs, Sonnenschein, London 1880; 2ed. 1881; 3ed. 1883; 4ed. 1886; 5ed. 1893. Translation of *Lehrbuch der Botanik*, 3ed. Leipzig 1879, edited by Sydney Howard Vines.

3072. [PRATT, *Miss* Anne] *The field, the garden, and the woodland.* 16mo, xii+324 pp, 1 col. pl., text figs, Knight, London 1838; 2ed. 1841; 3ed. 12mo, 1847 signed; ed. 1853. By a lady.

3073. —— *Flowers and their associations.* 8vo, xii+409 pp, 4 col. pls, Knight, London 1840; ed. 12mo, 1846; 2ed. 1847.

3074. —— *The pictorial catechism of botany.* 16mo, vii+230 pp, text figs, Suttaby, London 1842; ed. 1846.

3075. —— *Chapters on common things of the sea-side.* 12mo, ix+345 pp, 1 pl., text figs, SPCK, London 1850; ed. 1853; 1856.

3076. —— *Wild flowers of the year.* 12mo, London [1851]; ed. 1940.

3077. —— *The green fields and their grasses.* 8vo, 96 pp, text figs, London 1852; ed. [ ].

3078. —— *Our native songsters.* 16mo, vi+350 pp, 73 col. pls, SPCK, London 1852. Later editions.

3079. —— *Wild flowers.* 16mo, 2 vols, 192 col. pls, SPCK, London 1852. Later editions.

3080. —— *Wild flowers.* Roy. 4to, 96 col. pls with letterpress, SPCK, London 1853. A 4to version of No. 3079. Each plate was also sold separately.

3081. —— *Flowering plants and ferns of Great Britain.* Roy. 8vo, 5 vols, 238 col. pls, SPCK, London [1850–57]; ed. 1861; ed. 6 vols 1873 title *Flowering plants, grasses, sedges and ferns . . .*; ed. 1886; ed. 1891; ed. 4 vols, 1899–1905. See also No. 3082.

3082. —— *Ferns of Great Britain and their allies, the club mosses, pepperworts and horsetails.* 8vo, iv+164 pp, 41 col. pls, SPCK, London [1855]; 2ed. [1862]; ed. [1871]. This is Vol. 6 of No. 3081 with a separate title.

3083. —— *The poisonous, noxious and suspected plants of our fields and woods.* 8vo, xii+208 pp, 43 col. pls, SPCK, London [1857]. Also available as separate leaves.

3084. — British grasses and sedges. 8vo, viii+136 pp, 35 col. pls, SPCK, London [1859].

3085. —— *Grasses, sedges and ferns of Great Britain.* Roy. 8vo, col. pls, SPCK, London 1860; ed. 1866. Also forms Vol. 6 of No. 3081.

3086. —— *Haunts of the wild flowers.* Sm. 8vo, 320 pp, 8 col. pls, Routledge, London 1863; ed. 1866. Books for the Country.

3087. PRATT, Samuel Jackson *Pity's gift: a collection of interesting tales to excite the compassion of youth for the animal kingdom.* 12mo, viii+147 pp, text figs, Longman & Newbery, London 1798; 2ed. 1798; 3ed. 1801; 4ed. 1807; 5ed. 1810; 6ed. 1816; ed. Belfast 1823.

*A preliminary flora of the Stoneyhurst district.* 1886. *see* Gerard, J.

3088. PRENTIS, Walter *Notes on the birds of Rainham; including the district between Chatham and Sittingbourne.* Sm. 8vo, 92 pp, Gurney & Jackson, London 1894.

3089. PRESTON, Thomas Arthur *Flora of Marlborough; with notices of the birds.* 8vo, xxiv+129 pp, 1 map, Van Voorst, London 1863; 2ed. 1870–76.

3090. —— *The flowering plants of Wilts, with sketches of the physical geography and climate of the County.* 8vo, lxix+435 pp, 1 pl., 1 map, privately printed [Devizes 1888].

3091. *A pretty book of beasts.* 32mo, 28 pls, Luckman & Suffield, Coventry [c.1810].

3092. *A pretty book of birds.* 32mo, 28 pls, Luckman & Suffield, Coventry [c.1810].

3093. *A pretty book of pictures for little masters and misses, or,*

*Tommy Trip's history of beasts and birds.* 32mo, Newbery, London 1753; 2ed. 1754; 5ed. 1760; 8ed. 1765; 9ed. 1767; 10ed. 1769; 13ed. 1778; 14ed. 1787; ed. Glasgow 1802 title *The natural history of four footed beasts.*

3094. PRICE, Sarah *Illustrations of the Fungi of our fields and woods.* Roy. 4to, 2 ser., x+x pp, 20 col. pls, Reeve, London 1864, 65.

3095. PRICHARD, James Cowles *Researches into the physical history of man.* 8vo, iv+558 pp, Arch, London 1813; 2ed. 2 vols 1826 title . . . *of mankind*: 3ed. 5 vols 1836–47; 4ed. Vol. 1 1851.

Prince, – *see* Baker, T. B. L.

*Principia botanica: or, a concise and easy introduction to the sexual system of Linnaeus.* 1787. *see* Darwin, R. W.

3096. PRIOR, Richard Chandler Alexander *On the popular names of British plants.* 8vo, xxvi+250 pp, Williams & Norgate, London 1863; 2ed. 1876; 3ed. 1879.

3097. PRITCHARD, Andrew *The microscopic cabinet of select animated objects.* 8vo, 246 pp, 10 col.+2 plain pls, text figs, Whittaker, London 1832.

3098. —— *The natural history of animalcules.* 8vo, 196 pp, 7 pls, Whittaker, London 1834; ed. 1841 title *A history of Infusoria*; 1842; 1845; ed. 1852 title *A history of infusorial animalcules*; 4ed. 1861 title *A history of Infusoria*; ed. of plates alone [c.1837] title *Coloured impressions of engravings from The natural history of animalcules.*

3099. —— *A list of two thousand microscopic objects; with remarks on the circulation in animals and plants.* 12mo, 36 pp, Whittaker, London 1835; 2ed. 1842.

3100. —— *Notes on natural history selected from The microscopic cabinet.* 12mo, iv+93 pp, 10 col. pls, Whittaker, London 1844; ed. 1854 title *Notes on aquatic microscopic subjects.* See No 3097.

—— *see* Goring, C. R.

*Prodromus Lepidopterorum Britannicarum.* 1802. *see* Haworth, A. H.

3101. PRYOR, Alfred Reginald *Notes on a proposed reissue of the Flora of Hertfordshire.* 8vo, 14 pp, Hertford 1875.

3102. —— *A flora of Hertfordshire.* 4to or 8vo, lix+588 pp, 2 col.+1 plain maps, Gurney & Jackson, London 1887. The 4to issue was of 50 copies for the Watford and Hertfordshire Natural History Society.

3103. PULTENEY, Richard *A general view of the writings of Linnaeus.* 8vo, iv+425 pp, Payne, London 1781; 2ed. 4to 1805.

3104. —— *Historical and biographical sketches of the progress of botany in England.* 8vo, 2 vols, Cadell, London 1790.

3105. —— *Catalogues of the birds, shells, and some of the more rare plants of Dorsetshire.* Fol., 92 pp, 1 portr, for the compiler and his friends, London 1799; ed. 1813. Reprinted from Hutchins, John *The history and antiquities of the County of Dorset*, 2ed. 1796–1815.

3106. PURCHAS, William Henry & Ley, Augustin *editors A flora of Herefordshire.* 8vo, xi+xxxvii+549 pp, 3 pls, 1 map, Jakeman & Carver for the Woolhope Naturalists' Field Club, Hereford 1889.

3107. PURTON, Thomas *A botanical description of British plants in the Midland Counties, particularly those in the neighbourhood of Alcester.* 8vo, 2 vols, ix+795 pp, 8 col. pls, Ward printed, Stratford-upon-Avon 1817; Vol. 3 title *An appendix to . . .*, 2 pts, xiv+575 pp, 30 col. pls, Longman for the author, London 1821.

3108. PYCRAFT, William Plane *The courtship of animals.* 8vo, Hutchinson, London 1894; 2ed. 1913.

3109. —— *The story of bird life.* 18mo, 244 pp, 1 pl., text figs, Newnes, London 1900. Library of Useful Stories.

Pye-Smith, Phillip Henry *see* Smith, P. H. P.

## Q

3110. QUATREFAGES DE BREAU, Jean Louis Armand de *The rambles of a naturalist on the coasts of France, Spain, and Sicily* [*Channel Islands*]. 8vo, 2 vols, Longman, London 1857. Translation of *Souvenirs d'un naturaliste*, 2 vols, Paris 1854, by Henry Lawson.

3111. —— *Metamorphoses of man and the lower animals.* 8vo, xx+284 pp, Hardwicke, London 1864. Translation of *Physiologie comparée. Metamorphoses de l'homme et des animaux*, Paris 1862, by Elise C. Otté.

3112. Quekett, John Thomas *A practical treatise on the use of the microscope.* 8vo, xx+464 pp, 9 pls, text figs, Baillière, London 1848; 2ed. 1852; 3ed. 1855. Library of Illustrated Standard Scientific Works, Vol. VI.

3113. —— *Lectures on histology delivered at the Royal College of Surgeons of England, 1850–51, 1851–52.* 8vo, 2 vols, text figs, Baillière, London 1852, 54.

3113a —— *Catalogue of the contents of the Museum. Part 1. Plants and invertebrate animals in the dried state.* 4to, xliii+514 pp, Royal College of Surgeons, London 1860.

Quelet, Lucien *see* Cooke, M. C.

3114. *Queries proposed to gentlemen in several parts of Great Britain in hope of obtaining from the answers, a better knowledge of its antiquities and natural history.* 8vo, 15 pp, London 1754.

*The question concerning the sensibility . . . of insects.* 1837. *see* Badham, C. D.

3115. QUIGGIN, John *Quiggin's illustrated guide . . . through the Isle of Man. (The natural history by E. Forbes).* 12mo, 130 pp, pls, 1 map, text figs, Quiggin, Douglas 2ed. 1839; 1840; 1841; 1842; 1845; 3ed. 1847; 1848; 4ed. 1852; 5ed. 1856; 6ed. 1858. First ed. 1836, by a resident, does not contain the natural history.

# R

R., G. *The three barriers.* 1861. *see* Rorison, G.

3116. RAFF, Georg Christian *A system of natural history adapted for the instruction of youth, in the form of a dialogue.* 12mo, 2 vols, pls, Johnson, London 1796. Translation of *Naturgeschichte für Kinde*, Gottingen 1781, edited by R. H.

3117. RAINEY, George *An experimental inquiry into the causes of the ascent and descent of sap.* 12mo, viii+47 pp, 2 pls, Pamplin, London 1847.

3118. —— *On the mode of the formation of the shells of animals, of bone, and of several other structures.* 8vo, xii+152 pp, text figs, Churchill, London 1858.

3119. RALFS, John *The British phaenogamous plants and ferns.* 8vo, xvi+208 pp, Longman, London 1839.

3120. —— *The British Desmideae.* Roy. 8vo, xxii+226 pp, 34 col.+1 plain pls, Reeve, London 1848; reprint 1962.

3121. RALPH, Thomas Shearman *Elementary botany.* 8vo, iv+64 pp, 20 col. pls, Pamplin, London 1849; ed. 1865 title *The young botanist.*

3122. —— *Icones carpologicae; or figures and descriptions of fruits and seeds. Leguminosae.* 4to, 48 pp, 40 col. pls, Pamplin, London 1849. All publ.

Ram, William *see* Dodoens, R.

*Rambles among the Channel Islands.* [c.1850]. *see* Lukis, F. C.

3123. *Rambles through the fields of nature by a father and his son.* 12mo, viii+194 pp, Richard Phillips, London 1801. Preface dated 1810.

Ramsey, D. *see* Linnaeus, C.

3124. RATTRAY, James *A botanical chart; or, a concise introduction to the Linnaean sytem of botany.* 8vo, s.sh., Glasgow [1835]; 2ed. [1838].

3125. RAVENSHAW, Thomas Fitz-Arthur Torin *A new list of the flowering plants and ferns growing wild in the County of Devon.* Sm. 8vo, viii+93+vii pp, Bosworth, London 1860; ed. with suppl. 1872.

3126. —— *Botany of North Devon.* 8vo, 110 pp, Ilfracombe [1874]. Reprinted from Stewart's *North Devon handbook.*

3127. [RAY, John] *Catalogus plantarum circa Cantabrigium nascentium.* 8vo, 182+103 pp, Nealand, Cambridge 1660; issue Martin Allestry & Dicas, London 1660; appendix [16]pp, 1663; 2ed. of appendix 1685.

3128. —— *Catalogus plantarum Angliae, et insularum adjacentium.* 8vo, 358 pp, Martyn, London 1670; 2ed. 1677.

3129. —— *A collection of English words not generally used . . . with catalogues of English birds and fishes.* 8vo, 148 pp, Burrell, London 1674; 2ed. 12mo, 1691; ed. 1737; ed. 1768; ed. 1874. The 18c eds contain *A collection of English proverbs* as well.

3130. —— *Methodus plantarum nova.* 8vo, 166 pp, 1 pl., Faithorne, London 1682; issue Amsterdam 1682; 2ed. 1703; ed. 1710; ed. 1733 [Tubingen]; reprint 1962.

3131. —— *Historia plantarum.* Fol., 3 vols, 1 pl., Vols 1 & 2 Faithorne, Vol. 3 Smith & Walford, London 1686–1704; issue 1693–1704 all vols Smith & Walford; L.P. A catalogue with

50, or 72 in later issues, plates is contained in No. 2976.

3132. —— *Fasciculus stirpium Britannicarum, post editum Plantarum Anglia catalogum observatarum.* 8vo, 27 pp, Faithorne, London 1688.

3133. —— *Synopsis methodica stirpium Britannicarum.* 8vo, 317 pp, 2 pls, Smith, London 1690; 2ed. 1693; 3ed. 1724; issue 2 vols 1724; facsimile of 3ed. 1973.

3134. —— *The wisdom of God manifested in the works of the creation.* 8vo, 249 pp, Smith, London 1691; 2ed. 1692; 3ed. 1701; 4ed. 1704; 5ed. 1709; 6ed. 1714; 7ed. 1717; 8ed. 1722; 9ed. 1727; 10ed. 1735; 11ed. 1743; 12ed. 12mo, Glasgow 1750; 13ed. 1756; ed. 1758; 12ed. 1759; ed. 1762; ed. 12mo, Glasgow 1768; ed. 12mo 1777; ed. 12mo, Edinburgh 1798; ed. 12mo. 1827; ed. 1844–46.

3135. —— *Synopsis methodica animalium quadrupedum et serpentini generis.* 8vo, 336 pp, 1 portr., Smith & Walford, London 1696.

3136. —— *Joannis Raii de variis plantarum methodis dissertatio brevis.* 8vo, 48 pp, Smith & Walford, London 1696.

3137. —— *Methodus insectorum: seu Insecta in methodum aliqualem digesta.* 8vo, 16 pp, Smith & Walford, London 1705; 2ed. 1708; ed. 1710 in *Historia insectorum* No. 3138.

3138. —— *Historia insectorum . . . cui subjungitur appendix de scarabaeis Britannicis autore M. Lister.* 4to, xv+400 pp, Churchill, London 1710.

3139. —— *Joannis Raii Synopsis methodica avium & piscium.* 8vo, 2 pts, 198+166 pp, 4 pls, Innys, London 1713.

3140. —— *Philosophical letters between the late learned Mr. Ray and several of his ingenious correspondants, natives and foreigners. To which are added those of Francis Willughby.* 8vo, 376 pp, Derham, London 1718.

—— *see* Wilkins, John.

—— *see* Willughby, F.

3141. RAY SOCIETY *Reports of progress in zoology and botany, 1841, 1842.* 8vo, viii+44+348+104+xx pp, Ray Society, London 1845. Ray Society Publ. No. 1 for 1844.

3142. —— *Reports and papers on botany.* 8vo, viii+493 pp, 7 pls, Ray Society, London 1846. Ray Society Publ. No. 6 for 1845.

3143. —— *Reports on zoology, 1843, 1844.* 8vo, viii+596 pp, Ray Society, London 1846. Ray Society Publ. No. 11 for 1847.

——, Publications of the, *see*

Alder, J. & Hancock, A. *British nudibranchiate Mollusca.* 1845–55, 1910.
Allman, G. J. *Gymnoblastic hydroids.* 1871–72.
Baird, W. *British Entomostraca.* 1850.
Blackwall, J. *Spiders of Great Britain and Ireland.* 1860–64.
Boulenger, G. A. *Tailless batrachians of Europe.* 1897–98.
Bowerbank, J. S. *British Spongiadae.* 1864–72.
Brady, G. S. *British free and semi-parasitic Copepoda.* 1878–80.
Brown, R. *Miscellaneous botanical works.* 1866–68.
Buckler, W. *Larvae of British butterflies and moths.* 1886–1901.
Buckton, G. B. *British Aphides.* 1876–83.
Cameron, P. *British phytophagous Hymenoptera.* 1882–93.
Carpenter, W. B. *Foraminifera.* 1862.
Darwin, C. R. *Cirripedia.* 1851–54.
Douglas, J. W. & Scott, J. *British Hemiptera.* 1865.
Forbes, E. *British naked-eyed Medusae.* 1848.
Hofmeister, *W. On the germination of the higher Cryptogamia.* 1862.
Huxley, T. H. *Oceanic Hydrozoa.* 1859.
Leighton, W. A. *British angiocarpous lichens.* 1851.
Lubbock, *Sir* John *Bart Collembola and Thysanura.* 1873.
McIntosh, W. C. *British annelids.* 1873–1923.
Meyen, F. J. F. *Geography of plants.* 1846.
Michael, A. D. *British Oribatidae.* 1884–88.
Nitzsch, C. L. *Pterylography.* 1867.
Parker, W. K. *Shoulder girdle and sternum in the Vertebrata.* 1868.
Steenstrup, J. J. S. *On the alternation of generations.* 1845.
Williamson, W. C. *Recent British Foraminifera.* 1858.

3144. READ, Robert H. *The birds of the lower Brent valley.* 8vo, 28 pp, [Ealing 1896]. Reprinted from *Rep. Trans. Ealing nat. Hist. Soc.*, 1896.

*The reason why.* [1856]. *see* Philp, R. K.

*The reason why . . . General science.* 1857. *see* Philp, R. K.

*The reason why . . . Natural history.* 1860. *see* Philp, R. K.

3145. RECLUS, Jean Jacques Élisée *The ocean, atmosphere, and life.* 8vo, 2 sections, maps, text figs, Chapman, London 1872; 1873; 2ed. 1875; ed. 1877. Translation of *L'Ocean, l'atmos-*

*phere, la vie*, Paris 1889, by Bernard Bolingbroke Woodward.

3146. *Recreations in natural history; or, popular sketches of British quadrupeds.* 4to or 8vo, xvi+367 pp, 24 pls, Clarke, London 1815. 4to plates are on India paper. Plates by Luke Clennell.

3147. REES, George *Gwersi mewn Llysileueg.* 8vo, 148 pp, Aberystwyth 1896.

3148. REEVE, Henry *Disputatio inauguralis de animalibus hieme sopitis.* 8vo, Edinburgh 1803.

3149. —— *An essay on the torpidity of animals.* 8vo, viii+152 pp, Longman, London 1809.

3150. REEVE, Lovell Augustus *Conchologia systematica; or complete system of conchology.* 4to, 2 vols, 301 col. pls, 1 tab., Longman, London 1841, 42. Pls 1–300, 114*.

3151. —— *Conchologia iconica; or, illustrations of the shells of molluscous animals.* 4to, 20 vols, col. pls, Reeve, London 1843–78. Completed by G. B. Sowerby (II) from 1865.

3152. —— *Initiamenta conchologica; or, elements of conchology.* Roy. 8vo, 2 vols, 62 col. pls, text figs, Reeve, London [1846–] 1860. Pls A–Q, 1–46.

3153. —— *The land and freshwater mollusks indigenous to, or naturalized in, the British Isles.* 8vo, xx+275 pp, 1 portr., 1 map, text figs, Reeve, London 1863.

—— *see* Catlow, Agnes.

3154. REEVES, J. A. *Does it rise from the roots?* 8vo, 82 pp, 1 pl., privately printed, ?London 1890.

3155. REID, Clement *Origin of the British flora.* 8vo, vii+191 pp, London 1899.

3156. REID, Hugo *Science of botany.* 18mo, Glasgow 1837; ed. Edinburgh 1840.

3157. REID, Thomas Mayne *Quadrupeds; what they are and where they are found.* Sm. 8vo, viii+167 pp, 24 pls, Clarke, London [1860]; ed. 1867; ed. 1872.

3158. —— *The naturalist in Siluria.* 8vo, iii+240 pp, 1 pl., text figs, Sonnenschein, London 1889.

3159. REID, William *List of the Lepidoptera of Aberdeenshire & Kincardineshire.* 8vo, 35 pp, Sonnenschein, London [1891].

3160. RELHAN, Richard *Richardi Relhan . . . Flora Cantabrigiensis exhibens plantas agro Cantabrigiensi indigenas.* 8vo, 490 pp, 7 pls, University Press, Cambridge 1785; suppl. 39 pp, 1786; 2nd suppl. 36 pp, 1788; 3rd suppl. 44 pp, 1793; ed. 1802; 3ed. 1820.

3161. —— *Heads of a course of lectures on botany.* 8vo, 8 pp, Archdeacon printed, Cambridge 1787.

3162. *Remarkable insects.* [Containing the honey bee, the fly, the ant, the spider, the gall insect]. 16mo, vi+32+32+32+32+32 pp, text figs, RTS, London 1842, [ ]. The parts available separately.

*Remarks on the Linnaean orders of insects.* 1828. *see* Turner, J. A.

3163. [RENNIE, James] *The menageries.* 8vo, 3 vols, text figs, Nattali & Bird, London 1829–40.

3164. [——] *Insect architecture.* 12mo, xii+420 pp, text figs, Knight, London 1830; 2ed. 1830; ed. 1845; ed. 1857 signed; ed. 1888. Library of Entertaining Knowledge. Ed. of 1857 is combined with *Insect miscellanies.*

3165. [——] *Insect transformations.* 12mo, xii+420 pp, text figs, Knight, London 1830. Library of Entertaining Knowledge.

3166. [——] *The architecture of birds.* 12mo, xi+392 pp, text figs, Knight, London 1831; ed. 1844 title *Bird architecture*; ed. 1858. Library of Entertaining Knowledge.

3167. [——] *Insect miscellanies.* 12mo, xii+414 pp, text figs, Knight, London 1831; ed. 1857 signed. Library of Entertaining Knowledge. Ed. of 1857 is combined with *Insect Architecture.*

3168. —— *Alphabet of insects for the use of beginners.* 12mo, 108 pp, text figs, Orr, London 1832; 2ed. 1834.

3169. —— *A conspectus of the butterflies and moths found in Britain.* Sm. 8vo, xxxvii+287 pp, Orr, London 1832.

3170. —— *Alphabet of zoology, for the use of beginners.* 12mo, xiv+154 pp, text figs, Orr, London 1833.

3171 [——] *The domestic habits of birds.* 12mo, xvi+379 pp, text figs, Knight, London 1833. Library of Entertaining Knowledge.

3172. —— *The handbook of plain botany; or, Linnaean lessons on common plants.* 16mo, 170 pp, text figs, Orr, London 1834; ed. 2 vols 1845; 2ed. 1848; ed. 1857; ed. 1869.

3173. [——] *The faculties of birds.* 12mo, xvi+338 pp, text figs, Knight, London 1835. Library of Entertaining Knowledge. Vol. 3 of birds with index to the three.

3174. —— *Alphabet of the sciences for the use of beginners.* 12mo, 4 vols, text figs, Orr, London 1836. A combination volume.

3175. —— *Bird miscellanies: illustrative of the habits and faculties of birds.* 12mo, London 1847.

3176. [——] & [WESTWOOD, John Obadiah] *The natural history of insects.* 12mo, 2 vols, text figs, Murray, London 1829, 35; 2ed. 1830–38. The Family Library, Vols VII & LI.

3177. RENNIE, Robert *Essay on the natural history and origin of peat moss.* 8vo, essay 1–2, vii+233 pp, for the author, Edinburgh 1807; essays 3–9, 16, pp. 235–665, for the author, Edinburgh 1810.

*Report of a committee . . . to consider the rules by which the nomenclature of zoology may be established.* 1842. *see* Strickland, H. E.

3178. REYNOLDS, Sidney Hugh *The vertebrate skeleton.* 8vo, xvi+559 pp, text figs, University Press, Cambridge 1897. Cambridge Natural Science Manuals. Biological Series.

3179. RHIND, William *Studies in natural history; exhibiting a popular view of the most striking and interesting objects of the material world.* Sm. 8vo, iv+247 pp, text figs, Oliver & Boyd, Edinburgh 1830; 2ed. 1833.

3180. —— *A catechism of zoology. A natural history of the animal kingdom.* 12mo, Oliver & Boyd, Edinburgh 1832.

3181. —— *A catechism of botany.* 12mo, 76 pp, Oliver & Boyd, Edinburgh 1833.

3182. —— *Excursions illustrative of the geology and natural history of the environs of Edinburgh.* Sm. 8vo, vi+146 pp, 1 col. pl., 1 map, text figs, Anderson, Edinburgh 2ed. 1836.

3183. —— *Elements of zoology.* 8vo, Oliver & Boyd, Edinburgh 1839.

3184. —— *A history of the vegetable kingdom.* 4to, x+711 pp, 12 pls, 1 portr., text figs, Blackie, Glasgow [1840–41]; 1852; 1855; 1860; 1862; 2ed. 1868; 1874; 1877.

3185. RICHARD, Achille *New elements of botany.* 8vo, xvi+758+xxiii pp, 7 pls, 1 tab., Byrne, Dublin 4ed. 1829. Translation of *Nouveaux élémens de botanique*, 4ed. Paris 1828, by P. Clinton.
*another translation Elements of botany and vegetable physiology.* 8vo, xxviii+563 pp, text figs, Blackwood, Edinburgh 1831, by William Macgillivray.

3186. RICHARD, Louis Claude Marie *Observations on the structure of fruits and seeds.* 8vo, xx+100 pp, 6 pls, Harding, London 1819. Translation of *Démonstrations botaniques, ou analyse du fruits*, Paris 1808, by John Lindley.

3187. RICHARDSON, *Sir* John *et al. The museum of natural history.* Imp. 8vo, 2 vols, 78 col.+58 plain pls, text figs, Mackenzie, Glasgow [1859–62]. Later editions, all undated.

3188. RIDEWOOD, Walter G. *Illustrated natural history of animals.* Imp. 8vo, col. pls, Tuck, London [1900].

3189. RIDLEY, Marian S. *A pocket guide to British ferns.* Sm. 8vo, 96 pp, Bogue, London 1881.

3190. [RILEY, George] *The beauties of creation; or a new moral system of natural history.* 18mo, 2 vols, 2 pls, text figs, for the author, London 1790; 2ed. 5 vols, 1793 unsigned. See also *A new moral system of natural history*, 1790, No. 2790, of which this is an extension.

3191. RIMMER, Richard *The land and freshwater shells of the British Isles.* Sm. 8vo, xxxii+208 pp, 11 pls, Bogue, London 1880; issue Allen, London [?1880]; issue Grant, Edinburgh [?1880]; issue Edinburgh 1907 title *Shells of the British Isles land and freshwater.*

3192. RIPPINGHAM, John *Natural history according to the Linnaean system explained by familiar dialogues in visits to the London museum.* 12mo, 4 vols, 32 pls, N. Hales, London 1817; 2ed. 1817. Juvenile Library.

3193. RITSON, *Mrs* Anne *Spring flowers.* 18mo, London 1832.

*The Riverside natural history.* 1888. *see* Kingsley, J. S.

3194. *Robert and William, or the beauties of nature.* Sm. 8vo, Butler, Northampton 1838.

3195. ROBERTS, Charles *The naturalist's diary, a day book of meteorology, phrenology, and rural biology.* 8vo, xlvi pp, Sonnenschein, London [1886].

Roberts, *Hon.* Francis *see* Morton, C.

3196. ROBERTS, George *Topography and natural history of Lofthouse and its neighbourhood; with the diary of a naturalist and rural notes.* Sm. 8vo, 2 vols, 10 pls, text figs, Vol. 1 Bogue, London, Vol. 2 for the author, Leeds, 1882, 85.

3197. [ROBERTS, Mary] *The wonders of the vegetable kingdom displayed in a series of letters.* 8vo, 242 pp, 1 pl., Whittaker, London 1822; 2ed. 1824.

3198. [——] *The conchologist's companion.* Sm. 8vo, xii+251 pp, text figs, Whittaker, London 1824; ed. 1834 signed; ed. 1835.

3199. [——] *Sister Mary's tales in natural history.* 18mo, text figs, Parker, London 1834; 7ed. 1850. Attribution uncertain.

3200. —— *The sea-side companion.* Sm. 8vo, viii+240 pp, text figs, Whittaker, London 1835.

3201. —— *Wild animals; their nature, habits, and instincts; with incidental notes on the regions they inhabit.* 8vo, text figs, Parker for SPCK, London 1836.

3202. —— *The progress of creation*. Sm. 8vo, xii+285 pp, 13 pls, Smith Elder, London 1837; 3ed. 1838; 1846.
3203. —— *Flowers of the matin and even song*. 8vo, xvi+192 pp, 4 col. pls, Clarke, London 1845.
3204. —— *Voices from the woodlands; or history of forest trees, ferns, mosses, and lichens*. Roy. 16mo, viii+368 pp, 20 col. pls, Reeve, London 1850. Popular Natural History Series.
3205. —— *A popular history of the Mollusca*. Roy. 16mo, ix+396 pp, 20 col. pls, Reeve, London 1851. Popular Natural History Series.

Robertson, David *see* Stebbing, T. R. R.

3206. ROBINSON, Louis *Wild traits in tame animals, being some familiar studies in evolution*. 8vo, vii+329 pp, pls, text figs, Blackwood, Edinburgh 1897.
3207. ROBINSON, Philip Stewart *Noah's Ark; or mornings in the zoo*. 8vo, xi+416 pp, Sampson Low, London 1882.
3208. —— *Birds of the wave and woodland*. Roy. 8vo, 224 pp, 17 pls, text figs, Isbister, London 1894.
3209. —— *In garden, orchard and spinney*. Sm. 8vo, 288 pp, Isbister, London 1897.
3210. ROBINSON, Thomas *New observations on the natural history of this world of matter, and this world of life*. 8vo, 222 pp, 2 pls, Newton, London 1696.
3211. —— *An essay towards a natural history of Westmorland and Cumberland*. 8vo, 2 pts, 95+118 pp, Freeman, London 1709.
3212. ROBSON, John Emmerson *A series of botanical labels for labelling herbaria*. 8vo, 255 pp, Bell, London 1873.
3213. —— & GARDNER, John *A list of British Lepidoptera and their named varieties*. 8vo, 46 pp, Kempster, London [c.1890]. Printed on one side.
3214. ROBSON, Stephen *The British flora . . . to which are prefixed the principles of botany*. 8vo, xx+330 pp, 5 pls, Blanchard, York 1777.
3215. ROBSON, Thomas *Birds of the Derwent valley*. 8vo, 64 pp, Dent printed, Consett 1896.
3216. RODD, Edward Hearle *A list of British birds, as a guide to the ornithology of Cornwall, especially in the Land's End district*. 8vo, 42 pp, Simpkin, London 1864; 2ed. 1869.
3217. —— *The birds of Cornwall and the Scilly Islands*. 8vo, lvi+320 pp, 1 portr., 1 map, Trübner, London 1880.
3218. RODWAY, James *The story of forest and stream*. Sm. 8vo,

vii+202 pp, 1 pl., text figs, Newnes, London 1897. Library of Useful Stories.

3219. ROEBUCK, William Denison *Locusts in Yorkshire*. 8vo, 22 pp, Brown, Huddersfield 1876. Reprinted from *The Naturalist*, 1876.

—— *see* Clarke, W. E.

3220. ROESEL VON ROSENHOF, Augustin Johann *A collection of curious insects.* Sq. 4to, t.p.+12 pls, Hemmerich [London c.1760]. Copied from plates in *Der monatlich-herausgegebenen Insekten Belustigung*, Nürnberg [1740–] 1746–61.

3221. [ROGER, Charles] *Botanizing excursions in the high and low lands by Professor Balfour's pupils during the summer of 1877*. Sm. 8vo, 80 pp, portr., for the author, Dundee 1877. By No. 86.

3222. ROGERS, Bertram Mitford Heron *editor British Association . . . handbook to Bristol and the neighbourhood*. 8vo, 237 pp, 1 col. map, British, Association, Bristol 1898.

3223. ROGERS, William Moyle *Handbook of the British Rubi*. 8vo, xiv+111 pp, Duckworth, London 1900.

3224. ROGET, Peter Mark *Animal and vegetable physiology considered with reference to natural theology*. 8vo, 2 vols, text figs, Pickering, London 1834; 2ed. 1834; 3ed. 1840; 4ed. 1867; 4ed. 1870. Bridgewater Treatise No. V.

3225. ROLLESTON, George *Forms of animal life*. Roy. 8vo, clxviii+268 pp, 12 pls, 1 tab., Clarendon, Oxford 1870; 2ed. 1888. Clarendon Press Series.

3226. —— *Scientific papers and addresses*. 8vo, 2 vols, pls, 1 portr., text figs, Clarendon, Oxford 1884.

3227. ROMANES, George John *The scientific evidences of organic evolution*. 8vo, Ross-shire Philosophical Society, Inverness 1877; ed. vi+88 pp, Macmillan, London 1882. London ed. is Nature Series.

3228. —— *Animal intelligence*. Sm. 8vo, xiv+520 pp, text figs, Kegan Paul, London 1882; 2ed. 1882; 3ed. 1883; 4ed. 1886; 5ed. 1889; 6ed. 1896; 7ed. 1898; 8ed. 1904; 9ed. 1910. International Scientific Series Vol. XLI.

3229. —— *Mental evolution in animals, with a posthumous essay on instinct by Charles Darwin*. 8vo, 411 pp, 2 pls, text figs, Kegan Paul, London 1883; 1885; 1887.

3230. —— *Jelly-fish, star-fish and sea-urchins; being a research on primitive nervous systems*. Sm. 8vo, vi+323 pp, text figs,

Kegan Paul, London 1885; 2ed. 1885. International Scientific Series, Vol. L.

3231. —— *Darwin, and after Darwin*. Sm. 8vo, 3 vols, portrs, text figs, Longman, London 1892–97; 2ed. Vol. 1 1893; 2ed. 1897; 3ed. 1901; 1905.

3232. —— *An examination of Weismannism*. Sm. 8vo, ix+221 pp, Longman London 1893.

3233. —— *Essays*. Sm. 8vo, 253 pp, Longman, London 1897.

3234. ROMER, A. *Anecdotal and descriptive natural history*. 8vo, 2 pts, vi+184 pp, 8 col. pls, text figs, Groombridge, London 1872.

3235. RONALDS, Alfred *The fly-fisher's entomology*. 8vo, x+115 pp, 20 col. pls, Longman, London 1836; 2ed. 1839; 3ed. 1844; 1849; 5ed. 1856; 6ed. 1862; 7ed. 1868; 8ed. 1877; 10ed. 1901; ed. 4to, Liverpool 1913 with fisherman's flies; ed. 1921.

3236. ROOKE, Hayman *Descriptions and sketches of some remarkable oaks in the park at Welbeck*. 4to, 33 pp, 10 pls, White, London 1790.

3237. —— *A description of the great oak in Salcey forest, in the County of Northampton*. 8vo, 8 pp, 1 pl., Tupman, Nottingham 1797.

3238. —— *A sketch of the ancient and present state of Sherwood Forest*. 8vo, 30 pp, 4 pls, Tupman, Nottingham 1799.

3239. ROOTSEY, Samuel *Syllabus of a course of botanical lectures*. 8vo, Bristol 1818.

3240. ROPER, Freeman Clark Samuel *The flora of Eastbourne*. Sm. 8vo, xliv+165 pp, 1 map, Van Voorst, London 1875.

3241. [RORISON, Gilbert] *The three barriers: notes on . . . Darwin's Origin of species*. 8vo, 180 pp, text figs, Blackwood, Edinburgh 1861. By G.R.

3242. ROSCOE, *Mrs* Margaret *Floral illustrations of the seasons*. 4to, 55 pls with letterpress, Havell, London 1829; ed. 1831.

Rose, Hugh *The elements of botany*. 1775. *see* Linnaeus, C.

3243. ROSENBERG, *Miss*. *Museum of flowers*. 8vo, 5 numbers, London 1845–46.

3244. ROSS, Charlotte M. W. *A manual of cryptogamic botany*. 8vo, 77+v pp, for the author, Battersea 1894.

3245. ROSSITER, William *A first book of botany*. 12mo, Allman, London 1866 [=1865]; 2ed. 1867; 4ed. 1874.

3246. ROSTAFINSKI, Josef Tomasz *Contributions to Mycologia Britannica. The Myxomycetes of Great Britain*. 8vo, iv+96 pp, 24 pls, London 1877. Translation of the relevant parts of *Sluzowce (Mycetozoa) monographia*, Paris 1874–75. The adaptation is by M. C. Cooke.

3247. ROTHERHAM, John *The sexes of plants vindicated in a letter to Mr. William Smellie.* 8vo, 43 pp, Edinburgh 1790.
—— *see* Fyfe, A.

Rothesay Royal Aquarium *see* Barker, E. E.

3248. ROUSSEAU, Jean Jacques *Letters on the elements of botany . . . and twenty-four additional letters* [by Thomas Martyn, F.R.S.]. 8vo, xxiii+503 pp, 1 chart, White, London 1785; 2ed. 1787; 3ed. 1791; 4ed. 1794; 5ed. 1796; 6ed. 1802; 7ed. 1807; 8ed. 1815. Translated from relevant parts in *Collection complete des oeuvres*, Geneva 1782. The 38 col. pls in No. 2507 are often found bound with this work.

3249. ROWDEN, Frances Arabella *A poetical introduction to the study of botany*. 4to, 1 col. & 6 plain pls, White, London 1801; 2ed. 8vo, 1812; 3ed. 1818.

3250. ROWE, Joshua Brooking *A catalogue of the mammals, birds, reptiles and Amphibia . . . of Devon*. 8vo, 50 pp, Van Voorst, London 1863. Reprinted from *J. Plymouth nat. Hist. Soc.*

3251. ROWE, Samuel *A perambulation of the antient and royal forest of Dartmoor, and the Venville precincts*. 8vo, iv+292 pp, 10 pls, 1 map, Rowe, Plymouth 1848; 2ed. 1856.

3252. ROWLEY, George Dawson *editor Ornithological miscellany*. 4to, 3 vols, 89 col.+40 plain pls, 3 portrs, 3 maps, Trübner, London 1875–78.

Royal Botanical Society *see* Sowerby, S. M.

3253. ROYAL COLLEGE OF SURGEONS OF ENGLAND *Invitation to surgeons and naturalists for donations to the Museum . . . also, directions for preserving animals*. 4to, 42 pp, the College, London 1826; ed. 1835 title *Directions for collecting and preserving animals*.

3254. —— *Catalogue of the contents of the Museum*. 4to, 2 vols, the College, London 1830–31.

3255. —— *Descriptive and illustrated catalogue of the physiology series of comparative anatomy*. 4to, 5 vols, the College, London 1833–40; 2ed. Vol. 1 1852; ed. 3 vols 1900–07.

3256. —— *Descriptive catalogue of the specimens of natural history in spirit. Vertebrata*. 4to, xxii+148 pp, Taylor & Francis printed, London 1859.
—— *see*
Cobbold, T. S. *Catalogue of Entozoa*. 1866.

Flower, *Sir* William H. *Synopsis of the Museum*. 1862.

—— *Catalogue of the osteology of vertebrates. Pt. 1 Man*. 1879.

—— *ditto. Pt. 2 Mammalia*. 1884.

Hillman, W. A. *Catalogue of the Mammalia and birds in store-bottles*. 1841.

Home, *Sir* Everard *Lectures on comparative anatomy*. 1814–28.

—— & Blizard, *Sir* William *Arrangement of the Hunterian Collection*. 1813.

Owen, *Sir* Richard *Synopsis of the arrangement of the preparations*. 1845.

—— *Catalogue of the osteological series*. 1853.

Quekett, J. T. *Catalogue of plants and invertebrate animals in the dry state*. 1860.

Sharpe, R. B. *Catalogue of the osteology of vertebrates. Pt. 3 Aves*. 1891.

3257. ROYAL GEOGRAPHICAL SOCIETY *Hints to travellers*. 8vo, the Society, 4ed. London 1878; 5ed. 1883; 6ed. 1889; 7ed. 1893; 8ed. 1901; 9ed. 1906; 11ed. 1935. The first three editions appeared in the *Journal* 1854 & 1864 and in the *Proceedings* 1872.

*The Royal natural history*. 1881. *see* Wylde, J.

*The Royal natural history*. 1893–96. *see* Lydekker, R.

3258. *The Royal portfolio of natural history pictures*. 4to, no text, 264 col. pls on 150 leaves, [London c.1897]. A collection of the plates from Allen's Naturalist's Library and Lloyd's Natural History.

*Rudiments of conchology; designed as a familiar introduction to the science*. 1826. *see* Venning, *Miss* M. A.

3259. *Rudiments of zoology*. 8vo, Chambers, Edinburgh [?1836]; ed. 1855.

3260. RUDLER, Frederick William *On natural history museums; with suggestions for the formation of a central Museum in Wales*. 8vo, 20 pp, London 1876.

A rural D.D. *see* Esdaile, D.

3261. RUSKIN, John *Love's meinie. Lectures on Greek and English birds*. 8vo, lectures 1 & 2, 83 pp, Allen, Keston 1873; lecture 3 & appendix, ix+85–195 pp, Allen, Orpington 1881. 1881 has t.p. and preliminaries for whole work, called Vol. 1, all publ.

3262. —— *Prosperina, Studies on wayside flowers*. 8vo, 2 vols, pls, text figs, Allen, Orpington [1875–]1879–86. Vol. 2 consists of pts VII–X only, all publ.

3263. RUSSEL, Alexander *The Salmon*. 8vo, viii+248 pp, Edmonston & Douglas, Edinburgh 1864.

3264. [RUSSELL, Anna] *Catalogue of the plants of Newbury*. 8vo, Newbury 1839.

Russell, C. *see* Gurney, J. H. *the son*.

3265. RUTTER, John *Delineations of the north-western division of the County of Somerset*. 8vo, xxiv+349 pp, 13 pls, 1 portr., 1 map, text figs, by the author, Shaftesbury 1829.

3266. RUTTY, John *An essay towards a natural history of the County of Dublin*. 8vo, 2 vols, 4 pls, 1 map, tabs, Sleater for the author, Dublin 1772.

3267. RYE, Bertram George *A handbook of the British Macro-Lepidoptera*. 8vo, col. pls, London 1895. Pts i–iv, all publ.

3268. RYE, Edward Caldwell *British beetles*. Sm. 8vo, xv+280 pp, 16 col. pls, text figs, Reeve, London 1866; 2ed. 1889; 1890. Crown Series.

Rymsdyk, Anthony van *see* Rymsdyk, J. van.

3269. RYMSDYK, Jan van & RYMSDYK, Anthony van *Museum Britannicum, being an exhibition of a great variety of antiquities and natural curiosities*. Fol., xvi+84 pp, 30 pls, text figs, Moore for the authors, London 1778; 2ed. 1791.

# S

S., M. *Guide to the Gardens of the Royal Botanical Society*. 1900. *see* Sowerby, M. S.

S., R. S. *The complete guide to Clevedon*. 1879. *see* Short, R. S.

3270. SACHS, Ferdinand Gustav Julius von *Text-book of botany, morphological and physiological*. Roy. 8vo, xii+858 pp, text figs, Clarendon, Oxford 1875; 2ed. 1882; ed. pt 2 only 1887 title *Outlines of classification and special morphology of plants*. 1887 ed. is by K. E. von Goebel. Translation of *Lehrbuch der Botanik*, Leipzig 1873, by Alfred William Bennett.

3271. —— *Lectures on the physiology of plants*. 8vo, xiv+836 pp, text figs, Clarendon, Oxford 1887. Translation of *Vorlesungen über Pflanzen-Physiologie*, Leipzig 1882, by Henry Baldwin Ward.

3272. SADLER, John *Narrative of ramble among the wild flowers on the Moffat Hills . . . in 1857*. 8vo, 64 pp, 5 pls of specimens, Muir, Moffat 1858.

3273. —— *Wild flowers and ferns of the Bridge of Earn*. 8vo, 56 pp, Edinburgh 1863.

3274. —— *Notice of Salix Sadleri and Carex frigida*. 8vo, Edinburgh 1874.

3275. —— *Botanical trip to Ben Nevis in . . . 1876*. 12mo, 12 pp, London 1877. Reprinted from *The Gardeners' Chronicle*.

—— *see* Balfour, J. H.

St Albans, *Viscount*. *see* Bacon, F.

3276. ST BRODY, Gustavus Ornano *The flora of Weston and its immediate neighbourhood*. 8vo, viii+174 pp, Weston-super-mare 1856.

3277. ST CLAIR, George *Darwinism and design; or, creation by evolution*. Sm. 8vo, xii+359 pp, Hodder & Stoughton, London 1873.

3278. ST JOHN, Charles William George *Short sketches of the wild sports and natural history of the Highlands*. 8vo, vi+281 pp, Murray, London 1846; 1847; 1849; ed. 1878; 1881; 1893; ed. 1919; ed. 1927. Home and Colonial Library, Vol. XVIII. The title varies.

3279. —— *A tour in Sutherlandshire, with extracts from the field books of a sportsman and naturalist*. 8vo, 2 vols, 9 pls, Murray, London 1849; 2ed. Edinburgh 1884; ed. 1891 title *A sportsman and naturalist's tour in Sutherlandshire*. See also No. 3280.

3280. —— *Natural history and sport in Moray*. 8vo, xlvii+336 pp, 1 pl., Edmonston & Douglas, Edinburgh 1863; ed. 1878; ed. 1882; ed. 1919. Incorporates parts of No. 3279.

3281. ST JOHN, John Seymour *Larva collecting and breeding: a handbook to the larvae of the British Macro-Lepidoptera and their food plants*. Sm. 8vo, iv+162 pp, Wesley, London 1890.

3282. ST JOHN, Percy Bolingbroke *The young naturalist's book of birds: anecdotes of the feathered creation*. 12mo, x+168 pp, 5 pls, Rickerby, London 1838; ed. 16mo [1843].

3283. SAINT-PIERRE, Jacques Henri Bernardin de *Studies of*

*nature*. 8vo, 5 vols, 1 map, Dilly, London 1796; 2ed. 3 vols 1799; 5ed. 4 vols 1809. Translation of *Études de la nature*, 5ed. Paris 1791–92, by Henry Hunter.
*another translation*. 12mo, 4 vols, Cundee, London 1807.
*abridged editions*. 8vo, xvi+448 pp, Myers, London 1798; ed. Vernor & Hood, London 1799; ed. 12mo, Dublin 1800; ed. 12mo, London [?1820].

3284. —— *Harmonies of nature, being a sequel to Studies of nature*. 8vo, 3 vols, 1 portr., Baldwin Craddock & Joy, London 1815. Translation of *Harmonies de la nature*, Paris 1815, by W. Meeston.

3285. SAINTER, Joseph Denby *The jottings of some geological . . . botanical, ornithological and zoological rambles around Macclesfield*. 8vo, viii+157 pp, 29 pls, 1 col. map, Swinnerton & Brown, Macclesfield 1878.

3286. SALISBURY, Richard Anthony *The generic characters in the English botany collated with those of Linné*. 8vo, iv+34 pp, Hatchard, London, 1807. 100 copies.

3287. —— *The genera of plants . . . a fragment containing part of Liriogamae*. 8vo, vi+143 pp, Van Voorst, London 1866.

Salisbury, W. *Hortus siccus gramineus*. 1802–06. *see* Curtis, W.

3288. SALMON, John Drew *Flora of Surrey; or, a catalogue of the flowering plants and ferns found in the County*. 8vo, xiv+367 pp, 2 maps in pocket, London 1863. Compiled by Brewer, James Alexander.

3289. SALMON, Nathaniel *Antiquities of Surrey . . . with some account of the present state and natural history of the County*. 8vo, 204 pp, for the author, London 1736.

*Salmonia: or days of fly fishing*. 1828. *see* Davy, *Sir* Humphry.

3290. [SALTER, James] *A catalogue of the rarities to be seen at Don Saltero's coffee-house in Chelsea*. 8vo, London 1729. Many later editions. By Don Saltero.

3291. SALTER, John Henry *List of the birds of Aberystwyth and neighbourhood*. 8vo, 19 pp, University of Wales Scientific Society, Aberystwyth [1900].

Salter, John William *see* Johnson, C. P.

3292. SALTER, Thomas Bell *A short account of the botany of Poole*. 8vo, 27 pp, Sydenham printed, Poole 1839. Reprinted from Sydenham, J. *History of Poole*, 1839.

Saltero, Don *pseudonym. see* Salter, J.

3293. SAMOUELLE, George *The entomologist's useful compendium; or, an introduction to the knowledge of British insects*.

8vo, 496 pp, 12 col. pls, Boys, London 1819; ed. 1824; 2ed. 1836. Plates often in duplicate, coloured and plain.

3294. —— *A nomenclature of British entomology*. 8vo, 44 pp, Longman, London 1819. Printed on one side.

3295. —— *General directions for collecting and preserving exotic insects and Crustacea*. 12mo, 70 pp, 4 col. pls, Longman, London 1826.

3296. —— *The entomological cabinet; being a natural history of British insects*. 12mo, 2 vols, 156 col. pls, Longman, London [1832–] 1833, 34; 2ed. 1 vol. 1841.

3297. SAMUELSON, James & HICKS, John Braxton *Humble creatures. I, The earthworm and the common housefly*. 8vo, viii+78 pp, 8 pls, Van Voorst, London 1858; 2ed. 1860.

3298. SANDEMAN, George *Problems of biology*. Sm. 8vo, 213 pp, Sonnenschein, London 1896.

3299. SANDERS, Samuel John W. *Brief notes on structural botany*. 12mo, Educational Trust Company, London [1870].

3300. [SANDHAM, *Mrs* Elizabeth] *Conversations on natural history*. Sm. 8vo, 128 pp, Baker, Southampton [?1815]. By the author of *The twin sisters*.

3301. SANDS, J. *Out of the world; or life in St Kilda*. 8vo, 92 pp, 15 pls, Maclachlan & Stewart, Edinburgh [1876]; 2ed. 1878 [=1877].

3302. SATCHELL, Thomas *Provisional index to a glossary of fish names*. 8vo, 12 pp, privately printed, London [1879].

3303. SAUNDERS, Edward *A catalogue of British Hemiptera-Heteroptera*. 8vo, 13 pp, London 1876; ed. Perth 1908 title *Catalogue of British Hemiptera*. 1908 edition has Homoptera by Edwards, J.

3304. —— *Synopsis of British Hemiptera-Heteroptera*. 8vo, 3 pts, [156] pp, 1 pl., Janson, London 1876. Reprinted from *Trans. ent. Soc. Lond.*, 1875–76. Original pagination is retained.

3305. —— *Catalogue of the British Hymenoptera, (Aculeata)*. 8vo, Milne Tannahil & Methven printed, Perth 1890; ed. 1902.

3306. —— *The Hemiptera-Heteroptera of the British Islands*. Roy. or Demy 8vo, vi+350 pp, 31 col.+1 plain pls, Reeve, London 1892. Demy copies have only the one plain plate.

3307. —— *The Hymenoptera Aculeata of the British Islands*. Roy. or demy 8vo, viii+391 pp, 51 col.+3 plain pls, Reeve, London 1896. Demy copies have only the three plain plates.

—— *see* Newbery, E. A.

—— *see* Saunders, G. S.

3308. SAUNDERS, George Sharp & SAUNDERS, Edward *Natural history of Reigate and its vicinity. List of land and fresh-water Mollusca*. 8vo, 8 pp, Newman for the Holmesdale Natural History Club, London [1861].

3309. SAUNDERS, Howard *A list of British birds, revised to April 1887*. 8vo, 22 pp, London 1887; ed. 1892 title . . . *revised to July 1892*; ed. 1905 title . . . *revised to July 1905*.

3310. —— *An illustrated manual of British birds*. 8vo, xl+754 pp, 3 col. maps, text figs, Gurney & Jackson, London [1888–] 1889; 2ed. [1897–] 1899; 3ed. 1927. Based on Yarrell, W. *A History of British birds*, 4ed. 1871–85.

3311. SAUNDERS, Samuel *A short and easy introduction to scientific and philosophical botany*. 8vo, viii+107 pp, White, London 1782.

3312. SAUNDERS, William Wilson, SMITH, Worthington George & BENNETT, Alfred William *Mycological illustrations*. 8vo, 2 pts, 48 pls with letterpress, Van Voorst, London 1871, 72. All publ.

Saville-Kent, William *see* Kent, W. S.

3313. SAWYER, Frederick Ernest *The birds and mammals of Sussex*. Sm. 8vo, 11 pp, Infield for the author, Brighton [1875]. Reprinted from *The Brighton Daily News*.

3314. —— *An old boy's natural history reminiscences*. 8vo, 12 pp, King & Thorne, Brighton 1886.

3315. SAXBY, Henry Linckmyer *The birds of Shetland*. Roy. 8vo, xv+398 pp, 8 tinted pls, Maclachlan & Stewart, Edinburgh 1874.

Scarabaeus *pseudonym see* Badham, C. D.

*The Scarborough natural historians*. 1829. *see* Cole, J.

*Scenes of industry displayed in the bee-hive and the ant-hill*. [?1829]. *see* Johnstone, *Mrs* C. I.

3316. SCHACHT, Hermann *The microscope in its special application to vegetable anatomy and physiology*. Sm. 8vo, 1 pl., text figs, Highley, London 1853; 2ed. 1855. Library of Science and Art. Section II, Natural History.

3317. SCHARFF, Robert Francis *The history of the European fauna*. Sm. 8vo, vii+364 pp, text figs, Scott, London 1899. Contemporary Science Series, No. XVII.

Schenk, Heinrich *see* Strasburger, E.

3318. SCHERREN, Henry *Ponds and rock pools*. Sm. 8vo, 208 pp, 1 pl., text figs, RTS, London 1894; 1900; 1906; 1919.

3319. —— *Popular history of animals for young people*. 8vo, vii-

+376 pp, 13 col. pls, text figs, Cassell, London 1895; 1896.

3320. —— *Through a pocket lens*. 8vo, 192 pp, 1 pl., text figs, London 1897; 1904.

Schimper, Andreas Franz Wilhelm *see* Strasburger, E.

3321. SCHLEGAL, Herman *Essay on the physionomy of serpents*. 8vo, vii+254 pp, 2 pls, 1 map, Edinburgh 1843. Translation of *Essai sur la physiognomie des serpens*, Amsterdam 1837, by Thomas Stewart Traill.

3322. SCHLEIDEN, Matthias Jacob *The plant, a biography*. 8vo, viii+365 pp, 5 col. + 1 plain pls, text figs, Baillière, London 1848; 2ed. 1853. Translation of *Die Pflanzen und ihr Leben*, Leipzig 1848, by Arthur Henfrey.

3323. —— *Principles of scientific botany; or, botany as an inductive science*. 8vo, viii+616 pp, 6 pls, text figs, Longman, London 1849. Translation of *Die Botanik als inductive Wissenschaft*, 2ed. 1845–46, by Edwin Lankester.

3324. SCHMEIL, Franz Otto *Text-book of zoology*. 8vo, 3 pts, xvi+493 pp, text figs, Black, London 1900–01. Translation of *Lehrbuch der Zoologie*, Leipzig 1898, by Rudolph Rosenstock.

3325. SCHMIDT, Eduard Oskar *The doctrine of descent and darwinism*. Sm. 8vo, vi+334 pp, text figs, King, London 1875; 2ed. 1875; 5ed. 1883; 7ed. 1887; 8ed. 1905. International Scientific Series, Vol. XII. Translation of *Descendenzlehre und Darwinismus*, Leipzig 1884.

3326. —— *The Mammalia in their relation to primeval times*. Sm. 8vo, xi+308 pp, text figs, Kegan Paul, London 1885; 2ed. 1894. International Scientific Series, Vol. LIV. Translation of *Säugethiere in ihrem Vërhaltniss zur Vorwelt*, Leipzig 1884.

3327. SCHOEDLER, Friedrich Carl Ludwig *The book of nature*. Sm. 8vo, 2 pts, xii+578 pp, text figs, Griffin, London 1851; 2ed. 1854; 3ed. 1855; 5ed. 1859; ed. 1862 title *Treasury of science*; ed. [1872]. Translation of *Das Buch der Natur*, Brunswick, 5ed. 1850, by Henry Medlock.

3328. SCHOUW, Joakim Frederik *The earth, plants and man*. Sm. 8vo, 402 pp, 1 map, Bohn, London 1852; 1859. Bohn's Scientific Library. Translation of *Natur-Skildringer*, Copenhagen 1839–42, by Arthur Henfrey.

3329. SCHUBERT, Gotthilf Heinrich von *Natural history of the animal kingdom for the use of young people*. Fol., 3 pts, pls, SPCK, London 1889. Also issued in 3 parts; 1. Mammalia; 2.

Birds; 3. Reptiles fish etc. Translation of *Die Geschichte der Natur*, Erlangen, by William Forsell Kirby.

3330. SCHWANN, Theodor *Microscopial researches into the accordance in the structure and growth of animals and plants*. 8vo, xx+268 pp, 6 pls, Sydenham Society, London 1847. Translation of *Microscopische Untersuchungen*, Berlin 1839, by Henry Smith.

*Science Ladders*. *see* Bell, Nancy, *Mrs* Arthur.

3331. *Science lectures for the people*. 8vo, 11 series in 9 vols, text figs Heywood, Manchester 1866–80; ed. 6 vols 1883–85. There are several reissues of parts of this series. The individual contributions are signed.

*The scientific tourist through Ireland*. 1818. *see* Walford, T.

3332. SCLATER, Philip Lutley *Zoological sketches . . . made for the Zoological Society*. Fol., 50 col. pls with letterpress, Graves, London 1861. The plates are by Wolf, J. and it is often listed under his name.

3333. —— *Zoological sketches . . . second series*. Fol., 50 col. pls with letterpress, Graves, London 1867. The plates are by Wolf, J. and it is often listed under his name.

3334. —— *On the present state of our knowledge of geographical zoology*. 8vo, 49 pp, London [1876]. British Association address.

—— *Guide to the gardens of the Zoological Society*. 1859. *see* Zoological Society of London.

—— *see* Sclater, W. L.

3335. SCLATER, William Lutley & SCLATER, Philip Lutley *The geography of mammals*. 8vo, xviii+335 pp, 8 col. maps, text figs, Kegan Paul, London 1899. Ch. 1–7 by W. L. S. reprinted from *Geogr. J.*, 1894–97. Ch. 8 by P. L. S. reprinted from *Proc. zool. Soc. Lond.*, 1897.

3336. SCOFFERN, John B. *Outlines of botany*. 8vo, vi+297 pp, text figs, Kent, London [1857]. Cassell's Educational Course.

3337. SCOTT, Dukinfield Henry *Introduction to structural botany*. 8vo, 2 pts, 1 pl., text figs, Black, London 1894, 96; pt. 1 2ed. 1894; 3ed. 1895; 4ed. 1897; 5ed. 1899; 6ed. 1902; 6ed. 1906; 8ed. 1912; 11ed. 1927; 12ed. 1933; pt. 2 2ed. 1897; 3ed. 1899; 4ed. 1904; 5ed. 1907; 6ed. 1912; 7ed. 1917; 10ed. 1932; 12ed. 1955.

Scott Elliot, George Francis *see* Elliot, C. F. S.

Scott, John *see* Douglas, J. W.

Scully, Reginald William *see* Colgan, N.

3338. *Sea-anemones; or, tanks and their inhabitants*. 8vo, Simpkin, London 1856.

3339. SEAGER, Herbert West *editor Natural history in Shakespeare's time*. 8vo, viii+358 pp, text figs, Elliot Stock, London 1896; reprint 1972.

3340. [SEALEY, Alfred Forbes] *A classified list of the names and latin synonyms of the British birds*. 8vo, vi+27 pp, Webb, Cambridge 1853. Printed on one side.

*Sea-side pleasures*. 1853. *see* Gosse, P. H.

*Sea-weeds*. [c.1855]. *see* Johns, C. A.

3341. *The seasons displayed: or an account of the various productions of nature.* 12mo, 72 pp, text figs, Pratt, Coventry [?1820].

3342. SEDGWICK, Adam *A student's text book of zoology*. Roy. 8vo, 3 vols, text figs, Sonnenschein, London 1898–1909; ed. 1927–32.

3343. SEEBOHM, Henry *A history of British birds, with coloured illustrations of their eggs*. Roy. 8vo, 3 vols & atlas vol., 68 col. pls. Porter for the author, London 1883–85; ed. 1883–85 [=1896]; issue of atlas 1904.

3344. —— *Classification of birds; an attempt to diagnose the subclasses . . . of existing birds*. Roy. 8vo, xi+53 pp, Porter, London 1890; suppl. viii+28 pp, 1895.

3345. —— *On the geographical distribution of British birds: appendix – brief summary . . . arranged for labels*. Roy. 8vo, vi+39+[72] pp, Porter, London 1893. Appendix printed on one side. Text reprinted in No. 3343, 1896.

3345a —— *On the classification of British birds*. Roy. 8vo, Porter, London 1895. Reprinted in No. 3343, 1896.

3346. —— *Coloured figures of the eggs of British birds, with descriptive notices.* Roy. 8vo, xxiv+304 pp, 60 col. pls, Pawson & Brailsford, Sheffield 1896. Plates 1–59, 58*. Text largely by Sharpe, R. B.

3347. SEELEY, Harry Govier *The fresh-water fishes of Europe.* Roy. 8vo, x+444 pp, 1 pl., text figs, Cassell, London 1886.

3348. SEEMAN, Berthold Carl *The British ferns at one view.* 8vo, 2 pp, 1 col. pl., Van Voorst, London [1860].

3349. SELBY, Prideaux John *Illustrations of British ornithology.* 4to, 19 numbers, Constable, Edinburgh 1821–34; ed.

Vol. 1 only 8vo, 1825; ed. 2 vols 8vo, 1833. This is the text for No. 3350.

3350. —— *Illustrations of British ornithology*. Fol., 2 vols, 214 col. + 4 plain pls, Vol. 1 Constable, Vol. 2 Lizars, Edinburgh [1818–] 1834; issue 1834; issue 1841. Rarely may be as many as 219 col. pls.

3351. —— *Pigeons.* Sm. 8vo, 228 pp, 30 col. pls, 1 portr., Lizars, Edinburgh 1835; ed. [1845–46]. Jardine's Naturalist's Library Vol. XI.

3352. —— *A history of British forest trees.* Roy. or demy 8vo, xx+450 pp, text figs, Van Voorst, London [1841–]1842. Natural History of the British Isles Series.

3353. SELENKA, Emil *A zoological pocket-book: or, a synopsis of animal classification*. 8vo, viii+237 pp, Griffin, London 1890. Translation of *Zoologische Taschenbuch*, 3ed. Erlangen [ ], by James Richard Ainsworth Davis.

3354. [SELLON, Gertrude] *Short stories about animals for little children, in words of one syllable.* 8vo, 92 pp, col. pls, text figs, Griffith & Farran, London 1882; ed. 4to 1892 signed.

3355. SELWYN, *Miss* Amelia *A key or familiar introduction to the science of botany.* 12mo, 122 pp, 3 pls, Souter, London 1824. Perhaps a pseudonym for Stubs, *Miss* Amelia.

3356. SEMPER, Carl Gottfried *The natural conditions of existence as they affect animal life.* Sm. 8vo, xvi+472 pp, text figs, Kegan Paul, London 1881; 2ed. 1882; 3ed. 1883; 4ed. 1889; 5ed. 1899. International Scientific Series, Vol. XXXI. Translation of *Die Natürlichen Existence bedingungen,* Leipzig 1880.

3357. SEMPLE, Charles Edward Armand *Aids to botany.* 12mo, Baillière, London 1877; ed. [1883].

Senilis, Johannes *pseudonym see* Nelson, J.

3358. *The senses. Containing the eye, the tongue, the hand, the ear, the sense of smell*. 16mo, iv+160 pp, col. text figs, RTS, London [ ].

*A series of botanical labels for the herbarium*. 1841. *see* Cowell, M. H.

3359. *A series of thirty studies from nature drawn on stone*. Fol., 30 pls, Ackermann, London 1820.

Seton, Ernest Evan Thompson *see* Thompson, E. E. S.

3360. SETON, George *St Kilda past and present*. 8vo, xv+346 pp, 11 pls, 2 maps, Blackwood, Edinburgh 1878.

3361. SHARP, David *A revision of the British species of Homalota*. 8vo, 182 pp, Janson, London 1869. Reprinted from *Trans. ent. Soc. Lond.*, 1869.

3362. —— *Catalogue of British Coleoptera*. 8vo, 37 pp, Janson, London 1871; 2ed. 1883.

3363. —— *The object and method of zoological nomenclature*. 8vo, 39 pp, Janson, London 1873.

3364. —— & FOWLER, William Weeks *Catalogue of British Coleoptera*. 8vo, 46 pp, Reeve, London 1893.

3365. SHARPE, Richard Bowdler *Young collector's handbook of British birds*. Sm. 8vo, 32 pp, text figs, Sonnenschein, London [1883]. Young Collector Handbook Series.

3366. —— *Birds in nature*. 4to, 78 pp, 39 col. pls, text figs, Sampson Low, London 1888.

3367. —— *Catalogue of the specimens illustrating the osteology and dentition of the vertebrated animals. Part III. Aves*. 8vo, lvii+460 pp, Royal College of Surgeons, London 1891. For Parts I & II see Flower, *Sir* William H.

3368. —— *A review of the recent attempts to classify birds*. 8vo, 90 pp, 1 col. + 2 plain pls, text figs, 2nd International Ornithological Congress, Budapest 1891.

3369. —— *A handbook to the birds of Great Britain*. Sm. 8vo, 4 vols, 128 col. pls, Shaw, London 1894–97; issue 1896–97. First issue is Allen's Naturalist's Library, second is Lloyd's Natural History.

3370. —— *A chapter on birds . . . rare British visitors*. Sm. 8vo, xii+124 pp, 18 col. pls, SPCK, London 1895.

3371. —— *Sketch book of British birds*. Sm. 4to, xx+255 pp, 1 col. pl., text figs, SPCK, London 1898.

3372. —— *Wonders of the bird world*. 8vo, xvi+399 pp, pls, text figs, Gardner Darton, London [1898].

—— *see* Seebohm, H.

3373. SHAW, C. *Illustrations of ferns for amateurs. Part 1*. 8vo, 20 pp, 20 col. pls, Wolverhampton [1878]. All publ.

3374. SHAW, G. F. *The panorama of nature, or a comprehensive view of natural history and botany, comprising the whole of Goldsmith's History of the earth*. 4to, xii+812 pp, 78 col. + 2 plain pls, London 1819.

3375. SHAW, George *Speculum Linnaeanum . . . or, Linnaean*

*zoology*. 4to, 8 col. pls with letterpress, J. Davis printed, London 1780. All publ.

3376. —— *Musei Leveriani explicatio, anglica et latina. Museum Leverianum, containing select specimens from the Museum of the late Sir Ashton Lever*. 4to, 2 vols, 72 col. pls, Parkinson, London 1792, 96. Vol. 2 has pp. 1–48, all publ.

3377. —— *General zoology, or systematic natural history*. 8vo, 14 vols in 28, 1217 pls, Arch, London 1800–26.

3378. —— *Zoological lectures, delivered in the years 1806 and 1807 at the Royal Institution*. Roy. or demy 8vo, 2 vols, 167 pls, Kearsley, London 1809. L.P. a few col. copies. Pls 1–163, 36*, 132*, 159*, 159**.

3379. —— *The naturalist's pocket book*. 12mo, 127 pp, 13 col. pls, London 1813.

—— *see* Miller, J. J. F.

3380. —— *The panorama of nature, or a comprehensive view of natural history and botany*. 4to, 80 col. pls, London 1817.

3381. —— *A select cabinet of natural history; with an historical account of the silkworm, and an elegant method of obtaining very exact and pleasing representations of plants*. 32mo, iv+100 pp, 26 col. pls, Wright, London 1820. Plate numbers erratic.

Shaw, James *see* Wallace, R.

3382. ——, NODDER, Frederick P. & NODDER, Richard P. *Vivarium naturae, sive rerum naturalium icones. The naturalist's miscellany*.. 4to, 24 vols, 1064 col. + 1 plain pls, Nodder, London [1789–] 1790–1813. Vols 1–12 by G. S. & F. P. N., Vols 13–24 by G. S. & R. P. N. Continued as *The zoological miscellany*, by Leach, W. E. A few copies on large paper.

3383. SHAW, Simeon *Nature displayed in the heavens and on the earth*. 8vo, 6 vols, 51 col. + 199 plain pls, text figs, Whittaker for *Sir* Richard Phillips, London 1823. The plates also published in folio title *Atlas of nature*. Derived from Pluche, N. A. *Spectacle de la nature*.

3384. *Shells and their inmates*. 16mo, xvi+214 pp, 1 col. pl., text figs, RTS, London 1841.

3385. SHEPPARD, John *On trees, their uses and biography*. 12mo, 162 pp, Jackson, London 1848.

3386. SHERBORN, Charles Davies *An index to the generic and trivial names of animals, described by Linnaeus, in the 10th and 12th editions of his Systema naturae*. 8vo, viii+108 pp, Dulau, London 1899. Manchester Museum Handbook, No. 9.

3387. SHIELD, Richard *Practical hints respecting moths and butterflies*. 12mo, 194 pp, Van Voorst, London 1856.

3388. SHIPLEY, *Sir* Arthur Everett *Zoology of the Invertebrata*. 8vo, viii+458 pp, text figs, Black, London 1893.
—— *see* Harmer, S. F.

3389. SHOBERL, Frederic *The natural history of quadrupeds*. 16mo, 2 pts, 64 pls, Harris, London 1834; 2ed. [c.1843]. Little Library, Vols 13 & 14.

3390. —— *The natural history of birds*. 16mo, viii+275 pp, 13 pls, Harris, London 1836. Little Library, Vol. 18.

3391. SHORE, Thomas William *Elementary practical biology. Vegetable*. 8vo, viii+173 pp, Churchill, London 1887.

3392. *A short and easie method to give children an idea or true notion of celestial and terrestrial beings: and to teach them the names of most things that are useful and necessary to human life Méthodé courte et facile* . . . 8vo, 38 pp, text figs, J. Nutt printed, in the Savoy [London] 1710. Text in French & English.

3393. *Short chapters on natural history*. 16mo, 207 pp, text figs, SPCK, London [c.1875].

3394. *A short description of sixty-four beasts, birds, fishes, & insects*. 16mo, 88 pp, 16 pls, Harris, London 1818.

3395. *A short history of birds and beasts*. 32mo, 24 pp, text figs, Houlston, Wellington [c.1830].

3396. *A short introduction to natural history intended for the rising generation*. 16mo, text figs, Crosby, London 1807.

*A short history of insects . . . a pocket companion to those who visit the Leverian Museum*. 1797. *see* Fenn, *Lady* Eleanor.

*Short stories about animals*. 1882. *see* Sellon, G.

*Short studies from nature*. [c.1885]. *see* Dallas, W. S. *and others*.

3397. [SHORT, R. S.] *The complete guide to Clevedon . . . with notes on the natural history*. 8vo, viii+253 pp, 8 pls, 1 map, Clevedon [1879]. By R. S. R.

3398. SHUCKARD, William Edward *Essay on the indigenous fossorial Hymenoptera*. 8vo, xii+259 pp, 1 col. + 3 plain pls, 1 chart, for the author, London 1837.

3399. —— *Elements of British entomology: Part 1*. 8vo, 240 pp, 1 pl., text figs, Baillière, London 1839. All publ.

3400. —— *The British Coleoptera delineated*. 8vo, vii+83 pp, 94 pls, Crofts, London 1840; ed. 1861. Plates by Spry, William J. and the work is often known as Spry & Shuckard.

3401. —— *British bees*. Sm. 8vo, vi+371 pp, 16 col. pls, Reeve, London 1866; ed. [ ]. Crown Series.

—— *see* Swainson, W.

3402. SHUFELDT, Robert Wilson *The myology of the raven (Corvus corax sinuatus), a guide to the study of the muscular system in birds*. 8vo, xix+343 pp, text figs, Macmillan, London 1890.

3403. SHUTE, Daniel Kerfoot *A first book in organic evolution*. Sm. 8vo, xvi+285 pp, 12 col. pls, text figs, Kegan Paul, London 1899.

3404. SIBBALD, *Sir* Robert *Nuncius Scoto-Britannus, sive adminitio de atlante Scotica*. Fol., 15 pp, Lindsay, Edinburgh 1683. A duplicate of pp. 1–15 of No. 3405 with variant title leaf.

3405. —— *Scotia illustrata, sive prodromus historiae naturalis*. Fol., 15+102+114+56 pp, 22 pls on 20 leaves, portr., Kniblo for the author, Edinburgh [1683–]1684; English ed. London 1684 title *Scotland illustrated*; ed. 1727. Pages 1–15 comprise unsold copies of No. 3404 and are present in all copies.

3406. —— *Phalainologia nova; sive observationes de rarioribus quisbusdam balaenis in Scotia littus nuper ejectis*. 4to, 105 pp, Redi, Edinburgh 1692; ed. London 1773.

3407. —— *Auctarium Musaei Balfourniani, e Musaeo Sibbaldiano*. 8vo, Edinburgh Academy, Edinburgh 1697.

3408. —— *A letter from Sir R---- S---- to Dr. A. Pitcairn*. Sm. 8vo, Edinburgh 1709.

3409. —— *The history, ancient and modern of the Sheriffdoms of Fife and Kinross . . . with an account of the natural products of the land and waters*. Fol., 164 pp, 1 pl., 2 maps, Watson for the author, Edinburgh 1710; ed. Cupar 1803.

3410. —— *Vindiciae Scotiae illustratae sive Prodromi naturalis historiae Scotiae, contra prodromomastiges sur larva libelli De legibus historiae naturalis latentes*. Fol., 30 pp, Edinburgh 1710.

3411. SIBLY, Ebenezer *An universal system of natural history*. 8vo, 4 vols, 267 col. pls, for the proprietor, London [1794–1904]; 2ed. 14 vols, [1794–1810] title *A genuine and universal . . .* Translation of Gmelin, F. of Linnaeus, C. *Systema naturae*, with additional matter. Only Vols 1–3 are by S., the rest by the editors of the *Encyclopaedia Londiniensis*, i.e. Wilks, J.,

Jones, J. & Jones, G. Engraved title to Vol. 1 *Magazine of natural history* etc. S's name on engraved title only. There is an issue of the plates of the first seven vols, without text or printed title page, and without S's name on engraved title.

3412. SIBTHORP, John *Flora Oxoniensis; exhibens plantas in agro Oxoniensi sponte crescentes*. 8vo, xxiv+422 pp, Sheldonian, Oxford 1794.

3413. SIDDALL, John D. *Classification of the vegetable kingdom*. Fol., s. sh., Chester [1886].

3414. —— & BRADY, Henry Bowman *Catalogue of British recent Foraminifera*. 8vo, 10 pp, Griffith printed, Chester 1879.

3415. SIEBOLD, Carl Theodor Ernst von *On a true parthenogenesis in moths and bees*. 8vo, vii+110 pp, 1 pl., Van Voorst, London 1857. Translation of *Wahre Parthenogenesis bei Schmetterlingen und Bienen*, Leipzig, 1856, by William Sweetland Dallas.

3416. —— *On tape and cystic worms*. 8vo, Sydenham Society, London 1857. Translation of *Ueber die Band-und-Blasenwürmer*, Leipzig 1854, by Thomas Henry Huxley.

3417. —— & STANNIUS, Hermann *Comparative anatomy. Vol. 1 Anatomy of the Invertebrata*. 8vo, xiv+470 pp, Trübner, London 1854. Translation of *Lehrbuch der vergelichenden Anatomie*, 2 vols, Berlin 1848, 46, by Waldo Irving Burnett. Vol. 1 all publ., Boston, Mass printed.

3418. *Sights in all seasons. Spring. (Summer, Autumn, Winter.)* 16mo, 4 pts, 372 pp, 1 col. pl., text figs, RTS, London [1844–?45]. The parts available separately.

3419. SILVER, Alexander *Outlines of elementary botany*. 18mo, 380 pp, Renshaw, London 1866 [=1865]; ed. 1877 [=1876].

3420. SIM, George *The food of fishes, both in fresh and salt water*. 8vo, 50 pp, Clowes, London 1883. Great International Fisheries Exhibition, Vol. XI.

3421. SIMEON, Cornwall *Stray notes on fishing and natural history*. Sm. 8vo, xx+263 pp, 1 pl., text figs, Macmillan, Cambridge 1860; ed. London 1863.

Simmons, Samuel Foart *see* Person, Claude.

3422. SIMMS, George E. *Butterfly and moth collecting; where to search and what to do*. Sm. 8vo, 116 pp, text figs, Upcott Gill, London [1892].

3423. *Simple studies in natural history. Birds.* 12mo, [36] pp, col. text figs, Darton, London [c.1830].

3424. SIMPSON, Archibald *Hieroglyphica animalium terrestrium, volatilium, natatilium, reptilium, insectorum, vegetivorum, metallorum, lapidum*. 4to, 3 pts, Finlason, Edinburgh 1622–24.

3425. SIMSON, James *Contributions to natural history, and papers on other subjects*. 8vo, Edinburgh 1875; 2ed. London 1880.

3426. SINCLAIR, George *Hortus Gramineus Woburnensis: or an account of the results of experiments on the produce and nutritive qualities of different grasses*. Fol., vi+lxvi+316 pp, 123 pls of grass specimens and 35 seeds, M'Millan printed, London 1816; ed. 8vo 60 col. pls, 1824; 2ed. 1825; 3ed. 1826; 4ed. [1838]; 5ed. [1869].

*Sister Mary's tales in natural history*. 1834. *see* Roberts, Mary

*A sister's stories*. 1833. *see* Martin, Selina

Skepper, Edmund *see* Henslow, J. S.

Skertchly, Sydney Barber Josiah *see* Miller, S. H.

3427. SKETCH, Sally *pseudonym An alphabetical arrangement for little naturalists*. 16mo, t.p.+28 col. pls, Harris, London 1821.

*Sketches from nature; or, hints to juvenile naturalists*. 1830. *see* Traill, *Mrs* C. P.

3428. *Sketches in natural history. History of the Mammalia*. Sm. 8vo, 6 vols, text figs, C. Cox, London 1849. Knight's Monthly Volume.

*Sketches of the physiology of vegetable life*. 1811. *see* Jackson, *Mrs* M. E.

3429. SKRIMSHIRE, Fenwick *A series of essays introductory to the study of natural history*. 12mo, 2 vols, Johnson, London 1805.

3430. SKUSE, Frederick A. Askew *British stalk-eyed Crustacea and spiders*. Sm. 8vo, 128 pp, 1 pl., text figs, Sonnenschein, London 1887. Young Collector Series.

3431. SLACK, Henry James *Marvels of pond-life*. Sm. 8vo, xix+197 pp, 7 col. pls, text figs, Groombridge, London 1861; 2ed. 1871; 3ed. 1878; 4ed. 1880; 1889; ed. 1891.

3432. SLADEN, Frederick William Lambart *The humble bee, its life history and how to domesticate it*. 8vo, 35 pp, 1 col. pl., text figs, the author, Ripple Court near Dover 1892; ed. 1912. First ed. stencilled.

3433. —— *A list of wild bees (Anthophila) of the Deal and Dover district*. 8vo, 11 pp, text figs, the author, Ripple Court near Dover 1897. Stencilled.

3434. SLANEY, Robert Aglionby *An outline of the smaller British birds; intended for the use of ladies and young persons*. Sm. 8vo, 143 pp, text figs, Longman, London 1832; 2ed. 1833.

3435. SMART, Thomas Gregory *Birds on the British list, their title to enrolement considered especially with reference to the British Ornithological Union's list*. 8vo, xxiv+148 pp, London 1886. For the work referred to see No. 481.

3436. SMEE, Alfred *Instinct and reason: deduced from electrobiology*. 8vo, xxxiv+320 pp, 7 col.+3 plain pls, text figs, Reeve, London 1850.

3437. SMELLIE, William *The philosophy of natural history*. 4to, 2 vols, Vol. 1 Heirs of Charles Elliot, Vol. 2 Bell & Bradfute, Edinburgh 1790, 99; ed. 8vo, 2 vols, Dublin 1790. Vol. 1 not so-called. Dublin ed. is contents of Vol. 1 only.

3438. SMILES, Samuel *Life of a Scotch naturalist: Thomas Edward. Appendix – Selections from the fauna of Banffshire*. 8vo, xix+438 pp, 6 pls, 1 portr., text figs, Murray, London 1876; 3ed. 1877; ed. 1891.

3439. —— *Robert Dick: baker of Thurso, geologist and botanist*. 8vo, xx+436 pp, 13 pls, 1 portr., text figs, Murray, London 1878.

3440. SMITH, Alfred Charles *The birds of Wiltshire*. 8vo, xxv+588 pp, Porter, London 1887.

Smith, Annie Lorraine *see* Crombie, J. M.

3441. SMITH, Cecil *The birds of Somersetshire*. 8vo, x+643 pp, Van Voorst, London 1869.

3442. —— *The birds of Guernsey and the neighbouring islands, Alderney, Sark, Jethou, Herm*. Sm. 8vo, xix+223 pp, Porter, London 1879.

3443. SMITH, Charles Hamilton *Introduction to Mammalia*. Sm. 8vo, 313 pp, 31 col. pls, 1 portr., Lizars, Edinburgh 1842; issue [1845–46]; ed. 1858. Jardine's Naturalist's Library, Vol. XXV, 2nd issue Vol. XV, Mammalia Vol. I.

3444. SMITH, *Mrs* Charlotte *A natural history of birds intended chiefly for young persons*. 12mo, 2 vols, 22 col. pls, Johnson, London 1807; ed. 2 vols in one, 1819.

3445. SMITH, Christopher Parker *The moss flora of Sussex; together with notes on the structure and reproduction of mosses*. 8vo, 16 pp, Brighton & Sussex Natural History Society, Brighton 1870.

3446. SMITH, Edward *Structural and systematic botany*. Sm. 8vo, text figs, Houlston & Stoneman, London 1854–56.

Reprinted from Orr, W. S. *Circle of the sciences*, Vol. II & III.
—— *see* Ansted, D. T.

3447. —— & DALLAS, William Sweetland *A system of natural history: being a structural and classified arrangement of plants and animals*. Sm. 8vo, 2 vols, text figs, Houlston & Stoneman, London 1855. This is Orr, W. S. *Circle of the sciences. Organic nature*. Vols II & III with an additional title page.

3448. [SMITH, Frederick] *List of specimens of British animals in the collection of the British Museum. Part VI: Hymenoptera Aculeata*. 12mo, 134 pp, British Museum, London 1851.

3449. [——] *List of specimens of British animals in the collection of the British Museum. Part XIII. Nomenclature of Hymenoptera*. 12mo, 74 pp, British Museum, London 1853.

3450. —— *Catalogue of British Hymenoptera in the British Museum. Part 1. Apidae – Bees*. 12mo, 252 pp, 11 pls, British Museum, London 1855; 2ed. 8vo, (*Pts. 1. & 2. Andrenidae and Apidae*), 1876; issue 1891, title *Catalogue of British bees*.

3451. —— *Catalogue of the British fossorial Hymenoptera, Formicidae, and Vespidae in the British Museum*. 12mo, 236 pp, 6 pls, British Museum, London 1858.

3452. —— *A catalogue of the British Hymenoptera; Aculeata.* 8vo, viii+44 pp, Napier for the Entomological Society of London, London 1858; ed. 1871.

3453. SMITH, Gerard Edwards *A catalogue of rare and remarkable phaenogamous plants, collected in south Kent.* 8vo, viii+76 pp, 6 col. pls, London 1829.

3454. SMITH, Henry *Flora Sarisburiensis; a repository of English botany.* Roy. 8vo, viii+67 pp, 30 col. pls, Brodie & Dowding printed, Salisbury [1817]. Nos 1–5, all publ.

3455. SMITH, *Sir* James Edward *English botany; or coloured figures of British plants.* Roy. 8vo, 36 vols, 2592 col. pls, Sowerby, London 1790–1814; general index 1814; suppl. 5 vols, 406 col. pls, 1831 [–1866]; 2ed. 12 vols [1832–] 1835–46; 3ed. 7 vols [1847–] 1849–54; 3ed. 12 vols, 1863–72, 86; suppl. [1891–] 1892; issue 12 vols 1899; suppl. 1902. First suppl. Vol. 5 is unfinished. 2ed. is known as the small edition. 3ed. first issue is flowering plants only. The original text is only signed J.E.S. from Vol. IV onwards. The work is usually listed under Sowerby, James, the illustrator and publisher.

3456. —— *Syllabus of a course of lectures on botany*. 8vo, iv+71 pp, for the author, London 1795.

3457. —— *Tracts relating to natural history.* 8vo, xiv+312 pp, 6 col.+1 plain pls, Davis, London 1798.
3458. —— *Flora Britannica.* 8vo, 3 vols, White, London 1800–04.
3459. —— *Compendium florae Britannicae.* 8vo, viii+122 pp, White, London 1800; ed. 1816; 3ed. 1818; 4ed. 1825; 5ed. 1828; 6ed. 1829; English ed. 1829 title *A compendium of the English flora*; 2nd English ed. 1836; ed. 1844.
3460. —— *An introduction to physiological and systematical botany.* 8vo, xxiii+533 pp, 15 col. pls, Longman, London 1807; 2ed. 1809; 3ed. 1814; 4ed. 1819; 5ed. 1825; 6ed. 1827; 7ed. 1833 title *An introduction to the study of botany*; ed. 1836; ed. 1838.
3461. —— *A grammar of botany illustrative of the artificial as well as the natural classification.* 8vo, xiv+242 pp, 21 col. pls, Longman, London 1821; 2ed. 1826. Later editions in No. 3460.
3462. —— *The English flora.* 8vo, 5 vols, Longman, London 1824–36; 2ed. of Vols I–IV, 1828–30. Vol. V, pt. 1, Mosses, is by Hooker, W. J.; Vol. V, pt. 2. Fungi, is by Berkeley, M. J. This vol. is also Vol. II of Hooker, W. J. *British flora* q.v. See also Cowell, M. H.

—— *A compendium of the English flora.* 1829. *see* No. 3459.

3463. —— *Memoir and correspondance.* [With *a review of the modern state of botany.*] 8vo, 2 vols, Longman, London 1832. The review reprinted from *Encyclopaedia Britannica*, 6ed. 1816–24, Suppl. Vol. 2.

—— *An introduction to the study of botany.* 7ed. 1833. *see* No. 3460.

3464. SMITH, John *of Kew. Ferns, British and foreign.* 8vo, xii+412 pp, 1 pl., text figs, Kew 1866; 2ed. 1878; ed. 1896.
3465. —— *Domestic botany. An exposition of the structure and classification of plants.* 8vo, xii+547 pp, 16 col. pls, Reeve, London 1871; ed. 1883.
3466. —— *Historia filicum; an exposition of the nature, number, and organography of ferns.* 8vo, xiv+429 pp, 30 pls, Macmillan, London 1875.
3467. SMITH, John *of Kilwinning The botany of Ayrshire.* 8vo, 107 pp, Ardrossan 1896.
3468. SMITH, Phillip Henry Pye *Catalogue of the preparations of comparative anatomy in the Museum of Guy's hospital.* 8vo, 294 pp, London 1874.
3469. SMITH, Thomas *The wonders of nature and art.* 12mo, 12 vols, London 1803–04.

3470. —— *The naturalist's cabinet.* 8vo, 6 vols, 60 pls, Cundee, London 1806–07; ed. 12mo, 6 vols, 60 pls, Cundee, London 1806–07. Title pages of Vol. 1 of both eds call for 4 vols only. 8vo of 47 numbers; 12mo of 60 numbers.

3471. SMITH, William *A synopsis of British Diatomaceae.* Roy. 8vo, 2 vols, 7 col.+62 plain pls, Van Voorst, London 1853–56.

3472. —— *List of British Diatomaceae in the collection of the British Museum.* 12mo, 55 pp, British Museum, London 1859.

Smith, William G. *see* Hogg, R.

—— *see* Saunders, W. W.

3473. SMITH, William Ramsay & NORWELL, J. Stewart *Illustrations of zoology, invertebrates and vertebrates.* 4to, 70 pls with letterpress, Pentland, Edinburgh 1889; 2ed. [ ].

3474. SMITH, Worthington George *Mushrooms and toadstools; how to distinguish easily the difference between edible and poisonous Fungi.* 8vo, 64 pp, 2 col.+2 plain pls, Hardwicke, London 1867; 2ed. 1875; 3ed. 1876; 4ed. 1879. The plates and text available separately.

3475. —— *Clavis agaricinorum; an analytical key to the British Agaricini.* 8vo, 40 pp, 6 col. pls, Reeve, London 1870.

3476. —— *Supplementum to Outlines of British fungology.* [by M. J. Berkeley]. 8vo, xii+386 pp, Reeve, London 1891.

3477. —— *Guide to Sowerby's models of British Fungi in the Department of Botany, British Museum (Natural History).* 8vo, 82 pp, text figs, British Museum (Natural History), London 1893; 1903; ed. 1908.

—— *see* Fitch, W. H.

3478. SMITHSON, T. Spencer *Pond life. (Algae, diatoms etc.).* Sm. 8vo, 93 pp, text figs, Sonnenschein, London 1890. Young Collector Series. Later editions.

3479. SNAITH, W. A. & FIELD, William *Elementary botany for elementary classes.* 16mo, Heywood, Manchester 1873; ed. 1877; 3ed. 1878.

3480. SNELGROVE, Edward *Object lessons in botany from forest, field and garden.* Sm. 8vo, 3 vols, Jarrold, London 1894–1900; appendix on correlation of studies with object lessons, 214 pp, 1897. The work is divided into Standards 1–6 and teachers' aid.

3481. SNOOKE, William Drew *Flora Vectiana.* 8vo, 35 pp, London 1823.

Solander, Daniel Carl *see* Ellis, J.

3482. SOLE, William *Menthae Britannicae: being a new botanical arrangement of all the British mints.* Fol., viii+55 pp, 24 pls, Cruttwell printed, Bath 1798.

*Some particulars relating to the history of Epsom.* 1825. *see* Pownall, H.

3483. SOMERVILLE, Alexander *County and Vice-County divisions of the British Isles (for biological purposes).* Fol., s.sh., Glasgow [1898].

3484. SOMERVILLE, Mary *On molecular and microscopic science.* Sm. 8vo, 2 vols, 8 tinted pls, text figs, Murray, London 1869.

A son of the marshes *pseudonym (edited by* Mrs J. A. Owen). *see* Jordan, D.

3485. SONNTAG, C. O. *A pocket flora of Edinburgh and the surrounding district.* Sm. 8vo, xii+246 pp, 1 map, Williams & Norgate, London 1894.

Souter's Catechisms *see* Irving, C.

3486. SOUTH, Richard *The Entomologist synonymic list of the British Lepidoptera.* 8vo, vi+40 pp, West Newman, London 1884. Printed on one or both sides.

South Kensington Museum *see* Murray, Andrew.

3487. SOUTHALL, John *A treatise of Buggs; showing when and how they were first brought into England.* 4to, xii+44 pp, Roberts, London 1730; 2ed. 1730; 2ed. 8vo, Ipswich 1793 title *A treatise on the Cimex lectularius; or bed bug.*

3488. SOUTHWELL, Thomas *The seals and whales of the British seas.* 4to, vi+128 pp, pls, 2 tabs, text figs, Jarrold, London 1881.

—— *see* Lubbock, R.

—— *see* Stevenson, H.

3489. SOWERBY, Charles Edward *An illustrated catalogue of British plants, arranged according to the natural orders.* 12mo, 56 pp, 48 pls, London 1841. Nos 1–6, all publ.

3490. SOWERBY, George Brettingham I. *The conchological illustrations.* 8vo, [116] pp, 200 col. pls, London [1832–] 1841–42.

3491. —— *editor Thesaurus conchyliorum: or monographs of genera of shells.* Roy. 8vo, 5 vols, 530 col. pls, Sowerby, London [1842–] 1847–87. Plate numbering is irregular.

—— *see* Sowerby, James *the elder*

3492. SOWERBY, George Brettingham II *A conchological*

*manual.* 8vo, v+130 pp, 24 col. pls, Sowerby, London 1839; 2ed. 1842; 3ed. 1846; 4ed. 1852.

3493. —— *Descriptive catalogue of British chitons.* 8vo, Sowerby, London 1841.

3494. —— *Popular British conchology.* Roy. 16mo, xii+304 pp, 20 col. pls, Reeve, London 1854. Popular Natural History Series.

3495. —— *Foraminifera from the Colne tidal river, Essex.* 8vo, 1 pl. with letterpress, privately printed by the author, London [1856].

3496. —— *Popular history of the aquarium.* Roy. 16mo, xvi+327 pp, 20 col. pls, Reeve, London 1857; ed. 1865. Popular Natural History Series.

3497. —— *Companion to Mr. Kingsley's Glaucus.* 12mo, 48 pp, 12 col. pls, Macmillan, Cambridge 1858. For a combination volume of this and *Glaucus* see No. 2095.

3498. —— *Illustrated index of British shells . . . (Labels for all the recognized species)*, 4to, xv+[48] pp, 24 col. pls with letterpress, Sowerby, London 1859; 2ed. 1887. The label list sold separately with title *Six hundred labels for British shells.*

—— *see* Leach, W. E.

3499. SOWERBY, James *the elder. An easy introduction to drawing flowers according to nature.* Obl. 4to, 6 col. pls with letterpress, for the author, [London 1788]; 2ed. [1807] title *A botanical drawing book: or . . .*

—— *English botany.* 1790–1814. *see* Smith, *Sir* James E.

3500. —— *Coloured figures of English Fungi or mushrooms.* Fol., 3 vols, 400 col. pls, J. Davis printed (Vols 1 & 2), R. Wilks (Vol. 3), London [1795–] 1797–1809; suppl. 40 col. pls on 36 leaves, 1803–1809 [–1815]. Pls 1–400, suppl. 401–440. Suppl. all publ. No title leaf.

3501. —— *The British miscellany: or, coloured figures of new, rare, or little known animal subjects; many not before ascertained to be inhabitants of the British Isles.* 4to, 2 vols, 76 col. pls with letterpress, Sowerby, London [1804–] 1806; issue 1 vol. 1875. T.p. & index, Quaritch, London 1875. Vol. 2 pp 1–31, all publ.

3502. —— & SOWERBY, George Brettingham I. *The genera of recent and fossil shells.* 8vo, 3 vols, 263 col.+1 plain pls, Sowerby, London 1820–25 [–1834]. Title page and index Quaritch, London 1875. All publ.

3503. SOWERBY, James *the younger The mushroom and cham-*

*pignon illustrated . . . and distinguished from the poisonous Fungi which resemble them.* 4to, viii pp, 5 col. pls, London 1832.

3504. SOWERBY, John Edward *An illustrated key to the natural orders of British wild flowers.* 8vo, 42 pp, 9 col. pls, Van Voorst, London 1865. A key to Johnson, C. P. *The British wild flowers illustrated*, 1860.

——*The ferns of Great Britain.* 1855. *see* Johnson, C.

—— *The fern allies.* 1856. *see* Johnson, C.

—— *The grasses of Great Britain.* 1858. *see* Johnson, C.

—— *Wild flowers worth notice.* 1861. *see* Lankester, *Mrs* P.

3505. SOWERBY, J. G. *Rooks and their neighbours.* Roy. 8vo, 169 pp, text figs, Gay & Bird, London 1895; 2ed. 1895; 4ed. 1895.

3506. [SOWERBY, Sarah Mary] *Guide to the gardens of the Royal Botanical Society.* 8vo, 22 pp, London 1900. By M. S.

3507. SPALLANZANI, Lazzaro *An essay on animal reproductions.* 8vo, iv+86 pp, Becket, London 1769. Translation of *Prodrome di un opera da imprimersi sopra le riproduzioni animali*, Modena 1768, by Matthew Maty.

3508. —— *Dissertations relative to the natural history of animals and vegetables.* 8vo, 2 vols, 5 pls, London 1784; ed. 1789; ed. 1796; ed. 1797. Translation of *Dissertazioni di fisica animale e vegetabile*, Modena 1780.

3509. —— *Tracts on the nature of animals and vegetables.* 8vo, viii+394 pp, 6 pls, Edinburgh 1799; 2ed. 2 vols, 1803 title *Tracts on the natural history . . .* Translation of *Opusculi di fisica animale e vegetabile*, Modena 1776, by *Sir* John Graham Dalyell. First edition is an abridgement.

3510. SPARKES, John Charles Lewis & BURBIDGE, Frederick William *Wild flowers in art and nature.* 4to, iv+92 pp, 21 col. pls, text figs, Arnold, London [1893–94].

Specht, Friedrich *see* Vogt, C. C.

3511. *Specimen flora; or, British botany exemplified.* 8vo, Longman, London 1847.

*Spectacle de la nature: or nature delineated.* 1739. *see* Pluche, N. A. de la

*Spectacle de la nature: or nature display'd.* 1733. *see* Pluche, N. A. de la

3512. SPEEDY, Thomas *Craigmillar and its environs. With notices of the topography, natural history, and antiquities of the district.* 4to, xv+248 pp, 12 pls, text figs, Lewis, Selkirk 1892.

3513. [SPEIGHT, H.] *Through Airedale from Goole to Malham.* 8vo, lxiv+302 pp, 1 map, text figs, Leeds 1891. By Johnnie Gray.

3514. SPENCE, Robert F. [*The birds of Orkney.*] 4to, 280 pp, 1883. All printed, not published.

Spence, William *see* Kirby, W.

3515. SPENCER, Herbert *First principles. (A system of philosophy).* 8vo, xii+503 pp, Williams & Norgate, London 1862; 1863; 2ed. 1867; 3ed. 1870; 4ed. 1880; 5ed. 1884. Many later editions which form Vol. 1 of *A system of synthetic philosophy.*

3516. —— *The principles of biology.* 8vo, 2 vols, text figs, Williams & Norgate, London [1863–] 1864; 1867; ed. 1864–67 [2nd thou.]. Many later editions which form Vols 2 & 3 of *A system of synthetic philosophy.*

3517. —— *The factors of organic evolution.* 8vo, iv+76 pp, Williams & Norgate, London 1887. Reprinted with additions from *The Nineteenth Century*, Vol. 19, 1886.

3518. —— *The inadequacy of 'natural selection'.* 8vo, 69 pp, Williams & Norgate, London [1893]. Reprinted from *the Contemporary Review*, 1893.

3519. —— *A rejoinder to Professor Weismann.* 8vo, 29 pp, Williams & Norgate, London [1893]. Reprinted from *The Contemporary Review*, 1893.

3520. —— *Weismannism once again.* 8vo, 24 pp, Williams & Norgate, London 1894. Reprinted from *The Contemporary Review*, 1894.

—— *see* Collins, F. H.

3521. SPICER, Thomas Thevethan *The study of nature. A lecture delivered to the pupils of the Misses Bowes.* 8vo, London 1849.

3522. —— *Zoology. Notes of a lecture delivered at Crosby Hall.* 8vo, 16 pp, for the author, London 1853.

3523. SPIERS, William *Rambles and reveries of a naturalist.* 8vo, 256 pp, 1 pl., text figs, Kelly, London 1890.

3524. SPRENCEL, Kurt Polycarp Joachim *An introduction to the study of cryptogamous plants. In letters.* 8vo, viii+411 pp, 10 col. pls, White, London 1807. Translation of *Anleitung zur Kenntniss der Gewächse*, Pt 3, Halle 1804, by Carl Dietrich Eberhard König.

—— *see* Candolle, A. P. de

3525. SPRUCE, Richard *On Cephalozia (a genus of Hepaticae), its subgenera and some allied genera.* 8vo, vi+96 pp, for the author, Malton 1882.

Spry, William J. *see* Shuckard, W. E.

3526. STACKHOUSE, John *Nereis Britannica; continens species omnes fucorum in Insulis Britannicis crescentium; containing all the species of Fuci, natives of the British coasts.* Fol., xxxix+112 pp, 17 col. pls, appendix 2 11, 7 col. drawings, Hazard, Bath [1795–] 1801; ed. 4to, Oxford 1816. Portr. in some copies.

3527. STAINTON, Henry Tibbats *An attempt at a systematic catalogue of the British Tineidae and Pterophoridae.* 8vo, iv+32 pp, Van Voorst, London 1849; suppl. iv+38 pp, 1851.

3528. —— *A catalogue of the British Tineidae and Pterophoridae.* 8vo, 4 pp, Van Voorst, London 1849; additional catalogue, 2 pp, 1851. Printed on one side.

3529. —— *The entomologist's companion being a guide to the collection of Micro-Lepidoptera, and comprising a calendar of British Tineidae.* 12mo, iv+75 pp, Van Voorst, London 1852; 2ed. 1854.

3530. —— *Insecta Britannica. Lepidoptera: Tineina.* 8vo, viii+313 pp, 10 pls, Reeve, London 1854; suppls 1–3, 1854–56. Insecta Britannica Series. Supplements are preprints from *The Entomologists' Annual*, 1855–57.

3531. —— *List of British Tineina for labelling cabinets.* 8vo, Van Voorst for the author, London 1854. Printed on one side. Also available printed on both sides with title *List of British Tineina for interchange among collectors.*

3532. —— *List of specimens of British animals in the collection of the British Museum. Part XVI. Lepidoptera (completed).* 12mo, 199 pp, British Museum, London 1854. For previous parts see Stephens, J. F.

3533. —— *June: a book for the country in summer time.* 8vo, 141 pp, Longman, London 1856.

3534. —— *A manual of British butterflies and moths.* 12mo, 2 vols, text figs, Van Voorst, London [1856–] 1857–59; ed. 1867.

3535. —— *List of British Lepidoptera . . . arranged as in the Manual of British butterflies and moths.* 12mo, 2 pts, Van Voorst, London 1857–59. Some copies printed on one side.

3536. —— *British butterflies and moths.* Sm. 8vo, xii+292 pp, 16 col. pls, Reeve, London 1867; 2ed. [ ]. Crown Series.

3537. —— *assisted by* ZELLER, Philipp Christoph, DOUGLAS, John William, & FREY, Johann Friedrich Heinrich Konrad

*The natural history of the Tineina.* 8vo, 13 vols, 104 col. pls, Van Voorst, London 1855–73. Titles and text in English, French, German and Latin.

3538. STANLEY, Edward *A familiar history of birds: their nature, habits and instincts.* 16mo, 2 vols, text figs, SPCK, London 1835; 2ed. 1836; 3ed. 1840; 4ed. 1848; 6ed. 1854; ed. 1 vol. 1865; ed. 8vo 1880.

3539. STANNARD, Henry *Outdoor common birds.* Sm. 8vo, 215 pp, 1 pl., text figs, Warne, London [1873]; 1877; 1890.

3540. STARK, John *Elements of natural history.* 8vo, 2 vols, 8 pls, text figs, Blackwood, Edinburgh 1828.

3541. STARK, Robert Mackenzie *A popular history of British mosses.* Roy. 16mo, xvi+322 pp, 20 col. pls, Reeve, London 1854; 2ed. 1860; ed. [ ].

3542. —— *The marine aquarium: directions for its preparation and management.* 8vo, Edinburgh 1857.

—— *see* White, A.

3543. STAVELEY, *Miss* E. F. *British spiders.* Sm. 8vo, xv+280 pp, 16 col. pls, Reeve, London 1866; ed. [ ]. Crown Series.

3544. —— *British insects.* Sm. 8vo, xv+392 pp, 16 col. pls, text figs, Reeve, London 1871. Crown Series.

3545. STEAD, J. *Description and natural history of English song birds . . . to which is prefixed . . . a general history of the feathered tribe.* 12mo, xii+59 pp, 3 pls, for the author, London 1797.

3546. STEBBING, Thomas Roscoe Rede *Essays on Darwinism.* 8vo, vi+184 pp, Longman, London 1871.

3547. —— *Museums and our Museum.* 8vo, 19 pp, Torquay 1872.

3548. —— *The naturalist on Cumbrae: a true story of the life of David Robertson.* 8vo, x+398 pp, Kegan Paul, London 1891.

3549. —— *A history of Crustacea: recent Malacostraca.* Sm. 8vo, xvii+466 pp, 19 pls, text figs, Kegan Paul, London 1893.

3550. [STEELE, Francesca Maria] *The glory of the sea . . . with . . . a table of British shells.* 8vo, vi+212 pp, RTS, London [1887]. By Darley Dale *pseudonym.*

3551. [——] *Mr Mygale's hobby. A story about spiders.* [With a list of British species.] 8vo, 192 pp, 1 pl., text figs, RTS, London 1889. By Darley Dale [*pseudonym*] on cover. By the author of *The glory of the sea* on title page.

3552. [——] *Swallow-tails and skippers.* [With a list of British butterflies]. 8vo, 158 pp, 1 col. pl., RTS, London 1886. By Darley Dale *pseudonym.*

3553. STEELE, William Edward *Handbook of field botany, comprising the flowering plants and ferns indigenous to the British Isles.* 8vo, xiv+249 pp, 1 pl., McGlashan, Dublin 1847; 2ed. 1851.

3554. STEENSTRUP, Johan Japetus Smith *On the alternation of generation.* 8vo, viii+132 pp, 3 pls, Ray Society, London 1845. Ray Society Publ. No. 4 for 1843. Translation of *Ueber den Generationswechsel*, Copenhagen 1842, itself a translation of *Om Fortplannting og Udvikling gjennem vexlende Generations-Raekker*, Copenhagen 1842, by George Busk.

Steele-Elliott, J. *see* Elliott, J. S.

3555. STEGGALL, John *The pupil's introduction to botany.* 8vo, 174 pp, 8 pls, Highley, London 1829; 2ed. 1831 title *Elements of botany.*

3556. [STEP, Edward] *The fresh-water aquarium.* 8vo, 43 pp, text figs, Marshall & Japp, London 1881. By Weston, James *pseudonym.*

3557. —— *Plant life. Popular papers on the phenomena of botany.* 8vo, xii+216 pp, text figs, Marshall & Japp, London 1881; 2ed. 1883; 3ed. 1884.

3558. —— *Easy lessons in botany.* 8vo, 48 pp, Fisher Unwin, London 3ed. 1883.

3559. [——] *Stories and pictures of birds, beasts, and fishes.* 4to, 96 pp, text figs, Partridge, London [1886]. By Weston, James *pseudonym.* There is an edition in No. 3561.

3560. [——] *Stories and pictures of animal life, with a few words concerning plants.* 4to, 96 pp, text figs, Partridge, London 1889. By Weston, James *pseudonym.* There is an edition in No. 3561.

3561. —— *By sea-shore, wood and moorland: peeps at nature.* 8vo, 320 pp, text figs, Partridge, London 1891. See also Nos 3559 & 3560.

3562. —— *Wayside and woodland blossoms.* [First series.] 8vo, vi+173 pp, 128 col. pls, Warne, London 1895; 2ed. 1896. See also No. 3565.

3563. —— *By the deep sea. A popular introduction to the wild life of the British shores.* 16mo, 321 pp, 1 pl., text figs, Jarrold, London 1891; 3ed. 1896.

3564. —— *By vocal woods and waters.* 8vo, 254 pp, Bliss Sands, London 1894.

3565. —— *Wayside and woodland blossoms. Second series.* 8vo, xv+170 pp, 128 col. pls, text figs, Warne, London 1896. See also No. 3562. There are many later issues of these two volumes in The Wayside and Woodland Series.

3566. —— *Romance of wild flowers.* Sm. 8vo, 357 pp, 32 pls, Warne, London 1899.

3567. STEPHENS, James Francis *Illustrations of British entomology.* 8vo, 11 vols, 80 col. pls, Baldwin & Cradock for the author, London [1827–] 1828–35 [–1845]; suppl. v+32 pp, 15 col. pls, 1845; suppl. issue 1846. Made up of Mandibulata Vols I–VII, and Haustellata Vols I–IV. Plates Mandibulata I–XXXVI, XXXVIII–XL; Haustellata I*, II*, III*, IV, 5–8, IX, 10–41; suppl. XXXVII, XLI–XLVII, 42–48. All publ.

3568. —— *A systematic catalogue of British insects.* 8vo, 2 pts, xxxiv+416+388 pp, Baldwin & Cradock for the author, London 1829.

3569. —— *The nomenclature of British insects, being a compendious list of such species as are contained in the Systematic catalogue of British insects.* 8vo, 68 pp, Baldwin & Cradock for the author, London 1829; 2ed. pt. 1 (Mandibulata) 1833. 2ed. all publ.

3570. —— *An abstract of the indigenous Lepidoptera contained in the Verzeichniss bekanter Schmetterling by Hübner.* 8vo, 24 pp, London 1835. Reprinted from No. 3567, Vol. IV, 1834.

3571. —— *A manual of British Coleoptera or beetles.* Sm. 8vo, xii+443 pp, Longman, London 1839.

3572. —— *List of the specimens of British animals in the collection of the British Museum. Part V. Lepidoptera.* 12mo, xii+353 pp, British Museum, London 1850; 2ed. 1856. For the rest of this list see Nos 3532, 3573 & 3574.

3573. —— *List of the specimens of British animals in the collection of the British Museum. Part X. Lepidoptera (continued).* 12mo, xiii+120 pp, British Museum, London 1852. For the rest of this list see Nos 3532, 3572 & 3574.

3574. —— *List of the specimens of British animals in the collection of the British Museum. Part XII. Lepidoptera (continued).* 12mo, iv+54 pp, British Museum, London 1852. For the rest of this list see Nos 3532, 3572 & 3573.

3575. STEPHENS, William *Botanical elements: published for the use of the Botany School in the University of Dublin.* 8vo, 48 pp, Risk, Dublin 1727.

3576. STERLAND, W. J. *The birds of Sherwood Forest.* Sm. 8vo, ix+244 pp, 3 col.+1 plain pls, Reeve, London 1869. Crown Series.

*3577. —— *The handbook of natural history: Mammalia.* 8vo, London [1879].

3578. —— & WHITAKER, Joseph *Descriptive list of the birds of Nottinghamshire.* 8vo, 71 pp, Gouk, Mansfield 1879.

3579. STEVENSON, Henry *A list of the birds of Norfolk.* 8vo, 10 pp, Sheffield 1864. Reprinted from White's *History and directory of the County of Norfolk.*

3580. —— *The birds of Norfolk.* 4to, 3 vols, 5 col.+3 plain pls, portr., text figs, Van Voorst, London 1866–90. Vol. 3 completed by Southwell, Thomas.

3581. STEVENSON, John *Mycologia Scotica: the Fungi of Scotland.* 8vo, xx+443 pp, 1 map, Cryptogamic Society of Scotland, Edinburgh 1879.

3582. —— *Hymenomycetes Britannici: British Fungi (Hymenomycetes).* Sm 8vo, 2 vols, text figs, Blackwood, Edinburgh 1886.

3583. STEWART, Alexander *Nether Lochaber. The natural history, legends, and folk-lore of the West Highlands.* 8vo, xviii+417 pp, 1 pl., Paterson, Edinburgh 1883. Reprint of papers in *The Inverness Courier*, under pseudonym of Nether Lochaber.

3584. —— *'Twixt Ben Nevis and Glencoe: the natural history of the West Highlands.* 8vo, xv+384 pp, 1 portr., Paterson, Edinburgh 1885.

3585. [STEWART, Charles] *Elements of natural history: being an introduction to the Systema naturae of Linnaeus: comprising the characters of the whole genera, and most remarkable species, particularly of all those that are natives of Britain.* 8vo, 2 vols, 12 pls, Cadell, London 1801–02; 2ed. Edinburgh 1817 title *Elements of the natural history of the animal kingdom*, signed.

3586. STEWART, R. B. *Outlines of botany: a sketch of the Linnean arrangement of plants.* 8vo, Ridgway, London 1835.

3587. STEWART, Robert *Handbook of the Torquay flora.* Sm. 8vo, iv+187 pp, Croydon, Torquay 1860.

3588. STEWART, Samuel Alexander & CORRY, Thomas Hughes *A flora of the north-east of Ireland.* Sm. 8vo, xxv+331 pp, Macmillan, Cambridge 1888; 2ed. Belfast 1938. Suppl. to first edition in *Proc. Belfast Fld. Nat. Club*, 1895 by Praeger, Robert Lloyd. Second suppl., xii+129 pp, Belfast 1923.

3589. STILLINGFLEET, Benjamin *Miscellaneous tracts relating to natural history.* 8vo, xxx+230 pp, Dodsley, London 1759;

2ed. 1762 title . . . *to which is added the calendar of flora*; 3ed. 1775; 4ed. 1791. The tracts are translations of Linnaean dissertations. The calendar is No. 3590.

3590. —— *The calendar of flora, Swedish and English.* 4to, xxiii+45 pp, Dodsley, London 1761. Later editions are in No. 3589. The Swedish calendar is a translation of Berger, Alexander Maj *Calendarium florae*, Uppsala 1749, a Linnaean dissertation.

3591. —— *Literary life and select works.* Roy. 8vo, 2 vols, 17 col.+2 plain pls, 4 portrs, Longman, London 1811. Vol. 2 is in two parts. The *Literary life* alone reprinted 1818.

3592. STONE, John Frederic Matthias Harris *Interdependance of animals and plants.* 8vo, Bath 1879.

3593. STORER, John *The wild white cattle of Great Britain.* 8vo, xx+384 pp, 8 pls, text figs, Cassell, London [?1879]; ed. [1879].

3594. *Stories about animals.* 4to, 36 pp, col. text figs, Webb Millington, Leeds [ ].

*Stories of animals.* 1820, 1831. *see* Hack, Maria

3595. STORMONTH, James *A manual of scientific terms.* 8vo, xi+488 pp, Edinburgh 1879; 2ed. 1885; ed. 1903.

3596. STORRIE, John *The flora of Cardiff.* 8vo, v+129 pp, Cardiff 1886.

3597. STRASBURGER, Eduard *Handbook of practical botany for the botanical laboratory and private student.* 8vo, 425 pp, Sonnenschein, London 1887; 3ed. 1893; 4ed. 1897; 5ed. 1900; 6ed. 1908; 7ed. 1911; 8ed. 1925. Translation of *Das botanische Practicum*, Jena 1884, by William Hillhouse.

3598. ——, NOLL, Fritz, SCHENK, Heinrich & SCHIMPER, Andreas Franz Wilhelm *A text-book of botany.* 8vo, 632 pp, text figs some col., Macmillan, London 1898; 2ed. 1903; 3ed. 1908; 4ed. 1912; 5ed. 1921; 6ed. 1930. Translation of *Lehrbuch der Botanik*, Jena 1894; 2ed. 1895, by H. C. Porter.

3599. STRATTON, Frederick *Wild flowers of the Isle of Wight.* Obl. 12mo, 20 pp, 1 col. map, Newport I.O.W., 1900.

3600. [STRICKLAND, Miss Agnes] *Narratives of nature, and history book for young naturalists.* 12mo, xii+276 pp, 7 pls, Lacey, London [?1845]. By the author of *Sketches from nature* etc., which is by her sister, *Mrs* Catherine Parr Traill q.v.

Strickland, Catherine Parr *see* Traill, *Mrs* C. P.

3601. STRICKLAND, Charlotte & STRICKLAND, Juliana

Sabina *Select specimens of British plants.* Fol., 2 pts, 7 pp, 5 pls, London 1797, 1809. Text of part 2 by Shaw, G.

3602. STRICKLAND, Hugh Edwin *editor Report of a Committee appointed "to consider of the rules by which the nomenclature of zoology may be established on a uniform and permanent basis".* 8vo, 17 pp, British Association, London 1842; ed. Edinburgh 1863 title *Rules for zoological nomenclature*; ed. Dumfries 1865; ed. London 1878. The first edition was also printed in *Rep. Brit. Assoc. Adv. Sci.*, 1842; the 3rd in that for 1865.

3603. —— *Memoirs.* [With a selection of his scientific writings.] Roy. 8vo, cclxv+xvi+441 pp, 18 pls, 2 portrs, 1 map, Van Voorst, London 1858.

Strickland, Juliana Sabina *see* Strickland, Charlotte

*A stroll in a marsh, in search of wild flowers.* 1895. *see* Cooke, M. C.

3604. STROUD, T. B. *The elements of botany, physiological and systematical: to which is added a . . . dictionary.* 8vo, 257 pp, Delahoy for the author, Greenwich 1821.

3605. *The structure, physiology, and classification of plants.* 8vo, Reynolds, London [1873]. A text to accompany wall charts for schools, one of a set of 7. See also *A comprehensive view of the animal kingdom.*

Strugnell, W. Bishop *see* Witchell, C. A.

3606. STRUTT, Jacob George *Sylva Britannica; or portraits of forest trees.* Fol., 49 pls with letterpress, for the author, London 1822 [–1825]; ed. 1826; ed. Imp. 8vo [1830–36].

3607. —— *Deliciae sylvarum; or, grand and romantic forest scenery in England and Scotland.* Fol., no text, 8 pls with letterpress, London 1828 [–1829].

3608. [STUART, John, *Earl* of Bute] *A tabular distribution of British plants.* 4to, 57 pp, for the author, London 1780. See also No. 3609.

3609. —— *Botanical tables containing the different familys of British plants.* 4to, 9 vols, 566 col. pls, for the author, [London ?1785]. Contains much of No. 3608. Very few copies, ?12.

3610. STUBBS, Arthur G. *Hints on cleaning the smaller transparent species of British land Mollusca.* 8vo, Eastbourne 1900.

Stubs, *Miss* Amelia *see* Selwyn, A.

3611. STURM, Christoph Christian *Reflections on the works of God and of his providence, throughout all nature, for every day of the year.* 12mo, 3 vols, Edinburgh 1788; ed. 1791; 5ed. 1795; 6ed. 1798; 7ed. 1800; 8ed. 1802; ed. 1804; 10ed.

1810. Translation of *Considerations sur les oeuvres de Dieu*, The Hague 1786, itself a translation of *Betrachtungen über die Werke Gotes im Reich der Natur*, Halle 1784. Translated by a lady.

Apart from this first translation, there are very many editions and new translations, some direct from the German. The following have been noted:

[?1800] 8vo, 2 vols, London. Dr Balfour translator.
1801 12mo, 4 vols, Badcock, London. Adam Clarke translator.
1803 18mo, 3 vols, Wallis, London. The author of *The Adviser* translator.
1804 8vo, 2 vols, Brightly, Bungay. New ed.
1807 8vo, 2 vols, Brightly, Bungay. New ed.
1808 12mo, 4 vols, Cundee, London. F. Schoberl translator.
1808 12mo, 3 vols, Tegg, London.
1808 12mo, 3 vols, Scholey, London.
1808 12mo, 2 vols, London.
1809 8vo, 2 vols, London.
1809 24mo, 2 vols, London. Walker's Classics.
1810 12mo, 4 vols, London. Adam Clarke translator.
1811 8vo, 2 vols, London. Adam Clarke translator.
1813 8vo, 2 vols, London. Adam Clarke translator.
1813 8vo, 2 vols, Faversham.
1814 12mo, 2 vols, Crosby, London. New translation.
1816 8vo, 2 vols, Ellis, Glasgow.
1817 12mo, 2 vols, London. Walker's Classics. Dr Balfour translator.
1818 8vo, 2 vols, London. Adam Clarke translator.
1819 12mo, 3 vols, Goodsir, London.
1823 8vo, 2 vols, Whittaker, London.
1824 24mo, 2 vols, Dove, London.
1824 12mo, 4 vols, Whittingham, London. Sharpe's edition.
1824 8vo, 2 vols, Whittaker, London. Dr Balfour translator.
1826 8vo, 1 vol., Tegg, London.
1836 12mo, 1 vol., Bohn, London.
1837 8vo, 1 vol., Tegg, London.
1838 8vo, 1 vol., Virtue, London.
1849 8vo, 1 vol., Virtue, London.
1850 12mo, 1 vol., Bohn, London.
1850 12mo, 2 vols, Tegg, London. Adam Clarke translator.
1859 12mo, 2 vols, Tegg, London. Adam Clarke translator.

1860 12mo, 2 vols, Tegg, London. Adam Clarke translator.
1860 12mo, 1 vol., Bohn, London.
Abridgement, 1812, title *Beauties of Sturm*, Darton, London. New ed. by Eliza Andrews.
Abridgement, 1823, Rivington, London. By –. Hemet.

3612. STUTTARD, John *The butterfly, its history, development and attributes.* 12mo, 92 pp, Unwin, London 1890 [=1889].

3613. [SUCKLING, Horatio John] *Anti-Darwin: or some reasons for not accepting his hypothesis.* 8vo, viii+212 pp, text figs, privately printed, Twickenham 1884. ed. London 1887; ed. 1888. By the author of *Ceylon ancient and modern.*

3614. SUFFOLK, W. T. *On microscopical manipulation.* Sm. 8vo, xv+227 pp, 8 pls, text figs, Gillman, London 1870.

3615. SUNDEVALL, Carl Jacob *Sundevall's Tentamen*, 8vo, xiv+316 pp, 1 portr., London 1889. Translation of *Methodi naturalis avium disponendarum tentamen*, Stockholm 1872–73, by Francis Nicholson.

3616. SURR, Elizabeth *Sea-birds, and the lessons of their lives.* 16mo, text figs, Nelson, London [1876]; [1880].

3617. SWAINSON, Charles *Provincial names and folk-lore of British birds.* 8vo, x+243 pp, The Dialect Society, London 1885; issue 1886 title *The folk-lore and provincial names . . .*

3618. SWAINSON, William *The naturalist's guide for collecting and preserving all subjects of natural history and botany.* 8vo, viii+72 pp, 2 pls, Baldwin Cradock, London 1822; 2ed. 1822; 2ed. [=3ed.] 1822 [?=1835.]

3619. —— *Zoological illustrations.* [First series]. Roy. 8vo, 3 vols, 182 col. pls, Baldwin Cradock, London 1820–23; later issues of plates alone.

3620. —— *Zoological illustrations. Second series.* Roy 8vo, 3 vols, 136 col. pls, London 1829–33.

3621. —— *A preliminary discourse on the study of natural history.* Sm. 8vo, viii+462 pp, Longman, London 1834. Lardner's Cabinet Cyclopaedia, or as an independant work.

3622. —— *The elements of modern conchology.* 12mo, viii+62 pp, Baldwin & Cradock, London 1835.

3623. —— *On the natural history and classification of quadrupeds.* Sm. 8vo, viii+397 pp, text figs, Longman, London 1835; ed. 1845. Lardner's Cabinet Cyclopaedia.

3624. —— *A treatise on the geography and classification of animals.* Sm. 8vo, vii+367 pp, 1 portr., text figs, Longman, London 1835. Lardner's Cabinet Cyclopaedia.

3625. —— *On the natural history and classification of birds.* Sm. 8vo, 2 vols, text figs, Longman, London 1836–37. Lardner's Cabinet Cyclopaedia. Part 3 of No. 3626, title *Two centuries and a quarter of birds* is a suppl. and is sometimes bound with Vol. 2.

3626. —— *Animals in menageries.* Sm. 8vo, 3 pts, vi+373 pp, text figs, Longman, London 1838. Lardner's Cabinet Cyclopaedia. Part 3 is suppl. to No. 3625.

3627. —— *Flycatchers.* Sm. 8vo, 256 pp, 31 col. pls, portr., Lizars, Edinburgh 1838; ed. [1845–46]. Jardine's Naturalist's Library, Vol. XXI, 2ed. Vol. XIII.

3628. —— *On the natural history and classification of fishes, amphibians and reptiles, or monocardian animals.* Sm. 8vo, 2 vols, text figs, Longman, London 1838–39. Lardner's Cabinet Cyclopaedia.

3629. —— *On the habits and instincts of animals.* Sm. 8vo, vi+375 pp, text figs, Longman, London 1840. Lardner's Cabinet Cyclopaedia, or as an independant work.

3630. —— *Taxidermy; with the biographies of zoologists and notices of their works.* Sm. 8vo, 392 pp, text figs, Longman, London 1840. Lardner's Cabinet Cyclopaedia.

3631. —— *A treatise on malacology, or the natural classification of shells and shellfish.* Sm. 8vo, viii+419 pp, text figs, Longman, London 1840. Lardner's Cabinet Cyclopaedia.

3632. —— & SHUCKARD, William Edward *On the history and natural arrangement of insects.* Sm. 8vo, 406 pp, text figs, Longman, London 1840. Lardner's Cabinet Cyclopaedia.

*Swallow-tails and skippers.* [1866]. *see* Steele, F. M.

3633. SWAMMERDAM, Jan *Ephemeri vita; or, the natural history and anatomy of the ephermeron.* 4to, 44 pp, 8 pls, Faithorne, London 1681. Translation of *Ephemeri vita* [Dutch], Amsterdam 1675, by Edward Tyson.

3634. —— *The book of nature; or, the history of insects.* Fol., xx+236+153+lxiii pp, 53 pls, Seyffert, London 1758. Translation of *Bybel der natuure*, Leyden 1737–38, by Thomas Flloyd.

3635. SWAN, John *Speculum mundi: or, a glasse representing the face of the world.* 4to, 504 pp, text figs, Printers (Buck & Daniel) to the University, Cambridge 1635; 2ed. 1643; ed. 1643; 3ed. London 1665; 4ed. London 1670. Only the dedication is signed.

3636. SWANN, Harry Kirke *The birds of London.* Sm. 8vo, xv+136 pp, Sonnenschein, London 1893.

3637. —— *A concise handbook of British birds.* 8vo, viii+210 pp, Wheldon, London 1896.

3638. SWANTON, Ernest W. *A monograph of the British puff-balls and earth-balls.* 8vo, 6 pp, 1 pl., text figs, Economic & Educational Museum, Huddersfield 1896. Issued with *The Naturalists' Journal*, Vol. 5.

3639. SWAYNE, George *Gramina pascua: or a collection of . . . common pasture grasses.* Fol., [21] pp, 6 pls of specimens, Bonner for the author, Bristol 1790.

3640. SWAYSLAND, Walter *Familiar wild birds. First (fourth) series.* Sm. 8vo, 4 vols, 160 col. pls, Cassell, London 1883 [–1888]; ed. [ ].

3641. SWEDENBORG, Emanuel *The economy of the animal kingdom, considered anatomically, physically, and philosophically.* 8vo, 2 vols, text figs, Newbery, London 1845, 46. Translation of *Oeconomia regni animalis*, London 1740–41.

3642. SWEET, Robert *The British warblers. An account of the genus Sylvia.* 8vo, 24 pp, 6 col. pls, for the author, London 1823; issue 16 col. pls, 1823. Later editions in No. 242.

—— *see* Weddell, H.

3643. SWETE, Edward Horace *Flora Bristoliensis.* 8vo, xxvi+138 pp, 2 pls, 1 col. map, Hamilton Adams, London 1854.

3644. SWINTON, Archibald Henry *Insect variety: its propagation and distribution. Treating of the odours, dances, colours and music.* 8vo, x+323 pp, 7 col.+1 plain pls, text figs, Cassell, London [1880].

*Syllabus of the Mammalia.* [188–]. *see* Ewart, J. C.

*Sylva: or a discourse on forest trees.* 1664. *see* Evelyn, J.

3645. *The sylvan melodist; or, a cabinet of the more familiar British birds.* 16mo, xii+132 pp, text figs, Darton, London [?1842].

Syme, John Thomas Irvine *later* Boswell *see* Smith, *Sir* James E.

3646. SYME, Patrick *A treatise on British song birds.* 8vo or 12mo, vi+231 pp, 15 col. pls, Anderson, Edinburgh 1823.

3646a. SYMONDS, William Samuel *Records of the rocks; or, notes on the geology, natural history and antiquities of North and South Wales, Devon and Cornwall.* 8vo, xviii+433 pp, 1 pl., text figs, Murray, London 1872.

3647. SYMONS, Jelinger *Synopsis plantarum Insulis Britannicis indigenarum.* 12mo, vii+207 pp, White, London 1798.

*A system of anatomy and physiology.* 2ed. 1787. *see* Fyfe, A.
*A systematic catalogue of the . . . indigenous Mammalia and birds that are preserved in the British Museum.* 1816. *see* Leach, W. E.
*A systematic catalogue of British birds in the Manchester Museum of natural history. Aves Britannicae.* 1836. *see* Manchester Society of Natural History.

# T

3648. T., C. *editor. Chronicles of the seasons.* Sm. 8vo, 4 vols, text figs, Parker, London 1844.
Tabley, *Baron* de *see* De Tabley, *Baron*
*A tabular distribution of British plants.* 1780. *see* Stuart, J. *Earl of* Bute
3649. TALBOT, William *The birds of Wakefield.* 8vo, iv+33 pp, Huddersfield 1877. Reprinted from *The Naturalist*, Vols 1–2, 1875–76.
Tallis, John *Scripture natural history.* [1850, 51]. *see* Bicknell, W. I.
Tandon, Moquin *see* Moquin-Tandon, C. H. D. A.
3650. TATE, George *The history of the Borough, castle and Barony of Alnwick.* 4to, 2 vols, 21 pls partly col., Blair, Alnwick 1866–69.
3651. TATE, Ralph *Flora Belfastiensis. The plants around Belfast.* 8vo, xix+92 pp, Belfast 1863.
3652. —— *A plain and easy account of the land and freshwater mollusks of Great Britain.* Sm. 8vo, viii+244 pp, 11 col. pls, text figs, Hardwicke, London 1866.
3653. TAUVRY, Daniel *A new rational anatomy, containing an explication of the uses of the structure of the body of man, and some other animals, according to the rules of mechanicks.* 8vo, 301 pp, 21 pls, Midwinter, London 1701. Translation of *Nouvelle anatomie raisonnée*, Paris 3ed. 1698.
*Taxidermy.* 1820. *see* Bowdich, *Mrs* S.
3654. [TAYLOR, Charles] *Surveys of nature, historical, moral, and entertaining.* 4to, 2 vols in 6 pts, 73 col.,+86 plain pls, Taylor, London 1787 [–?1789]. By Fitzgerald, Francis *pseudonym.*

3655. TAYLOR, Jeffreys *The forest, or rambles in the woodland.* 16mo, 220 pp, 36 pls, Harris, London 1831; 2ed. 1832; 3ed. 1835. Little Library, Vol. 3.
3656. TAYLOR, John Ellor *Half hours at the sea-side; or recreations with marine objects.* Sm. 8vo, xii+266 pp, 1 pl., text figs, Hardwicke, London 1872; 2ed. 1875; 3ed. 1875; 4ed. 1880; ed. [1884]; ed. 1890; ed. Edinburgh 1907.
3657. —— *Half-hours in the green lanes: a book for a country stroll.* Sm. 8vo, Hardwicke, London 1873; 3ed. 1875; 4ed. 1877; 5ed. 1879; 6ed. 1884; 7ed. 1890; 8ed. 1891; 1898; 9ed. 1907.
3658. —— *The aquarium; its inhabitants, structure and management.* Sm. 8vo, xv+316 pp, text figs, Hardwicke, London 1876; 2ed. 1881; ed. 1884; ed. Edinburgh 1910.
3659. —— *editor Notes on collecting and preserving natural-history objects.* Sm. 8vo, viii+215 pp, text figs, Hardwicke, London 1876; ed. 1883; ed. 1896. Reprinted from *Science Gossip.*
3660. —— *Flowers; their origin, shapes, perfumes and colours.* 8vo, xxii+347 pp, 8 col. pls, text figs, Hardwicke, London 1878; 2ed. 1878; 3ed. Edinburgh 1906; 4ed. [1880].
3661. —— *Natural history rambles; mountain and moor.* Sm. 8vo, 256 pp, text figs, SPCK, London 1879. Natural History Rambles Series.
3662. —— *Natural history rambles. Underground.* Sm. 8vo, 256 pp, text figs, SPCK, London 1879. Natural History Rambles Series.
3663. —— *Nature's bypaths. A series of recreative papers in natural history.* Sm. 8vo, viii+408 pp, text figs, Bogue, London 1880; ed. 1884.
3664. —— *The sagacity and morality of plants.* Sm. 8vo, xi+311 pp, 1 col. pl., text figs, Chatto & Windus, London 1884; ed. 1891; ed. [1904].
3665. —— *The playtime naturalist.* Sm. 8vo, xv+287pp, 1 pl., text figs, Chatto & Windus, London 1889.
3666. [TAYLOR, John William] *Label list of British land and fresh-water shells.* 8vo, 16 pp, Taylor, Leeds [1863]. Printed on one side; also printed on both sides with title *Exchange list of . . .*
3667. —— *A monograph of the land and freshwater Mollusca of the British Isles.* Roy. 8vo, 4 vols, 77 col. pls, 1 portr., maps, text figs, Taylor, Leeds 1894–1921. Vol. 4, pts 1–3, all publ.
3668. TAYLOR, Joseph *Ornithologia curiosa; or, the wonders of*

*the feathered creation.* 18mo, 216 pp, 1 pl., Lackington, London 1807.

3669. —— *Arbores mirabiles: or, a description of the most remarkable trees, plants and shrubs, in all parts of the world.* 12mo, 192 pp, 6 pls, Darton, London 1812; issue [ ] title *The wonders of trees, plants and shrubs.*

3670. —— *Anecdotes of remarkable insects, selected from natural history and interspersed with poetry.* 12mo, 224 pp, 1 pl., text figs, Baldwin & Cradock, London 1817.

3671. —— *Naturales curiosae: curiosities in natural history.* 8vo, 160 pp, Allman, London 1819.

—— *see Curiosities for the ingenious.*

3672. TAYLOR, Silas *The history and antiquities of Harwich . . . and observations relating to natural history.* 4to, xxiv+464 pp, 14 pls, Davis & Green, London 1730; 2ed. 1732. The natural history by Samuel Dale.

Taylor, Thomas *see* Hooker, *Sir* William J.

3673. TEGETMEIER, William Bernhard *First lines in botany.* 8vo, text figs, Clarke, London [1849]. Later editions.

3673a —— *Information on common objects for the use of schools.* 8vo, 4ed. 1858. Later editions. A guide for a box of natural history specimens.

3673b —— *The classification of animals and vegetables after Cuvier and Decandolle.* Single large sheet, text figs, London [c.1860]. Also available on a roller.

3674. —— *Pallas's sand-grouse; its natural history, with a plea for its preservation.* 8vo, 23 pp, 1 col. pl., text figs, Cox, London 1888.

3675. TELL-TRUTH, T. *pseudonym Natural history of birds.* 12mo, 106 pp, 1 pl., text figs, Newbery, London 1770; 1778. Other printings.

3676. —— *Natural history of four-footed beasts.* 12mo, 1 pl., text figs, Newbery, London 1769; 1778; 1781. Other printings.

3677. TEMMINCK, Coenraad Jacob & YARRELL, William *A list of the birds of Great Britain according to the arrangement of Mr. C. J. Temminck and of Mr. Yarrell.* Fol., s.sh., Makeig, Crewkerne [c.1845]. The Temminck drawn from *Manuel d'ornithologie*, Amsterdam 1815; the Yarrell from *A history of British birds*, London 1837–43.

3678. [THEAKSTON, Michael] *A list of natural flies that are taken by trout, grayling & smelt in the streams of Ripon.* 12mo, x+129 pp, 8 pls, Harrison, Ripon 1853; ed. London 1862 title *British angling flies*; ed. [1883].

3679. THEOBALD, Frederick Vincent *An account of British flies (Diptera). Vol. 1.* 8vo, xix+215 pp, 4 pls, text figs, Elliot Stock, London [1891–] 1892 [1893]. All publ. For an earlier issue see Leigh, *Hon.* Mary C. E.

3680. —— *Insect life; a short account of the classification and habits of insects.* Sm. 8vo, xi+235 pp, text figs, Methuen, London 1896. University Extension Series.

3681. THOMAS, David Morgan *The day-book of wonders.* 8vo, 635 pp, Fisher Unwin, London 1897.

3682. [THOMAS, Michael Rogers Oldfield] *Directions for preparing small mammal skins.* 4to, 2 pp, British Museum (Natural History), London [1898].

3683. THOMÉ, Otto Wilhelm *Textbook of structural and physiological botany.* 8vo, ix+479 pp, 1 pl., text figs, Longman, London 1877; 2ed. 1878; 3ed. 1879; 4ed. 1882; 5ed. 1885; 7ed. 1891; 8ed. 1897. Translation of *Lehrbuch der Botanik*, Brunswick 1869, 4ed. 1875, by Alfred William Bennett.

3684. THOMPSON, Edward Pett *The note-book of a naturalist.* 12mo, xii+275 pp, Smith Elder, London 1845.

3685. —— *The passions of animals.* 8vo, viii+414 pp, Chapman & Hall, London 1851.

3686. THOMPSON, Ernest Evan Seton *Studies in the art anatomy of animals; being a brief analysis of the visible forms of the more familiar mammals and birds.* 4to, xii+96 pp, 49 pls, Macmillan, London 1896.

3687. THOMPSON, John *Botany displayed. No. 1 (2—4).* 4to, 4 pts, 10 pp, 12 col. pls, Thompson & Nunes, London 1798. All publ.

3688. THOMPSON, John Vaughan *A catalogue of the plants growing in the vicinity of Berwick-on-Tweed.* 8vo, xxiv+132 pp, 1 col. pl., White. London 1807.

3689. —— *Memoir on the Pentacrinus europaeus; a recent species discovered in the Cove of Cork.* 4to, 12 pp, 2 pls, King & Ridings, Cork 1827. Printed in 4to to stand with No. 2590.

3690. —— *Zoological researches and illustrations. No. I. (II—VI).* Roy. 8vo, 160 pp, 20 pls, text figs, King & Ridings, Cork 1829–34; facsimile 1968. All publ., being Vol. 1, pt. 1 (2).

3691. THOMPSON, William *The natural history of Ireland.* 8vo, 4 vols, 1 portr., Reeve, London 1849–56.

3692. THOMSON, Anthony Todd *Syllabus of a course of lectures on botany.* 8vo, 12 pp, Chelsea 1813.

3693. —— *Lectures on the elements of botany. Vol. 1.* 8vo, xxiii+688 pp, 10 pls, text figs, Longman, London 1822. All publ.

3694. THOMSON, *Sir* Charles Wyville *The depths of the sea.* 8vo, 13 pp, text figs, Royal Dublin Society, Dublin 1869.

3695. —— *The depths of the sea.* 8vo, 527 pp, 8 col. maps, text figs, Macmillan, London 1873; 2ed. 1874.

3696. —— *The voyage of the Challenger. The Atlantic, a preliminary account.* 8vo, 2 vols, 4 pls, 38 charts & diagrams some col., 1 portr., text figs, Macmillan, London 1877.

3697. THOMSON, *Sir* John Arthur *The study of animal life.* Sm. 8vo, xii+375 pp, text figs, Murray, London 1892; 2ed. 1892; 3ed. 1905; 4ed. 1917. University Extension Manuals.

3698. —— *The outlines of zoology.* Sm. 8vo, xiv+641 pp, 32 pls, Pentland, Edinburgh 1892; 2ed. 1895; 3ed. 1899; 4ed. 1907; 5ed. 1910; 6ed. 1914; 1919; 7ed. 1920; 8ed. 1929. Students' Manual Series.

3699. —— *Natural history of the year.* Sm. 8vo, 288 pp, text figs, Melrose, London [1896].

3700. —— *The science of life. An outline of the history of biology and its recent advances.* 8vo, x+246 pp, Blackie, London [1899].

—— *see* Geddes, *Sir* Patrick

3701. THOMSON, Robert *The natural history of a Highland parish Ardclach, Nairnshire.* 8vo, xv+304 pp, 1 pl., Bain, Nairn 1900.

3702. THOMSON, Spencer *Wanderings among the wild flowers: how to see and how to gather them.* Sm. 8vo, xix+318 pp, text figs, Groombridge, London 1854; 2ed. 1854; 3ed. 1856; 4ed. 1857 title *Wild flowers.* Later editions.

3703. —— *Wayside weeds, or botanical lessons from the lanes and hedgerows.* 8vo, xii+212 pp, Groombridge, London 1864; ed. 1872.

3704. THONNER, Franz *Analytical key to the natural orders of flowering plants.* 8vo, 151 pp, Sonnenschein, London 1895. Translation of *Amleitung zum Bestimmen der Familien der Phanerogamen*, Berlin 1891.

3705. THORNHILL, John *A fasiculus of thirty-five dried specimens of grasses, and . . . the seeds of each kind.* Fol., iv+16 pp, 35 11 of specimens, Gateshead 1806.

—— *see* Winch, N. J.
Thornton, C. G. *see* Wyatt, W.
3706. THORNTON, John Robert *New illustrations of the sexual system of C. von Linnaeus . . . A prize dissertation of the sexes of plants [by C. Linnaeus]* . . . And the temple of *flora*. Fol., 2 pts, 31 col.+91 plain pls, 28 portrs, text figs, Bensley for the author, London [1799–] 1807; ed. *Temple of flora* alone 1812. 1812 ed. is the smaller 'lottery' ed. Every copy of the first ed. differs in its plates.
3707. —— *Practical botany. Vol. 1.* 8vo, 90 pp, 85 pls, Baldwin, London 1807. All publ.
3708. —— *Alpha botanica; or, the first book that should be put into the hands of the botanical student.* 8vo, 89 pp, London 1810.
3709. —— *Botanical extracts: or philosophy of botany.* Fol., 4 vols, 169 pls, 31 portrs, text figs, White Johnson, London [1798–] 1810.
3710. —— *Elementary botanical plates, illustrative of the science of botany . . . but more particularly intended to illustrate Botanical extracts.* Fol., 160 pls, 17 portrs, London 1810.
3711. —— *A grammar of botany, containing an explanation of the system of Linnaeus, and the terms of botany.* 12mo, 45 col. pls, Phillips, London 1811; 2ed. 1814.
3712. —— *Outline of botany; or easy introduction to the science.* 8vo, 86 pp, 1 portr., 3 tab., London 1812; ed. 1812 as part of No. 3713.
3713. —— *Elements of botany.* Roy. 8vo, 2 vols, 176 pls, Whiting for the author, London 1812. Copies often have fewer plates. See No. 3712.
3714. —— *The British flora; or, genera and species of British plants.* 8vo, 5 vols, 342 pls, Whiting for the author, London 1812. Copies often have fewer plates.
3715. —— *Juvenile botany: being an easy introduction to that delightful science.* 8vo, v+307 pp, 15 pls, Sherwood, London 1818; ed. 1823 title *An easy introduction to the science of botany*; 2ed. 1826.
3716. THORPE, Charles *British marine conchology.* 12mo, 1+267 pp, 1 col.+7 plain pls, Lumley, London 1844.
*Thoughts on the study of natural history.* 1820. *see* Drummond, J. L.
3717. THRELKELD, Caleb *Synopsis stirpium Hibernicarum . . . being a . . . treatise on the native plants, especially such as grow spontaneously in the vicinity of Dublin.* 8vo, 26+[175]+60

pp, Davys, Dublin 1726; 1727; issue for the author, Dublin & London 1727.

*Through the copse. Another ramble after wild flowers.* 1895. *see* Cooke, M. C.

3718. TILER, William *The natural history of birds, quadrupeds, fishes, serpents, insects*. 12mo, text figs, Simpkin, London 1837; ed. 32mo, Halifax 1840.

3719. TIMBS, John *Strange stories of the animal world.* Sm. 8vo, xii+372 pp, 6 pls, text figs, Griffith & Farran, London 1866 [=1865]; 2ed. 1868.

3720. —— *Eccentricities of animal creation*. 8vo, 352 pp, 8 pls, Seeley, London 1869.

3721. —— *Curiosities of animal and vegetable life*. 12mo, Griffin, London 1872.

3722. TODD, Robert Bentley *editor The cyclopaedia of anatomy and physiology*. Roy. 8vo, 5 vols, text figs, Longman, London 1835–59. Vol. 5 is supplementary.

3723. TOMLINSON, Charles & TOMLINSON, Sarah, *Mrs* Charles *Lessons derived from the animal world*. 12mo, 2 vols, text figs, SPCK, London [1862].

3724. TOMLINSON, Sarah, *Mrs* Charles *First steps in general knowledge*. 16mo, 5 pts, text figs, SPCK, London [c.1860].
—— *see* Tomlinson, C.

*Tommy Trip's History of beasts and birds*. 1753. *see A pretty book of pictures*.

3725. *Tommy Trip's museum of birds*. 12mo, 2 parts, col. text figs, Harris, London [c.1820].

3726. *Tommy Trip's museum, or a peep at the quadruped race*. 12mo, 3 parts, col. text figs, Harris, London [?1821]; 1833.

Tonna, Charlotte Elizabeth *see* Phelan, *Mrs* C. E.

3727. TOPLIS, S. Grace *Jefferies's land: a history of Swindon and its environs*. 8vo, text figs, 1 map, Simpkin Marshall, London 1896.

3728. TOPSELL, Edward *A historie of foure-footed beasts*. Fol., 757 pp, text figs, Jaggard, London 1607; ed. 1658 in No. 3730.

3729. —— *The historie of serpents, or the second book of living creatures*. Fol., 315 pp, text figs, Jaggard, London 1608; ed. 1658 in No. 3730.

3730. —— & MOFFETT, Thomas *The history of four-footed beasts and serpents . . . whereunto is now added, the theater of insects*. Fol., 2 vols, 1130 pp, text figs, Sawbridge, London

1658. Pagination erratic; pp. 819–888 omitted. A combination edition of the English translation of No. 2624 with Nos 3728 & 3729.

3731. TOURNEFORT, Joseph Pitton de *The compleat herbal: or, the botanical institutions of Mr. Tournefort*. 4to, 2 vols, 268 pls, Bonwicke, London [1716–]1719–30. Plates 1–277; 22, 71 & 80 occur twice; 15 & 54–64 do not occur. Translation of *Institutiones rei herbariae*, Paris 1700–03, by John Martyn.

3732. —— *History of plants growing about Paris . . . with many additions, accommodated to the plants growing in Great Britain*. 8vo, 2 vols, Rivington, London 1732. Translation of *Histoire des plantes qui naissant aux environs de Paris*, 2ed., Paris 1725; translated and with additions by John Martyn.

*Tower menagerie*. [1829]. *see* Bennett, G. T.

Towndrow, Richard Francis *see* Edwards, W.

3733. TOWNLEY, Richard *A journal kept in the Isle of Man, giving an account of the . . . natural productions*. 8vo, 2 vols, Ware printed, Whitehaven 1791.

3734. TOWNSEND, Frederick *Notes on the flora of Hampshire*. 8vo, 15 pp, London 1879.

3735. —— *Flora of Hampshire, including the Isle of Wight*. 8vo, xxiv+524 pp, 1 col. & 1 plain pls, 1 col. map, Reeve, London 1883; 2ed. 1904; suppl. 1929.

3736. TOWNSON, Robert *Tracts and observations in natural history and physiology*. 8vo, ix+232 pp, 7 pls, for the author, London 1799.

3737. TOYNBEE, Henry *On the minuter inhabitants of the surface of the ocean. 8vo, 25 pp, 2 col. pls, London 1860.* Reprinted from *The Nautical Magazine*.

3738. TRADESCANT, John *Museum Tradescantianum: or, a collection of rarities. Preserved at South-Lambeth near London*. 8vo, 179 pp, 1 pl., 2 portrs, Grimmond printed, London 1656; ed. Oxford 1825. Ed. 1825 is Old Ashmolean Reprints No. 1; it omits pp. 74–178, the description of plants in Tradescant's garden.

3739. TRAHERNE, John Popkin *The habits of the salmon*. 8vo, v+163 pp, Chapman & Hall, London 1889.

3740. [TRAILL, Catherine Parr] *Sketches from nature; or hints to juvenile naturalists*. 12mo, xx+216 pp, 1 pl., Harvey & Darton, London 1830. By the author of *The young immigrants*. Often attributed to her sister *Miss* Agnes Strickland.

3741. TRAILL, George William *A monograph of the algae of the*

*Firth of Forth*. 4to, 16 pp, 4 pls of specimens, privately printed, Edinburgh 1885.

3742. TRAQUAIR, Ramsay Heatley *On popular delusions in natural history*. 8vo, 22 pp, Blackwood, Edinburgh 1896. Reprinted from *Trans. Edinb. Fld Nat. & Microscop. Soc.*, Vol. 3.

3743. [TRAVIS, William] *Catalogus plantarum circa Scarborough nascentium*. 4to, s. sh., [?Scarborough 1850].

*Treasures of the deep*. 1847. *see* Landsborough, D. *the father*.

*A treatise on zoology*. 1900–09. *see* Lankester, *Sir* Edwin R. *editor*.

Treitschke, Friedrich *see* Children, J. G.

Trembley, Abraham *see* Mortimer, C.

Trevor-Battye, Aubyn Bernard Rochfort *see* Battye, A. B. R. T.

3744. TRIMEN, Henry *Botanical bibliography of the British Counties.* 8vo, xxv pp, London [1874]. Revision of papers in *J. Bot.*, 1874.

3745. —— & DYER, *Sir* William Turner Thistleton *Flora of Middlesex*. Sm. 8vo, xli+428 pp, 1 map, Hardwicke, London 1869.

3746. TRIMMER, Kirby *Flora of Norfolk*. 8vo, xxvi+195 pp, Hamilton, London 1866; suppl. 73 pp, [1885].

3747. TRIMMER, Mary *A history of quadrupeds. Adapted to the capacities of youth and embellished with twenty-nine engravings on wood after Bewick.* 12mo, text figs, Tegg & Castleman, London 1803. ?Pseudonym.

—— *A natural history of the most remarkable quadrupeds, birds, fishes, serpents, reptiles, and insects*. 1825. *see* Boreman, T.

3748. TRIMMER, *Mrs* Sarah *An easy introduction to the knowledge of nature, and reading the holy scriptures*. 8vo, xiv+263 pp, for the author, London 1780; 6ed. 1789; 10ed. 1799; 11ed. 1802; 12ed. 1805; ed. [c.1820]; French ed., London 1788; ed. 1810; ed. 1817. *See also* Trimmer, *Miss* Sarah & Ventum, *Miss* Harriet.

3749. TRIMMER, *Miss* Sarah *Sequel to Mrs Trimmer's Easy introduction to the knowledge of nature*. 12mo, iv+229 pp, Rivington, London 1818.

Trip, Tommy *pseudonym. see A pretty book of pictures*; & Nos. 3725, 3726.

3750. TRIPP, F. E. *British mosses, their homes, aspects, structure and uses*. Roy. 8vo, xxii+235 pp, 39 col. pls, Bell & Daldy, London 1868; ed. 2 vols, 1874; 2 vols, 1888.

3751. TROUESSART, Édouard Louis *Microbes, ferments and molds*. Sm. 8vo, xi+314 pp, text figs, Kegan Paul, London 1886; 2ed. 1889; 3ed. 1894. Translation of *Les microbes, les ferments et les moisissures*, Paris 1886 [=1885]. International Scientific Series, Vol. LVII.

3752. TUCK, Julian George *Ornithology of Suffolk*. 8vo, 11 pp, Sheffield 1891. Reprinted from White's *History of Suffolk*.

3753. TUCK, William Henry *The avi-fauna of a Suffolk village, 1847–93*. Sm. 8vo, 7 pp, Groom, Bury St Edmunds [1894].

3754. TUCKER, Andrew G. C. *Ornithologia Danmoniensis; or, an history of the habits and economy of Devonshire birds. Vol. 1, No. 1 (2)*. 4to, lxxxviii pp, 6 col. + 1 plain pls, for the author, London 1809. Often with a duplicate set of plates, all plain. All publ.

3755. TUGWELL, George *A manual of the sea-anemones commonly found on the English coast*. Sm. 8vo, 123 pp, 6 col.+1 plain pls, Van Voorst, London 1856.

3756. —— *The North Devon hand-book . . . and an introduction to the natural history of the district*. 12mo, 252 pp, 14 pls, 1 map, Simpkin, London [1857]; 2ed. Ilfracombe [186–]; 4ed. [1877].

3757. TULK, Alfred & HENFREY, Arthur *Anatomical manipulation . . . in comparative anatomy and physiology, also an introduction to the use of the microscope*. Sm. 8vo, xvi+413 pp, text figs, Van Voorst, London 1844.

3758. [TUNSTALL, Marmaduke] *Ornithologia Britannica: seu avium omnium Britannicarum . . . catalogus*. Fol., 4 pp, text fig, Dixwell for the author, London 1771; ed. 8vo, 1880. 1880 ed. is Willughby Society Publication.

Tunstall Museum, Newcastle *see* Fox, G. T.

3759. TUPPER, James Perchard *An essay on the probability of sensation in vegetables; with . . . observations on instinct*. 8vo, xii+127 pp, White Cochrane, London 1811; 2ed. 1817.

3760. TURNBULL, Robert *Index of British plants according to the London catalogue (eighth edition)*. 8vo, 98 pp, Bell, London 1889. For the *Catalogue* see Watson, H. C. & Dennes, G. E.

3761. TURNBULL, William Patterson *Birds of East Lothain and a portion of the adjoining Counties*. 4to or Roy. 8vo, 48 pp, 1 col. pl., text figs, Murray for private circulation, Glasgow 1867. 4to 12 copies, 8vo 150. There is an earlier edition Sherman, Philadelphia [1863].

3762. TURNER, Dawson *A synopsis of the British Fuci.* 8vo, 2 vols, xlvi+400 pp, Yarmouth 1802. L.P.

3763. —— *Muscologiae Hibernicae spicilegium*. 8vo, xi+200+xiv pp, 16 col. pls, Black for the author, Yarmouth 1804. 350 copies.

3764. —— *Fuci, sive plantarum fucorum generi a botanicis ascriptarum icones, descriptiones et historia. Fuci: or, coloured figures and descriptions of the plants referred by botanists to the genus Fucus*. Roy. or demy 4to, 4 vols, 258 col. pls, Arch, London 1808–19. Roy. 4to 25 copies.

3765. —— & BORRER, William *Specimen of a Lichenographia Britannica; or an attempt at a history of British lichens*. 4to, ii+240 pp, for private circulation, Yarmouth 1839. All publ., c.30 copies.

3766. —— & DILLWYN, Lewis Weston [*List and questionnaire of British plants.*] 4to, 4 pp, [?London 1803]. This list has no title.

3767. —— & —— *The botanist's guide through England and Wales*. Sm. 8vo, 2 vols, xvi+804 pp, Phillips & Fardon, London 1805.

3768. [TURNER, James Aspinall] *Remarks on the Linnaean orders of insects, forming a short and familiar introduction to the study of entomology*. 4to, iv+40 pp, 7 col. pls, Longman, London 1828. By a member of the Manchester Natural History Society.

3769. TURNER, Richard *An easy introduction to the arts and sciences . . . eighth edition . . . with considerable additions and improvements; particularly an abridgement of natural history*. 12mo, 256 pp, 6 pls, text figs, Johnson, 8ed. London 1802; 9ed. 1803; 10ed. 1804; 13ed. 1809; 14ed. 1811; 16ed. 1814. Ed. 1 1783 and 2–7 do not contain the natural history.

3770. TURNER, Sharon *The sacred history of the world as displayed in the creation and subsequent events to the deluge*. 8vo, 3 vols, Longman, London 1832; 2ed. Vol. 1, 1832. Later editions.

3771. TURNER, William *Libellus de re herbaria novus*. 4to, 8 11, Bydell, London 1538; facsimile 1877; index 1878; facsimile 1966. Facsimile 1877 100 copies.

3772. —— *Avium praecipuarum, quarum Plinium et Aristotelum mentio est, brevis et succincta historia*. 8vo, [157] pp, Gymnicus, Cologne, 1544; ed. Cambridge 1823; ed. with translation, Cambridge 1903.

3773. —— *The names of herbes*. Sm. 8vo, 63 ll, Day & Seres, London [1548]; ed. 1881; facsimile 1966.

3774. —— *A new herball*. Fol., 2 pts, pt. 1 Mierdman printed, London; pt. 2 Birckman, Cologne 1551, 62; ed. Cologne 1568. Part 2 in 2 parts.

3775. TURTON, William *British fauna, containing a compendium of the zoology of the British Islands. Vol. 1.* 12mo, 230+vii pp, Evans printed, Swansea 1807. All publ.

3776. —— *A conchological dictionary of the British Islands*. 12mo, xxviii+272 pp, 28 col. pls, Booth, London 1819.

3777. —— *Conchylia Insularum Britannicarum. The shells of the British Islands systematically arranged*. 4to, xlvi+279 pp, 20 col. pls, Rodwell & Martin, Exeter 1822; ed. London [1830] title *Bivalve shells of the British Islands*; ed. London 1848 title *Conchylia dithyra Insularum Britannicarum.*

3778. —— *Enumeration of the marine shells found on the Devonshire coast*. 8vo, 1829. Reprinted from Carrington, N. T. *The Teignmouth, Dawlish & Torquay guide*, [1828].

3779. —— *A manual of the land and fresh-water shells of the British Islands*. Sm. 8vo, viii+150+16 pp, 10 col. pls, Longman, London 1831; ed. 1840; ed. 1842; ed. 1857.

3780. —— & KINGSTON, J. F. *The natural history of the district of Teignmouth, Dawlish and Torquay*. 12mo, unpaginated, Teignmouth [c.1810].

3781. TUTT, James William *Melanism and melanochroism*. 8vo, 66 pp, text figs, Sonnenschein, London 1891.

3782. —— *The British Noctuae and their varieties.* 8vo, 4 vols, Sonnenschein, London 1891–92; suppl. 4 vols, 1926–50.

3783. —— *Monograph of the Pterophorina of Britain*. 8vo, ii+161 pp, London [1891–93]; issue Hobson, Hartlepool [?1893]. Reprinted with additions from *The British Naturalist*, 1889–90.

3784. —— *Secondary sexual characters in Lepidoptera*. 8vo, xxiv pp, Blackheath 1892. The introduction to Vol. 3 of No. 3782, with new title leaf.

3785. —— *Random recollections of woodland, fen and hill*. 8vo, 183 pp, Sonnenschein, London [1893]; 2ed. 1895.

3786. —— *Woodside, burnside, hillside and marsh*. 8vo, v+241 pp, text figs, Sonnenschein, London 1894.

3787. —— *British butterflies*. 8vo, 469 pp, 10 col. pls, Gill, London 1896. Hedgerow and Woodland Series.

3788. —— *British moths*. Sm. 8vo, xii+368 pp, 12 col. pls, Routledge, London 1896; eds. [ ]. Country Books.

3789. —— *Insects and spiders, their structures, life-histories and habits*. Sm. 8vo, 116 pp, 18 pls, text figs, George Gill, London 1896. Practical Series of Object Lesson Books, Pt. 2.

3790. —— *Chats about British birds*. 8vo, 216 pp, text figs, Gill, London [1897]. Hedgerow and Woodland Series.

3791. —— *A natural history of the British Lepidoptera*. 8vo, 9 vols, I–V, VIII–X Sonnenschein, Vol. XI Eliot Stock, London 1899–1914. All publ. Vols. VIII–XI alternative title *A natural history of British butterflies,* Vols I–IV.

3792. TWAMLEY, Louisa Anne *later* Meredith *The romance of nature, or the flower seasons illustrated*. 8vo, 26 col. pls, Tilt, London 1836; 2ed. 1836; 3ed. 1836; ed. 1839.

3793. —— *Flora's gems*. Imp. 4to, 12 col. pls, Bogue, London [1837].

3794. —— *Our wild flowers familiarly described and illustrated*. 12mo, vi+312 pp, 12 col. pls, Tilt, London 1839; 2ed. 1839; ed. 1843.

3795. TWINING, Elizabeth *Illustrations of the natural orders of plants*. Fol., 2 vols, 169 pp, 160 col. pls, Cundall, London 1849, 55; ed. 8vo, 1868.

3796. —— *Short lectures on plants*. 16mo, xii+369 pp, Nutt, London 1858; ed. 1878 title *Lectures on plants*.

3797. —— *The plant world*. 8vo, 414 pp, 5 col. pls, Nelson, London 1866; ed. 1870.

3798. —— *A lecture on plants as water drinkers*. 8vo, London 1872.

*262 questions and answers; or, the child's guide to knowledge*. 1825. *see* Umphelby, *Miss* Frances.

3799. [TYAS, Robert] *Woodland gleanings, being an account of British forest trees*. Sm. 8vo, 188 pp, text figs, Tyas, London 1837; 1840; 2ed. 1853; 1859.

3800. —— *Favourite field flowers; or wild flowers of England*. 16mo, xi+196 pp, 12 col. pls, Houlston & Stoneman, London 1848; 2nd ser., ix+195 pp, 12 col. pls, 1850; ed. 1859, 60 title *The wild flowers of England; or, favourite field flowers*; ed. 1866.

3801. —— *The book of botany*. 32mo, vi+341+4 pp, 12 col. pls, Tyas, London [?1850].

3802. TYLOR, Alfred *Colouration in animals and plants*. Roy. 8vo, xiv+105 pp, 12 col.+11 plain pls, Alabaster Passmore printed, London 1886.

3803. TYNDALL, John *Fragments of science*. 8vo, 449 pp, Longman, London 1871; 2ed. 1871; 4ed. 1872; 5ed. 1876; 6ed. 2 vols 1879; 7ed. 1889; 8ed. 1892; 9ed. 1896; 1899; 1907. 6ed. and later contain No. 3804.

3804.—— *Address* [on evolution] *delivered before the British Association assembled at Belfast with additions*. 8vo, viii+65 pp, Longman, London 1874; ed. 1874. Later editions in No. 3803.

3805.—— *Essays on the floating matter in the air in relation to putrefaction and infection*. 8vo, xix+338 pp, text figs, Longman, London 1881; 2ed. 1883.

3806. TYRER, John *Taxidermy made easy, being plain and practical directions*. 12mo, xiii+63 pp, 1 pl., Burrill, Chatham 1844.

3807. [TYSON, Edward] *Phocaena, or the anatomy of a porpess, dissected at Gresham Colledge; with a preliminary discourse concerning anatomy*. 4to, pls, Tooke, London 1680.

## U

3808. ULLATHORNE, G. S. *A systematic list of British vertebrate animals*. Sm. 8vo, iv+36 pp, Durham 1861.

3809. ULLYET, Henry *Rambles of a naturalist round Folkestone, with occasional papers on the flora and fauna of the district*. Sm. 8vo, 157 pp, English, Folkestone 1880.

3810. [UMPHELBY, *Miss* Frances] *262 questions and answers; or, the child's guide to knowledge*. 24mo, 119 pp, S. Mills, Thetford 1825; 2ed. London 1828 title *The child's guide to knowledge*; 3ed. 1830; 4ed. 1832; 5ed. 1834. Many later editions to 60ed. 1900. By a lady; often attributed to her sister *Mrs* R. (Fanny) Ward.

3811. *Uncle Buncle's stories about beasts*. 12mo, 12 pp, 7 col. pls, Dean & Munday, London [c.1840].

Uncle Matt *pseudonym. see* Cooke, M. C.

Uncle Philip *pseudonym. see* Hawks, F. L.

3812. UNGER, Franz *Botanical letters to a friend*. 8vo, 116 pp, 1 pl., Highley, London 1853. Library of Science and Arts.

3813. *The universal powers of nature revealed . . . in a graphic outline of the history of the earth and animated nature*. 8vo, Atkinson, London 1856. By a philosopher.

University College of Wales (Aberystwyth) Scientific Society. *see* Salter, J. H.

3814. UNWIN, William Charles *Illustrations and dissections of the genera of British mosses*. 4to, [viii]+31 pp, 23 pls, pts i–xii G. P. Bacon, pt. xiii Sussex Advertiser, Lewes [1877–] 1878.

3815. URE, David *The history of Rutherglen and East Kilbride. Published with a view to promote the study of antiquity and natural history*. 8vo, vi+334 pp, 21 pls, Niven printed, Glasgow 1793.

3816. USSHER, Richard John & WARREN, Robert *The birds of Ireland; distribution, migration and habits*. Roy. 8vo, xxxi+419 pp, 1 col.+8 plain pls, 2 maps, text figs, Gurney & Jackson, London 1900.

## V

3817. VALENTINE, John Tristram Tristram *London birds and beasts*. Sm. 8vo, xii+319 pp, Cox, London 1895.

Van Beneden, Pierre Joseph *see* Beneden, P. J. van

Van Bruyssel, Ernest Jean *see* Bruyssel, E. J. van

Vandervelde, Emile *see* Demoor, J.

Van Huerck, Henri Ferdinand *see* Huerck, H. F. van

Van Rymsdyk, Jan & Andreas *see* Rymsdyk, J. & A. van

Varigny, Henry de *see* De Varigny, H.

3818. VASEY, George *Delineations of the ox tribe; or, the natural history of bulls, bisons and buffaloes*. 8vo, xvi+192 pp, 1 pl., text figs, Biggs, London 1851; issue 1857 title *A monograph of the genus Bos*.

3819. [VAUGHAN, Thomas] *A brief history, intermixed with variety of philosophical discourses . . . with refutation of such vulgar errors as our modern authors have omitted*. 8vo, Matthew Smelt, London 1669. By Philalethes, Eugenius *pseudonym*.

3820. VELLEY, Thomas *Coloured figures of marine plants, found on the southern coast of England . . . Plantarum maritimarum in oris Angliae australibus sponte crescentium icones pictae.* Fol., 9+8 pp, 6 col. pls, Hazard for the author, Bath 1795.

3821. VENABLES, Edmund *The Isle of Wight*. 8vo, vi+526 pp, Stanford, London 1860.

3822. *Venation in leaves, with specimens from nature illustrative of each class*. 4to, 250 pls with letterpress, London 1868. Nature printed.

3823. [VENNING, *Miss* Mary Ann] *Rudiments of conchology: designed as a familiar introduction to the science*. 12mo, vii+103 pp, 10 col. pls, Harvey & Darton, London 1826; 1828 title *Introduction to conchology*; 1837. Advertisement signed A.M.

3824. VENTUM, Harriet *Surveys of nature: a sequel to Mrs Trimmer's Introduction to the knowledge of nature . . . adapted to the capacities of children*. 12mo, vi+138 pp, Badcock, London 1802. *See also* Trimmer, *Mrs* Mary & Trimmer, *Miss* Mary.

3825. VERRALL, George Henry *A list of British Diptera*. 4to, 3 pts, 33+4+12 pp, London 1888; 2ed. 1901.

3826. —— *Catalogue of the Platypezidae of the European district with references and synonymy*. 8vo, 121 pp, Gurney & Jackson, London 1900. A separate issue of the catalogue in No. 3827 Vol. VIII.

3827. —— *British flies*. Roy. 8vo, 3 vols, 2 portrs, text figs, Vols V & VIII Gurney & Jackson, Vol. VI Cambridge University Press, London & Cambridge [1900–] 1901, 1909, 1961 [–1962]. Vol. VIII 1901, Vol. V 1909, Vol. VI (by Collin, James E.) 1961. All publ.

*Vertebrate fauna of Scotland. see* Buckley, T. E. & Brown, J. N. H. *editors*.

Verulam, *Baron see* Bacon, F.

*Vestiges of the natural history of creation*. 1844. *see* Chambers, R.

*View of nature in ascending regions*. 1853. *see* Emslie, J.

3828. VIGORS, Nicholas Aylward *An address delivered at the sixth and last anniversary meeting of the Zoological Club of the Linnean Society of London*. 8vo, 32 pp, Spottiswoode printed, London 1830.

—— *see* Baker, T. B. L.

3829. VINES, Sydney Howard *Student's text book of botany*. 8vo, x+430 pp, text figs, Sonnenschein, London 1894–95; ed. 1896; ed. 1902; ed. 1910.

3830. —— *An elementary text book of botany*. 8vo, xv+611 pp, text figs, Sonnenschein, London 1898; 2ed. 1905; 3ed. 1908.

3831. VIRCHOW, Rudolf Ludwig Carl *The freedom of science in the modern state.* [On darwinism]. 8vo, xvii+65 pp, Murray, London 1878. Translation of *Die Frieheit der Wissenschaft in modernen Staat*, Berlin 1877. See No. 1510 for a reply.

*The visible world*. 1792. *see* Comenius, J. A.

*The visit for a week*. 1794. *see* Peacock, Lucy.

3832. *A visit to the Tower, being an account of several birds and beasts*. 32mo, 16 pp, text figs, Kendrew, York [c.1820].

*A visit to the Zoological Gardens*. [1830]. *see* Bishop, J.

3833. VOGT, Carl Christoph & SPECHT, Friederich *The natural history of animals. (Class Mammalia – animals which suckle their young); in words and pictures*. Fol., 2 vols, 13 pls, text figs, Blackie, London [1887–88]. Translation of *Les mammifères*, Paris 1884, by George Goudie Chisholm.

3834. *The voice of flowers . . . character, habits . . . of many British plants*. 12mo, Broom, London 1871.

3835. *Voices of the garden, the woods, and the fields; or, the teachings of nature as seasons change*. Sm. 8vo, viii+412 pp, 4 col. pls, Nelson, London 1859; ed. 1860. By the author of *Success in life*, etc.

# W

3836. W., H. *Facts about flowers*. 4to, 12 col. pls, Christian Education Society, London 1871.

W., *Hon.* M. *The world of wonders revealed by the microscope*. 1858. *see* Ward, *Hon*. M.

W., *Hon*. M. & M., *Lady*. *Entomology in sport and entomology in earnest.* [1859]. *see* Ward, Hon. M. & Mahon, *Lady* J.

W., S. *The meadow queen.* [1836]. *see* Waring, *Miss* S.

W., T. W. *Curiosities of ornithology.* [1871]. *see* Wood, T. W.

3837. WADDELL, Cosslatt Herbert *Moss Exchange Club*

*catalogue of the British Hepaticae*. 8vo, 7 pp, Moss Exchange Club, London 1897.

3838. WADDY, Edith *A year with the wild flowers*. 16mo, viii+151 pp, 12 col. pls, text figs, Wesleyan Conference Office, London [1873].

3839. WADE, James *Systematic zoology. Grammar and synopsis of natural history*. Fol., [35] pp, for the author, Leeds 1842.

3839a WADE, Walter *Florae Dublinensis specimen*. Fol., 3 pls with letterpress, [?Dublin, 1789]. ?All publ. ?Prospectus only.

3840. —— *Catalogus systematicus plantarum indigenarum in Comitatu Dublinensi inventarum. Pt. 1.* 8vo, viii+275 pp, Dublin 1794. All publ.

3841. —— *Plantae rariories in Hibernia inventae; or habitats of some plants rather scarce and valuable found in Ireland: with Catalogus plantarum rariorum in Comitatu Galloviduae praecipue Cunnamara inventarum; or, a systematic account of the more rare plants principally found in the County of Galway; but more particularly in that part of it called Cunnemara*. 8vo, xiv+214+23 pp, 1 col. pl., Graisberry & Campbell printed, Dublin 1804. Reprint of 2 papers in *Trans. R. Dublin Soc*., 1804, 02.

3842. —— *Sketch of lectures on meadow and pasture grasses, delivered in the Dublin Society's Garden, Glasnevin*. 8vo, 55 pp, Graisberry & Campbell, Dublin 1808. Preprint from *Trans. Dublin Soc*., 6, 1810.

3843. —— *Salices, or an essay towards a general history of sallows*. 8vo, xxii+406+56 pp, 1 col. pl., Graisberry & Campbell printed, Dublin 1811.

3844. WAGNER, Moritz Friederich *The darwinian theory and the laws of migration of organisms*. 8vo, London 1873. Translation of *Die Darwinische Theorie, und das Migrationsgesetz der Organismen*, Leipzig 1868, by James L. Laird.

3845. WAITE, Frederick *Lepidoptera diurna; or, the British butterflies arranged after the modern system: Part the first*. 4to, 20 pp, 10 col. pls, Leeds 1832. ?All publ.

3846. WAKEFIELD, *Mrs* Priscilla *An introduction to botany in a series of familiar letters*. 12mo, xiii+184 pp, 12 pls, 1 tab., Newbery, London 1796; ed. Wogan, Dublin 1796; 2ed. 1798; 3ed. 1803; 5ed. 1807; 6ed. 1812; 7ed. 1816; 8ed. 1818; 9ed. 1823; 10ed. 1831; 11ed. 1841. For a supplementary work see No. 326.

3847. —— *Mental improvement; or, the beauties and wonders of nature and art, in a series of instructive conversations*. 12mo, 2 vols, Darton, London 1794; 2ed. 1795; 3ed. 1797; ed. 3 vols 1797–1800; 4ed. 1801; 5ed. 1804; 8ed. 1818.

3848. —— *Domestic recreation; or dialogues illustrative of natural and scientific subjects*. 12mo, 2 vols, pls, Darton, London 1805–07; ed. 1 vol. 1818.

3849. —— *Instinct displayed, in a collection of well-authenticated facts, exemplifying the extraordinary sagacity of various species of the animal creation*. 16mo, xvi+311 pp, Darton, London 1811; ed. 1812; 4ed. 1821; ed. [c.1825]; ed. [c.1835].

3850. —— *An introduction to the natural history and classification of insects; in a series of . . . letters*. 12mo, x+192 pp, 12 col. pls, Darton, London 1816.

—— *see* Hoare, Sarah.

3851. WALCOTT, John *Flora Britannica indigena: or plates of the indigenous plants of Great Britain*. 8vo, x+8+6+4+6 pp, 168 pls, 1 portr., Hazard printed, Bath 1778. A few copies have a plate 169.

3852. —— *Synopsis of British birds*. 4to, 2 vols, 255 col. pls with 59 11 of letterpress, Justins for the author, London 1789. Printed on one side of the paper. ?25 copies.

3853. WALFORD, Thomas *The scientific tourist through England, Wales, & Scotland*. 12mo, 2 vols, 8 col. pls, 4 maps, 1 tab., Booth, London 1818.

3854. [——] *The scientific tourist through Ireland.* 12mo, 34+[197] pp, 7 pls, 2 maps, Booth, London 1818. By an Irish gentleman.

3855. WALKER, Francis *Monographia chalciditum*. 8vo, 2 vols, Baillière for the author, London 1839.

3856. —— *Insecta Britannica. Diptera*. 8vo, 3 vols, 30 pls, Reeve, London 1851–56. Insecta Britannica Series, Vols I–III.

3857. —— *List of British Euplexoptera, Orthoptera, Thysanoptera, and Hemiptera*. 8vo, 55 pp, Newman, London 1860.

Walker, George *see* Boys, W.

3858. WALKER, John *of Edinburgh Institutes of natural history. Containing the heads of the lectures in national history . . . in the University of Edinburgh*. 8vo, 169 pp, Stewart Ruthven printed, Edinburgh 1792.

3859. —— *Essays on natural history and rural economy*. 8vo, ii+629 pp, University Press, Edinburgh 1808; ed. London 1812.

3860. WALKER, John *of London The bot fly of the ox (Oestrus ovis)*. 8vo, 15 pp, text figs, Jack, London [1886].
3861. WALKER, Richard *The flora of Oxfordshire, and its contiguous Counties . . . preceded by an introduction to botany*. 8vo, cxxxv+336 pp, 12 pls, 1 plan, Slatter, Oxford 1833.
3862. WALLACE, Alfred Russel *Contributions to the theory of natural selection*. 8vo, xvi+384 pp, Macmillan, London 1870; 2ed. 1871; 1875; 1891. 1891 ed. in No. 3867.
3863. —— *The geographical distribution of animals*. 8vo, 2 vols, 20 pls, 7 col. maps, Macmillan, London 1876.
3864. —— *Tropical nature and other essays*. 8vo, xiii+356 pp, Macmillan, London 1878; 1891. 1891 ed. is in No. 3867.
3865. —— *Island life*. 8vo, xvii+526 pp, 3 maps, text figs, Macmillan, London 1880; 2ed. 1892; 3ed. 1902; 1911.
3866. —— *Darwinism: an exposition of the theory of natural selection with some of its applications*. 8vo, xvi+494 pp, 1 portr., 1 map, text figs, Macmillan, London 1889; 2ed. 1889; 1890; 1905; 3ed. 1912.
3867. —— *Natural selection and tropical nature*. 8vo, xii+492 pp, text figs, Macmillan, London 1891. Almost entirely reprints of Nos 3862 & 3864.
3868. —— *Studies scientific and social*. 8vo, 2 vols, text figs, 1 map, Macmillan, London 1900.
3869. WALLACE, James *A description of the Isles of Orkney*. 8vo, 94+36 pp, 3 pls, Reid printed, Edinburgh 1693; ed. London 1700 title *An account of the Islands of Orkney*; ed. 1883.
3870. WALLACE, Robert *editor A country schoolmaster (James Shaw)*. Sm. 8vo, 488 pp, 8 pls, 1 portr., Edinburgh 1899.
3871. WALLICH, George Charles *Notes on the presence of animal life at vast depths in the sea*. 8vo, 38 pp, for private circulation, London 1860.
3872. —— *Deep sea researches on the biology of Globigerina*. 8vo, 75 pp, 1 pl., Van Voorst, London 1876.
3873. WALLIS, John *The natural history and antiquities of Northumberland: and so much of the County of Durham as lies between the rivers Tyne and Tweed, commonly called North Bishoprick*. 4to, 2 vols, Strahan for the author, London 1769.
3874. WALSINGHAM, Thomas De Gray *Baron* & DURRANT, John Hartley *Rules for regulating nomenclature with a view to secure strict application of the law of priority in entomological work* [Merton rules]. 8vo, 18 pp, Longman, London 1896.
3875. WALTON, John *On the British insects of the family*

*Curculionidae*. 8vo, 120 pp, 1 pl., London 1844 [=1852]. Reprints from *Ann. Mag. nat. Hist*., 1844–52.

3876. —— *List of British Curculionidae, with synonyms, in the collection of the British Museum*. 12mo, 46 pp, British Museum, London 1856.

3877. WARBURTON, George Egerton *Names and synonyms of British plants*. 12mo, 196 pp, Bell, London 1889.

Ward, Fanny, *Mrs* R. *see* Umphelby, *Miss* Frances.

3878. WARD, Harry Marshall *The oak: a popular introduction to forest botany*. 8vo, 175 pp, 2 pls, text figs, Kegan Paul, London 1892. Modern Science Series.

3879. WARD, James *British ornithology, or birds of passage*. 8vo, ix+288 pp, 2 pls, Maidstone 1871.

Ward, *Lady* Jane Mahon *see* Ward, *Hon.* Mary.

3880. [WARD, *Hon. Mrs* Mary] *A world of wonders revealed by the microscope. A book for young students*. Imp. 16mo, 56 pp, 14 col. pls, Groombridge, London 1858; 2ed. 1859. By *Hon. Mrs* W.

3881. [—— & WARD, *Lady* Jane Mahon] *Entomology in sport and entomology in ernest*. 8vo, 20+48 pp, 1 col. pl., col. text figs, Jerrard, London [1859]. By *Hon. Mrs* W. & *Lady* M.

3882. —— *Microscope teachings*. Sm. 8vo, 8 col. pls, text figs, Groombridge, London 1864; 3ed. 1869 title *The microscope*; 1870; 4ed. 1876.

3883. WARD, Nathaniel Bagshaw *On the growth of plants without open exposure to air*. 8vo, London 1836. Reprint of 'Improved method of transplanting living plants', *Companion to Bot. Mag*., 1:317–20.

3884. —— *On the growth of plants in closely glazed cases*. 8vo, vi+95 pp, Van Voorst, London 1842; 2ed. 1852.

3885. [——] *Aspects of botany*. [Kilkee, Co. Clare]. 8vo, Clapham [1864].

3886. WARD, Rowland *The sportsman's handbook to practical collecting, preserving, and artistic setting up of trophies and specimens*. Sm. 8vo, viii+103+xii pp, text figs, the author, London 1880; 2ed. 1882; 3ed. 1883; 4ed. 1887; 5ed. 1890; 6ed. 1891; 7ed. 1894; 8ed. 1899; 10ed. 1911; 12ed. 1923.

3887. WARD, Samuel *and others*. *A modern system of natural history*. 12mo, 12 vols, 117 pls, text figs, Newbery, London 1775–76. Later printings, and some sections were issued separately with new title pages, reading *Natural history of four-footed beasts* etc.

3888. [WARING, *Miss* S.] *The minstrelsy of the woods*. 12mo, xi+227 pp, 17 col. pls, Harvey & Darton, London 1832.

3889. [——] *The meadow queen; or, the young botanists*. 18mo, Harvey & Darton, London [1836]. By S. W.

3890. —— *The wild garland: or, prose and verse illustrative of English wild flowers and forest trees*. 12mo, 12 col. pls, Fry, London 1837; ed. 1843.

3891. WARMING, Johannes Eugenius Bülow *A handbook of systematic botany*. 8vo, xii+620 pp, text figs, Sonnenschein, London 1895. Translation of *Haandbog i den systematiske Botanik*, 2ed. Copenhagen 1884, by Michael Cresse Potter.

3892. WARNER, Richard *Plantae Woodfordienses: a catalogue of the more perfect plants growing spontaneously about Woodford*. 8vo, vii+238 pp, for the author, London 1771; additions, 8 pp, 1784. The additions by Foster, T. F.

3893. WARREN, George *The nature and properties of living animals*. 8vo, viii+144 pp, Longman, London 1828.

Warren, John Byrne Leicester *see* De Tabley, *Baron*

Warren, Robert *see* Ussher, R. J.

3894. WATERHOUSE, Charles Owen *Aid to the identification of insects*. 8vo, 2 vols, 189 col. pls, Janson, London 1880–90.

3895. WATERHOUSE, George Robert *A natural history of the Mammalia. Vol. 1, Marsupialia. (Vol. 2, Rodentia)*. 8vo, 2 vols, 22 col.+23 plain pls, text figs, Baillière. London 1846–48. All publ.

3896. —— *Catalogue of British Coleoptera*. 8vo, 4 pts, 117 pp, London 1858[–1861].

3897. —— *Pocket catalogue of British Coleoptera*. 8vo, 42 pp, Taylor, London 1861.

3898. WATERTON, Charles *Wanderings in South America . . . with . . . instructions for the perfect preservation of birds*. 4to, vii+326 pp, 1 pl., text figs, Mawman, London 1825; 2ed. 8vo 1828; 3ed. 1836; 4ed. 1839; ed. 1852; 6ed. 1866; ed. 1906; ed. 1925; ed. 1973.

3899. —— *An ornithological letter to William Swainson*. 8vo, Wakefield 1837.

3900. —— *Essays on natural history, chiefly ornithology*. [First series]. 8vo, Longman, London 1838; 3ed. 1839; 5ed. 1844; 8ed. 1851; 1861.

3901. —— *Essays on natural history, second series*. 8vo, cxlii+204 pp, 1 pl., Longman, London 1844; 2ed. 1844; 4ed. 1851; 1862.

3902. —— *Essays on natural history, third series*. 8vo, xlvii+290 pp, 1 portr., Longman, London 1857; 2ed. 1858.

3903. —— *Essays on natural history*. 8vo, vii+631 pp, 5 pls, 1 portr., Warne, London 1871 [=1870]; ed. [ ]. A one volume work based on Nos 3900–02.

3904. —— *The barn owl, a reprint of Waterton's essay*. 8vo, [12] pp, text fig., Society for the Protection of Birds, London 1895; 1902. Publ. No. 19.

3905. WATKINS, Charles R. W. *Principles and rudiments of mineralogy, botany, zoology*. 12mo, 32 pp, Hardwicke, London 1855. Nonsense.

3906. —— *Principles and rudiments of botany: delivered according an juliam system of arrangement and juliam method of classification*. Sm. 8vo, 114 pp, 2 tab., Partridge, London 1858. Nonsense.

3907. WATKINS, Morgan George *Pictures of bird life in pen and pencil*. Imp. 4to, 144 pp, 13 pls, text figs, Cassell, London 1882.

3908. —— *Gleanings from the natural history of the ancients*. 8vo, xiii+258 pp, Elliot Stock, London 1885; ed. 1896. Antiquary's Library.

Watkins, Watkin *see* Marchant, J. R. V.

3909. WATKINS, William *A treatise on forest trees*. 8vo, 46 pp, Rivington, London 1753.

3910. WATKINS, William *Directions for collecting, rearing and preserving British and foreign butterflies and moths*. 8vo, text figs, Crystal Palace Insectarium, Croydon 1884.

3911. WATSON, Frederick *The animal world display'd*. 8vo, xvi+302 pp, 18 pls, Baldwin, London 1754; 2ed. 1755.

3912. WATSON, Hewett Cottrell *Outlines of the geographical distribution of British plants belonging to the division of the Cotyledones*. 8vo, xv+334 pp, privately printed, Edinburgh 1832; ed. London 1835 title *Remarks on the geographical* . . .; 3ed. pt. 1, 1843 title *The geographical* . . . 3ed. all publ.

3913. —— *The new botanist's guide to the localities of the rare plants of Britain*. 8vo, 2 vols, xxiv+647 pp, Longman, London 1835–37.

3914. —— *Cybele Britannica; or, British plants and their geographical relations*. 8vo, 4 vols, 1 map, text figs, Longman, London 1847–59; suppl. pt. 1, 119 pp, for the author, 1860; 2nd suppl. iv+213 pp, privately printed, Thames Ditton 1872. The second suppl. is also a suppl. to No. 3915.

3915. —— *A compendium of the Cybele Britannica*. 8vo, 3 pts, vi+651 pp, 1 map, privately printed, London 1868–70. For suppl. see No. 3914.

3916. —— *Topographical botany ... the distribution of British plants*. 8vo, 2 pts, iv+740 pp, 1 col. map, privately printed, Thames Ditton 1873–74; 2ed. London 1883; suppl. 16 pp, 1905. 100 copies.

3917. —— & DENNES, George Edgar *The London catalogue of British plants*. 8vo, 16 pp, Botanical Society, London 1844; ed. [1845]; 2ed. fol., [1845]; ed. fol., 1847; ed. fol., 1848; ed. 1848; 3ed. 1850; 4ed. 1853; 5ed. 1857; 6ed. 1867; 7ed. 1874; 1877; 1881; 8ed. [1883]; 8ed. 1886; 9ed. 1895; 10ed. 1908; 11ed. 1925. See also No. 3760.

—— *see* Lees, F. A.

3918. WATSON, John *Sylvan folk. Sketches of bird and animal life in Britain*. Sm. 8vo, vi+286 pp, Unwin, London 1889.

3919. —— *Nature and woodcraft*. 8vo, vi+302 pp, 2 pls, Smith & Innes, London 1890.

3920. WATSON, John Selby *The reasoning power of animals*. 8vo, viii+471 pp, Reeve, London 1867.

3921. WATSON, Morrison *Observations in human and comparative anatomy*. 8vo, 78 pp, 2 pls, text figs, Oliver & Boyd, Edinburgh 1874. Mostly reprinted from *J. Anat.*

3922. WATTERS, John J. *The natural history of the birds of Ireland.* Sm. 8vo, xv+299 pp, McGlashan, Dublin 1853.

3923. —— *Scientific catalogue of the birds of Ireland (indigenous and migratory)*. 8vo, Dublin 1863.

3924. WATTS, William Marshall *A school flora*. 8vo, 113 pp, Warne, London 1878; ed. 1879. Several later printings.

Waugh, Richard *see* Winch, N. J.

3925. [WEAVER, Richard] *A systematic catalogue of British birds; intended as a guide to the ornithological department of the Birmingham Museum of Natural History.* 8vo, 7pp, Watton & Ashton printed, Birmingham 1832. Ascription uncertain.

3926. WEBB, Robert Holden & COLEMAN, William Higgins *A report of the progress made in the investigation of the flora of Hertfordshire*. 8vo, 14 pp, [Hertford 1843].

3927. —— & —— *Flora Hertfordiensis: or, a catalogue of the plants found in the County of Hertford*. 8vo, xlvi+290 pp, 1 col. map, Pamplin, London 1849; suppl. 20 pp, 1851; 2nd suppl. 23 pp. [1859].

3928. WEBSTER, Angus Duncan *British orchids*. 8vo, 103 pp, 8 pls, Bangor 1886; 2ed. 1898.

3929. —— *The flora of Kent*. 8vo, 24 pp, Bromley [1893].

3930. WEDDELL, H. & SWEET, R. *British botany. No. 1.* 4to, 9 pp, 2 col. pls, London 1831. All publ.

*A week at the Lizard*. 1848. *see* Johns, C. A.

3931. WEIR, Harrison William *Every day in the country*. 12mo, 8 col. pls with letterpress, text figs, Warne, London [1883].

3932. WEISMANN, Friedrich Leopold August *Studies in the theory of descent*. 8vo, 2 vols, xxxvi+729 pp, 8 col. pls, Sampson Low, London 1882. Translation of *Studien zur Descendenz-Theorie*, Leipzig 1875–76, by Raphael Meldola.

3933. —— *Essays upon heredity and kindred biological problems*. 8vo, x+435 pp, text figs, Clarendon, Oxford 1889; 2ed. 2 vols, 1891–92. In 2ed. vol. 1 is a reprint of the first edition; vol. 2 is new matter. Translations of Foreign Biological Memoirs Series, Vol. IV. Eight essays from various sources; no equivalent volume in German, edited by Edward Bagnall Poulton *et al*.

3934. —— *The germ-plasm: a theory of heredity*. Sm. 8vo, xxii+477 pp, text figs, Scott, London 1893. The Contemporary Science Series. Translation of *Das Keimplasma*, Jena 1892, by William Newton Parker & Harriet Rönnfeldt.

3935. —— *The effect of external influence upon development*. 8vo, 69 pp, Frowde, London 1894. Romanes Lecture, Oxford.

3936. WEISS, Frederick Ernest *Outline classification of the vegetable kingdom*. 8vo, 15 pp, Cornish, Manchester 1892. Manchester Museum Handbook, No. 4.

3937. WELLS, Herbert George *Text-book of biology*. Sm. 8vo, 2 pts, 47 pls, text figs, Clive, University Correspondance College Press, London [1891, 93]; ed. [1896]; 3ed. pt. 1, [1901]. Tutorial Series.

3938. —— & DAVIES, Arthur Morley *Text-book of zoology*. Sm. 8vo, 374 pp, text figs, Clive, University Correspondance College Press, London 1898. Several later editions.

3939. WELLS, William Charles *Two essays: one upon single vision with two eyes, the other on dew . . . with some observations on the causes of the differences in colour and form between the white and negro races of men*. 8vo, lxxiv+439 pp, Constable, Edinburgh 1818. *Essay on dew* first ed. 1814; *Essay on single vision* first ed. 1792; but the *Observations* only present in this edition.

Wernerian Club *see* Plinius Secundus, Caius.

—— *see* Ray, J. *Wisdom of God*, 1844–46.

3940. WESLEY, John *A survey of the wisdom of God in the creation*. 12mo, 2 vols, Bristol 1763; London 1764; ed. 5 vols, London 1809; ed. 3 vols, London 1836; ed. 1840. Edition of 1809 has title *A compendium of natural philosophy being . . .* Vol. IV is an abridged translation of Bonnet, Charles *La contemplation de la nature*, Amsterdam 1764. Vol. V contains translations of extracts from Duttens, Louis *Récherches sur l'origine des decouvertes attribuées aux modernes*, Paris 1766.

3941. WESTCOTT, Michael *The autobiography of a cabbage white butterfly*. 8vo, Wells, 1856.

3942. —— *The autobiography of a gossamer spider*. 8vo, Groombridge, London 1857.

3943. WESTELL, William Percival *Everyone's handbook of British breeding birds*. 12mo, 188 pp, text figs, Drane, London 1897; ed. 1899.

3944. —— *A year with nature*. Imp. 8vo, xvi+276 pp, text photographs, Drane, London 1900; 1901.

3945. WESTERMAIER, Maximilien *A compendium of general botany*. 8vo, x+299 pp, text figs, Chapman & Hall, London 1896. Translation from German, by Albert Schneider.

Weston, James *pseudonym. see* Step, E.

3946. WESTWOOD, John Obadiah *History of insects*. Sm. 8vo, 2 vols, text figs, Tegg, London 1828–29.

3947. —— *Address on the recent progress and present state of entomology. Read at the anniversary meeting of the Entomological Society, January 1835*. 8vo, 23 pp, Longman for the Society, London 1835.

3948. —— *The entomologist's text book*. 8vo, x+432 pp, 5 col. pls, text figs, Orr, London 1838. Reprinted from *The British Cyclopaedia of natural history*, 1835–37.

3949. —— *An introduction to the modern classification of insects*. 8vo, 2 vols, 1 col. pl., text figs, Longman, London 1839–40. Vol. 2 contains 'Synopsis of the genera of British insects'.

3950. —— *The butterflies of Great Britain and their transformations*. 8vo, xl+140 pp, 19 col.+2 plain pls, Orr, London [1854–] 1855. Later printings to 1887. An abridged version of No. 3951.

3951. —— & HUMPHREYS, Henry Noel *British butterflies and*

*their transformations*. 4to, 138 pp, 42 col. pls, Smith, London 1841; 2ed. [1845–] 1849.

3952. —— & —— *British moths and their transformations*. 4to, 2 vols, 124 col. pls, Smith, London 1843–45; 1849; ed. 1857 [–1858].

—— *see* Bate, C. S.

—— *see* Doubleday, E.

—— *see* Rennie, J.

3953. WHALE, William *Egham wild flowers*. 8vo, 40 pp, Egham [?1875].

3954. WHARTON, Henry Thornton *A list of British birds*. Sm. 8vo, lv+20 pp, Van Voorst, London 1877.

3955. WHATLEY, Richard *On instinct, a lecture before the Dublin Natural History Society . . . 1842*. Sm. 8vo, 32 pp, McGlashan, Dublin 1847.

3956. WHEELER, James *The botanist's and gardner's new dictionary . . . to which is prefixed an introduction to the Linnaean system of botany*. 8vo, viii+xxxi+480 pp, 2 pls, Strahan, London [1763].

3957. WHELDON, James Alfred *The York catalogue of British mosses*. 8vo, 16 pp, York [1889].

3958. WHELLER, Charles Gilbert *Illustrations of natural history and mineralogy*. 4to, 5 pls, Johnston, Edinburgh 1877. With a handbook for each plate.

Whitaker, Joseph *see* Sterland, W. J.

3959. WHITE, Adam *List of specimens of British animals in the collection of the British Museum. Part IV. Crustacea*. 12mo, iv+141 pp, British Museum, London 1850.

3960. —— *A popular history of Mammalia*. Roy. 16mo, viii+346+14 pp, 16 col. pls, Reeve, London 1850. Popular Natural History Series.

3961. —— *List of the specimens of British animals in the collection of the British Museum. Part VIII. Fish*. 12mo, xxiii+164 pp, British Museum, London 1851.

3962. [——] *List of specimens of British animals in the collection of the British Museum. Part XIV. Nomenclature of Neuroptera*. 12mo, 16 pp, British Museum, London 1853.

3963. [——] *List of specimens of British animals in the collection of the British Museum. Part XV. Nomenclature of Diptera. 1*. 12mo, 42 pp, British Museum, London 1853. All publ.

3964. [——] *List of specimens of British animals in the collection of the British Museum. Part XVII. Anoplura, Euplexoptera and Orthoptera.* 12mo, 17 pp, British Museum, London 1855.

3965. —— *A popular history of birds*. Roy. 16mo, viii+347 pp, 20 col. pls, Reeve, London 1855. Popular Natural History Series.

3966. —— *A popular history of British Crustacea*. Roy. 16mo, xii−358 pp, 20 col. pls, Reeve, London 1857. Popular Natural History Series.

3967. —— *Lessons from the natural history of animals*. Fol., 54 col. pls with letterpress, Stanford, London [ ]; 8ed. [1882].

3968. —— *Tabular view of Myriapoda, Arachnida, Crustacea, Annelida and Entozoa*. 4to, 4 pls, SPCK or Stanford, London [c.1860]. Diagrams of Natural History No. VII.

3969. —— *Tabular view of the orders and leading families of insects*. 4to, 4 pls, SPCK or Stanford, London [c.1860]. Diagrams of Natural History, No. VIII.

3970. —— *Heads and tales; or, anecdotes of quadrupeds and other beasts*. Sm. 8vo, xv+326 pp, 9 pls, Nisbet, London 1870; 2ed. 1870.

3971. —— & STARK, Robert Mackenzie *The instructive picture book, or progressive lessons from the natural history of animals and plants. Vol. 1. Lessons from the animal world*. Fol., 18 pp, 30 col. pls, Edmonston & Douglas, Edinburgh 1857; 3ed. 1859; 1860; 7ed. [1866]; 1877. For Vol. 2 see Yonge, *Miss* C. M.

3972. WHITE, Charles *An account of the regular gradation in man, and in different animals and vegetables.* 4to, xii+138, cxx-xix–clvi, 139–146 pp, 4 pls, Dilly, London 1799.

3973. WHITE, Francis Buchanan White *Fauna Perthensis: or contributions towards the knowledge of the animals inhabiting Perthshire. Part 1.* 4to, Perthshire Society for Natural Science, Perth 1871. All publ.

3974. —— *The flora of Perthshire.* Sm. 8vo, lix+407 pp, 1 portr., 1 col. map, Blackwood, Edinburgh 1898.

3975. [WHITE, Gilbert] *The natural history and antiquities of Selborne, in the County of Southampton*. 2 vols, 4to, v+468 pp, 9 pls, White, London 1789. No. pp. 441–42. Preface signed. Facsimile 1973.

3976. —— *The naturalist's calendar*. 4to or 8vo, 170 pp, 1 col. pl., White, London 1795.

—— Later editions of No. 3975 are very numerous and are summarized from Martin (1934) up to 1900, with additions.

There are many more after 1900, and the work is continuously in print. They usually contain No. 3976 at the end, although this and the antiquities are sometimes omitted. Editors are placed first, where given, followed by the publisher of the first edition by each editor.

1. J. Aiken & M. Markwick, White, 8vo, 2 vols, 1802 title *The works in natural history. . . .* The first collected edition, omits antiquities.
2. J. Mitford, White Cochrane, 4to 1813; large 4to, 2 vols, 1813; 8vo, 2 vols, 1813; 8vo, 2 vols, 1822; 8vo, 2 vols 1825.
3. *Sir* William Jardine, Constable, 12mo, Edinburgh 1829; 1830; 1832; 1833; 1836; 1853; 1879; 1880; 1882; 1887; 1890; 1891; 1891; 1894; 1894; 1895. Constable's Miscellany, Vol. XLV.
4. J. Rennie, Arch, 8vo, 1833.
5. *Lady* Dover, Hailes, 12mo, 1833; Tegg, London 1840; SPCK, 1842 and later issues. Arranged for young persons, i.e. bowdlerized.
6. T. Brown, Chambers, Edinburgh, 16mo, 1833; 1834; 1834; 1835; 1835; 1836; 1840; 1843; 1856; 1857; 1875; 1880; 1890. British Library. Vol. 1; The Golden Library.
7. E. Blyth, Orr & Smith, 12mo, 1836; 1850; 1858.
8. E. T. Bennett, Arch, Chiswick Press, 8vo, [1837].
9. L. Jenyns, Van Voorst, 8vo, 1843.
10. W. Jesse, Bohn, 8vo, 1849; 1851; 1854; 1861; 1867; 1870; 1872; 1881. Based on *Sir* William Jardine's edition. The Illustrated Library.
11. J. G. Wood, Routledge, 8vo, 1854; [1854]; 1857; 1859; 1860; 1864; 1869; 1872; 1879; 1893.
12. No editor, Bell & Daldy, 16mo, 1860; 1862. Pocket Volume Series.
13. J. E. Harting, Bickers, 8vo, 1875; 1876; 1880; 1884; 1887; 1888; 1889; 1891. Based on E. T. Bennett's edition.
14. F. Buckland, Macmillan, 4to or 8vo, 1875; 1876; 1879; 1880; 1883; 1887; 1891; 1894; 1897.
15. T. Bell, Van Voorst, Roy. or demy 8vo, 1877. Natural History of the British Islands Series. Roy. copies have plates on India paper.
16. G. C. Davies, Warne, 8vo, 1879; 1890. Chandos Library.
17. H. R. Haweis, Routledge, 16mo, 1885; 1886. Selections only. World Library.
18. R. Jefferies, Scott, 12mo, 1887; 1896. Camelot Classics.

19. H. Morley, Cassell, 18mo, 1887; 1892; 1897.
20. J. Burroughs, Macmillan, Sm. 8vo, 2 vols, 1895. English issues of USA edition.
21. G. Allen, John Lane, Roy. 8vo, 1900.
22. A. W. Pollard, Macmillan, Roy. 8vo, 1900. Library of English Classics.
23. R. B. Sharpe, Freemantle, 4to or 8vo, 1900.

3977. WHITE, John *An essay on the indigenous grasses of Ireland.* 8vo, xxix+156 pp, 2 col. pls, Graisberry & Campbell printed, Dublin 1808.

3978. WHITE, Robert *Nottinghamshire, Worksop, the Dukeries and Sherwood Forest.* Sm. 8vo, vii+338 pp, 1 map, text figs, Worksop 1875.

3979. [WHITE, T. Charters] *My microscope and some objects from my cabinet. A simple introduction to the study of "the infinitely little".* 8vo, 78 pp, 1 pl., Roper & Drowley, London 1887; 2ed. 1888. By a Quekett Club Man.

3980. WHITE, William Farren *Ants and their ways.* 8vo, xvi+279 pp, 1 pl., text figs, RTS, London [1883]; 2ed. 1895.

Whitehead, J. *see* Ashton-under-Lyne Linnaean Botanical Society.

3981. WHITING, James Edward *Where wild birds sing, being the notes of a naturalist.* Sm. 8vo, viii+158 pp, Mayle, London 1898.

3982. WHITLEY, H. *Natural history of British tits.* Sm. 8vo, 22 pp, 7 col. pls, London 1846. ?All publ, of a projected *Natural history of British birds.*

3983. WHITLOCK, F. B. *The birds of Derbyshire.* 8vo, vi+235 pp, 6 pls, 1 map, Bemrose, London 1893.

3984. —— *The migration of birds, a consideration of Herr Gätke's views.* 8vo, vi+140 pp, Porter, London 1897.

3985. WHITTOCK, N. *The art of drawing and colouring from nature, birds, beasts, fishes, and insects.* Obl. 4to, 96 pp, 24 col.+24 plain pls, Hinton, London 1829; 1830.

3986. *Who were the first miners?* 12mo, 72 pp, 1 col. pl., text figs, Nelson, London 1875.

3987. WIDDOWES, I. *Natural philosophy: or, a description of the world, namely of angels, of man, of the heavens, of the ayre, of the earth, and of the creatures in the whole world.* 4to, 64 pp, Bellamie, London 1621; 2ed. 1631.

3988. WIEDERSHEIM, Robert Ernst Eduard *Elements of the comparative anatomy of vertebrates.* 8vo, xxvi+345 pp, text figs some col., Macmillan, London 1886; 2ed. 1897; 3ed.

1907. Translation of *Grundriss der vergleichenden Anatomie der Wirbelthiere*, Jena 1884, by William Newton Parker.

3989. *Wild animals.* 16mo, 2 ser., pls, SPCK, London [1862]; ed. 1862.

3990. *Wild animals in freedom and captivity.* 8vo, xvi+440 pp, pls, Ward & Lock, London [1872].

3991. *Wild flowers.* 8vo, Chambers, London 1863. Library for Young People, Vol. 7.

3992. *Wild flowers.* 4to, 192 col. pls with letterpress, SPCK, London 1867.

3993. *Wild flowers and their teachings.* 16mo, 37 specimens, Binns & Goodwin, Bath, 1845; 2ed. 1848.

*Wild flowers of Dover* [?1845]. *see* Paley, F. A.

3994. *Wild flowers of the year.* Sm. 8vo, vi+192 pp, RTS, London [?1846]. Monthly Volume No. 3.

3995. *Wild flowers, showing where to gather them.* Sm. 8vo, iv+140 pp, text figs, Ward & Lock, London 1861; ed. [1880]. Indispensible Handy Books.

*The wild-fowl and sea-fowl of Great Britain.* 1895. *see* Jordan, D.

*The wild garland.* 1837. *see* Waring, *Miss* S.

3996. WILDMAN, Thomas *Treatise on the management of bees . . . to which are added the natural history of wasps and hornets.* 4to, xix+169 pp, 3 pls, Cadell for the author, London 1768; 2ed. 8vo 1770; 3ed. 1778.

3997. WILKES, Benjamin *Twelve new designs of English butterflies.* 4to, dedication pl.+12 col. pls, London 1741–42. Alternative title is *Bowles' New Collection of English moths and butterflies in twelve prints.* No printed text.

3998. [——] *The English moths and butterflies.* 4to, xxvi+63 pp, 120 col. pls, for the author, London 1747–60; ed. 1773 title *One hundred and twenty copper plates of English moths and butterflies*; 3ed. 1824.

3999. WILKINS, John *An essay towards a real character and a philosophical language: (An alphabetical dictionary).* Fol., 2 pts, 454+157 pp, 5 pls, 2 tabs, Gellibrand, London 1668. L.P. Contains Ray, J. 'Table of plants'.

4000. WILKINSON, Caroline Catherine *Lady. Weeds and wild flowers.* Sm. 8vo, x+421 pp, 12 col. pls, text figs, Van Voorst, London 1858.

4001. WILKINSON, Samuel James *The British Tortrices.* 8vo, vii+328 pp, 4 pls, Van Voorst, London 1859.

Wilks, J. *see* Sibly, E.

Wilks, Theodore C. *see* Woodward, B. B.

4002. WILLCOX, S. J. B. *Egg collector's dictionary of local names of British birds.* 16mo, 58 pp, Portsmouth 1894.

4003. WILLDENOW, Carl Ludwig *Principles of botany and of vegetable physiology.* 8vo, iv+508 pp, 1 col.+9 plain pls, Blackwood, Edinburgh 1805; ed. 1811. Translation of *Grundriss der Kräuterkunde*, Berlin 1792.

4004. WILLEMENT, E. E. *The conversations of little Herbert and his mother on zoology; or, the class Mammalia.* 8vo, v+90 pp, text figs, London 1848.

4005. WILLIAMS, Benjamin Samuel *Select ferns and lycopods: British and exotic.* Sm. 8vo, viii+343 pp, 15 pls, 48 full page text figs, for the author, London 1868; 2ed. 1873.

4006. WILLIAMS, Charles *of St John's Wood. Art in nature and science anticipated.* 12mo, xii+324 pp, 1 pl., Westley & Davis, London 1832.

4007. —— *The Surrey Zoological Gardens.* 32mo, 191 pp, text figs, Tilt, London [c.1835]. Hand-books for Children.

4008. —— *The Zoological Gardens, Regent's Park.* 32mo, vi+191 pp, text figs, Tilt, London [c.1835]. Hand-books for Children.

4009. [——] *Curiosities of animal life, with recent discoveries of the microscope.* Sm. 8vo, viii+184 pp, text figs, RTS, London [1848]; 1859.

4010. —— *Silver-shell, or the adventures of an oyster.* 8vo, vii+184 pp, 1 pl., text figs, Judd & Glass, London 1856; 2ed. 1857.

4011. —— *Pickings on the sea-shore; or, cliffs, sands, plants and animals . . . and the vivarium at home.* Sm. 8vo, ii+180 pp, text figs, Judd & Glass, London 1857.

4012. WILLIAMS, Charles *of Salisbury. The vegetable world.* 12mo, 288 pp, 1 pl., Ward, London 1833; 2ed. 1833.

4013. WILLIAMS, Frederick Newton *Provisional list of the orders and families of British flowering plants.* 8vo, 8 pp, [Brentford 1895]; 2ed. [1898].

4014. [WILLIAMS, J.] *The botanist's vade mecum.* 8vo, 56 pp, London 1855; 2ed. 1856. Reprinted from Balfour, J. H. *A manual of botany.*

4015. WILLIAMS, John *Faunula Grustensis: being an outline of the natural contents of the Parish of Llanwrst.* 12mo, 148 pp, Llanwrst 1830.

4016. WILLIAMS, Joseph William *The shell-collector's handbook for the field.* 8vo, xii+148 pp, text figs, Roper, London 1888.

4017. —— *Land and fresh-water shells.* Sm. 8vo, text figs, Sonnenschein, London 1888; 3ed. 1901. Young Collector Series.

4018. WILLIAMS, Robert *The history and antiquities of the town of Aberconwy . . . with notices of the natural history.* 8vo, 200 pp, Gee for the author, Denbigh 1835.

4019. WILLIAMSON, John *The British angler . . . an exact description of the several kinds of fish that are found in . . . Great Britain.* 12mo, viii+318 pp, 4 pls, Hodges, London 1740.

4020. WILLIS, John Christopher *A manual and dictionary of the flowering plants and ferns.* 8vo, 2 vols, University Press, Cambridge 1897; 2ed. 1904; 3ed. 1908; 1914; 4ed. 1919; 5ed. 1925; 6ed. 1931; 7ed. 1966.

4021. WILLIS, Thomas *De anima brutorum . . . exercitationes duae.* 4to, 400 pp, 8 pls, Sheldonian, Oxford 1672; ed. 8vo, London 1672; English ed. in *The remaining medical works*, Fol., London 1681–84.

4022. WILLS, George S. V. *Dictionary of botanical terms.* 12mo, 24 pp, Simpkin, London 1880.

4023. WILLSHIRE, William Hughes *The principles of botany.* 12mo, xii+232 pp, text figs, Highley, London 1840.

4024. WILLUGHBY, Francis *F. Willughbeii . . . Ornithologiae libri tres.* Fol., 307 pp, 77 pls, 2 tabs, Martyn, London 1676; issue 1676; English ed. 1678 title *The ornithology of Francis Willughby.* L.P. Facsimile of English ed. 1972.

4025. —— *F. Willughbeii . . . de historia piscium libri quatuor.* Fol., 343+30 pp, 187 pls, Sheldonian, Oxford 1686. Title page to plates dated 1685.

—— *see* Ray, J.

Willughby Society *see*

Leach, W. E., 1816.

Tunstall, M. 1771.

4026. WILMORE, J. H. *Lecture on birds of the neighbourhood of Queenwood (Hampshire), their nests and eggs.* 12mo, London 1874.

4027. WILSON, Andrew *Elements of zoology.* Sm. 8vo, xx+634 pp, text figs, Longman, London 1873.

4028. —— *The student's guide to zoology. A manual of the principles of zoological science.* Sm. 8vo, text figs, Churchill, London 1874.

4029. —— *Zoology: a description of types of animal structure.* 8vo, Chambers, London 1874. Elementary Science Manuals.

4030. —— *On the study of zoology and its place in medical and scientific education.* 8vo, Edinburgh 1875.

4031. —— *Sketches of animal life and habits.* Sm. 8vo, 208 pp, text figs, Chambers, London 1877; ed. [ ].

4032. —— *Leisure time studies chiefly biological.* Sm. 8vo, xv+381 pp, 1 pl., text figs, Chatto & Windus, London 1879; 3ed. 1884.

4033. —— *Illustrations of zoology and comparative anatomy.* 4to, 4 pls, Johnston, Edinburgh 1879. Pls numbered XXII–XXV. A handbook accompanies each plate. Also available on rollers.

4034. —— *Introduction to the study of flowers.* 8vo, 63 pp, Chambers, London 1880; 1891; 1893.

4035. —— *Chapters on evolution.* 8vo, text figs, Chatto & Windus, London 1882; 2ed. 1883; 3ed. 1886.

4036. —— *Leaves from a naturalist's note-book.* Sm. 8vo, 255 pp, Chatto & Windus, London 1882. Mayfair Library.

4037. —— *Wild animals and birds; their haunts and habits.* 4to, Cassell, London [ ]; ed. 1882.

4038. —— *Studies in life and sense.* Sm. 8vo, 354 pp, text figs, Chatto & Windus, London 1887.

4039. —— *Glimpses of nature.* 8vo, text figs, Chatto & Windus, London 1891.

—— *see* Allen, C. G. B.

4040. WILSON, James *Illustrations of zoology. Vol. 1.* Fol., ii+4+[138] pp, 36 col. pls, Blackwood, Edinburgh [1828–] 1831. All publ.

4041. —— *A treatise on insects, general and systematic, being the article Entomology.* 4to, 327 pp, 19 pls, Black, Edinburgh 1835. Reprinted from *The Encyclopaedia Britannica*, 7ed.

4042. —— *An introduction to the natural history of fishes; being the article Ichthylology.* 4to, pls, Black, Edinburgh 1837. Reprinted from *The Encyclopaedia Britannica*, 7ed.

4043. —— *The natural history of quadrupeds and whales; being the article Mammalia.* 4to, 17 pls, Black, Edinburgh 1837; 1838. Reprinted from *The Encyclopaedia Britannica*, 7ed.

4044. —— *An introduction to the natural history of birds; being the article Ornithology.* 4to, pls, Black, Edinburgh 1839. Reprinted from *The Encyclopaedia Britannica*, 7ed.

4045. —— *A voyage round the coasts of Scotland and the Isles.* 8vo, 2 vols, 3 pls, 1 map, Black, Edinburgh 1842.

4046. —— & DUNCAN, James *Entomologia Edinensis; or, a*

*description and history of the insects found in the neighbourhood of Edinburgh . . . Coleoptera.* 8vo, vi+351 pp, 2 pls, text figs, Blackwood, Edinburgh 1834. All publ.

4047. WILSON, John *A synopsis of British plants in Mr. Ray's method . . . with a botanical dictionary.* 8vo, 17+272 pp, 2 pls, text figs, Gooding printed, Newcastle upon Tyne 1744.

4048. WILSON, Owen S. *The larvae of British Lepidoptera and their food plants.* Roy. 8vo, xxxix+367 pp, 40 col. pls, Reeve, London [1872–] 1880.

Wilson, T. *pseudonym. see* Clark, Samuel

Wilson, William *Bryologia Britannica.* 3ed. 1855 *see* Hooker, *Sir* William & Taylor, T. *Muscologia Britannica.* 1818.

4049. WINCH, Nathaniel John *An essay on the geographical distribution of plants through the Counties of Northumberland, Cumberland and Durham.* 8vo, 52 pp, Newcastle upon Tyne 1819; 2ed. 1825.

4050. —— *Flora of Northumberland and Durham, addenda.* 4to, 9 pp, [Newcastle upon Tyne 1837]. The work itself is in *Trans. nat. Hist. Soc. Northumberland*, Vol. 2, 1831.

4051. —— *Contribution to the flora of Cumberland.* 4to, 17 pp, Newcastle upon Tyne 1833.

4052. ——, THORNHILL, John & WAUGH, Richard *The botanist's guide through the Counties of Northumberland and Durham.* 8vo, 2 vols, Hodgson printed, Vol. 1 Newcastle upon Tyne, Vol. 2 Gateshead upon Tyne 1805–07.

4053. WINCHESTER COLLEGE NATURAL HISTORY SOCIETY *A list of the macro-Lepidoptera and birds of Winchester and the vicinity.* 8vo, 32 pp, for the Society, Lincoln 1891.

4054. WINDSOR, John *Flora Cravoniensis, or a flora of the vicinity of Settle in Craven, Yorkshire.* 8vo, x+177 pp, for private distribution, Manchester 1873. Reprint of papers in *The Phytologist*, 1855–58.

4055. WINTRINGHAM, William Henry *British birds. Key to the present classification.* 8vo, 30 pp, 2 tabs, Grimsby News Co., Grimsby 1890.

4056. WISE, John Richard de Capel *The New Forest: its history and scenery.* 4to, viii+336 pp, 1 pl., 1 map, text figs, Smith Elder, London 1863 [=1862]; 2ed. 1863; 1867; 3ed. 18vo, 1880; 4ed. 1880; ed. 1883; 5ed. 1895.

4057. WISHART, R. S. *Self-educator in botany.* Sm. 8vo, 226 pp, Hodder, London 1900.

4058. WITCHELL, Charles A. *The evolution of bird song, with observations on the influence of heredity and imitation.* Sm. 8vo, x+253 pp, Black, London 1895.

4059. —— *Cries and call notes of wild birds.* Sm. 8vo, xi+84 pp, text music, Upcott Gill, London 1899.

4060. —— & STRUGNELL, W. Bishop *The fauna and flora of Gloucestershire.* Imp. 8vo, xxiv+301 pp, 10 pls, text figs, James, Stroud 1892.

*With the woodlanders and by the tide.* 2ed. 1893. *see* Jordan, D.

4061. WITHERBY, Harry Forbes *Forest birds: their haunts and habits.* 12mo, 98 pp, text figs, Kegan Paul, London 1894.

4062. WITHERING, William *A botanical arrangement of all the vegetables naturally growing in Great Britain.* 8vo, 2 vols, xcvi+838 pp, 14 pls, Swinney printed, Birmingham 1776; 2ed. 4 vols in 3 1787–93 title *A botanical arrangement of British plants*; 3ed. 1796 title *An arrangement of British plants*; 4ed. 1801 title *A systematic arrangement of British plants*; 5ed. 1812; 6ed. 1818; 7ed. 1830. For condensed editions see No. 4063.

4063. —— *A systematic arrangement of British plants . . . corrected and condensed . . . by W. Macgillivray.* 8vo, 391 pp, 10 pls, Dove, London 1830; 2ed. 1833; 3ed. 1835; 4ed. 1837; 5ed. 1841; 6ed. 1845; 7ed. 1848; 8ed. 1852; 9ed. 1855; 10ed. 1858; 11ed. 1863; 14ed. 1877. Title varies.

4064. WODARCH, Charles *An introduction to the study of conchology.* 8vo, xxiv+120 pp, 4 col. pls, Longman, London 1820; 2ed. 1822 title *Wodarch's Introduction . . .*; 3ed. 1825; 1827; 4ed. 1832.

4065. WOLF, Joseph *Feathered favourites . . . pictures of British birds from drawings by Joseph Wolf.* Roy. 8vo, 12 col. pls, Bosworth, London 1854.

—— *see* Elliot, D. G.

—— *see* Sclater, P. L.

4066. WOLLASTON, Thomas Vernon *On the varieties of species, with special reference to the Insecta.* Sm. 8vo, vi+206 pp, Van Voorst, London 1856.

Wolley, John *see* Newton, A.

4067. *Wonderful insects.* 4to, 6 col. pls, Groombridge, London [c.1850].

*The wonderful shape* [S.T.C. entry]. [?1521]. *see The noble life & nature.*

4068. *The wonders and beauties of creation, portrayed by Buffon.* Sm. 8vo, text figs, Gall, London 1875.

4069. *The wonders of animated nature.* 12mo, pls, Phillips, London 1810.

4070. *Wonders of nature.* 12mo, 12 col. pls, Nelson, London 1856.

4071. *The wonders of nature and art; being an account of whatever is most curious throughout the world.* 12mo, 4 vols, pls, Corbett, London 1750; ed. 3 vols, Newbery, London 1767; 2ed. 6 vols, 1768.

4072. *The wonders of nature and art. Comprising nearly three hundred of the most remarkable curiosities and phenomena in the known world.* 32mo, text figs, Milner, Halifax 1839.

4073. *Wonders of the animal world, birds. (Mammalia).* 16mo, 2 vols, text figs, SPCK, London 1847; 1848.

4074. *Wonders of the microscope; or, an explanation of the wisdom of the creator in objects comparatively minute.* 12mo, xii+118 pp, 5 pls, Tabart, London [1806]; 1811; [c. 1819]; 1823; ed. 16mo [c. 1839]. Plates full size after R. Hooke, *Micrographia.*

4075. *Wonders of nature.* 12mo, 24 11, 13 col. pls, Nelson, London 1856.

4076. *Wonders of organic life.* 12mo, RTS, London [1852]. Monthly Volume No. 79.

4077. *The wonders of the microscope photographically revealed . . . Olley's patent.* 8vo, 8 pp, 3 pls, London [1857–] 1861.

4078. *Wonders of the plant world.* Sm. 8vo, Nelson, London 1870.

4079. *Wonders of the waters.* [Containing the animalcule, the coral-maker, the sea-star, the lobster, the fish.] 16mo, 32+32+32+32+32 pp, text figs, RTS, London [1842]; [1853].

*The wonders of the vegetable kingdom displayed*, 1822. *see* Roberts, *Miss* M.

4080. WOOD, Albert *A list of land and freshwater shells found at Sutton Coldfield.* 8vo, 12 pp, Leeds 1897.

4081. WOOD, Charles Thorold *The ornithological guide.* 8vo, xviii+236 pp, Whitaker, London [1835].

4082. WOOD, Henry *A season among the wild flowers.* 8vo, 250 pp, text figs, Sonnenschein, London 1883.

4083. —— *Among the wild flowers. Vol. 1 Spring. (Vol. 2 Summer).* Sm. 8vo, 2 vols, text figs, Sonnenschein, London 1897. Young Collector Series.

4084. WOOD, John Bland *Flora Mancuniensis; or a catalogue of the flowering plants, the ferns and their allies of Manchester.* 8vo, ix+81 pp, Halifax 1840.

4085. WOOD, John George *The illustrated natural history.* 8vo, xxiv+524 pp, text figs, Routledge, London [1851–] 1853. Many later editions in 3 vols, to 20ed. 1894. Ed. 1 vol. [1923].

4086. —— *Sketches and anecdotes of animal life.* [First series]. Sm. 8vo, 427 pp, 8 pls, Routledge, London 1854; 2ed. 1855; 4ed. 1857. Later undated editions.

4087. —— *Sketches and anecdotes of animal life. Second series.* Sm. 8vo, 8 pls, Routledge, London 1855; 2ed. 1855; 4ed. 1857. Later undated editions.

4088. —— *My feathered friends. Anecdotes of bird life.* 8vo, xii+396 pp, 8 pls, text figs, Routledge, London 1856. Later editions.

4089. —— *Common objects of the country.* Sm. 8vo, iv+132 pp, 12 col. pls, Routledge, London 1857. Very many later editions up to 18ed. [c.1904].

4090. —— *The common objects of the sea shore; including hints for an aquarium.* Sm. 8vo, iv+132 pp, 12 col. pls, Routledge, London 1857. Very many later editions up to 17ed. 1912. Plates A–F, H–N.

4091. —— *Zoology. Mammalia.* 12mo, Routledge, London 1858. Educational Manuals.

4092. —— *Animal traits and characteristics.* Sm. 8vo, 2 ser., Routledge, London 1860; ed. 1867. ?Later editions of *Sketches and anecdotes*. No. 4087.

4093. —— *The boy's own book of natural history.* Sm. 8vo, v+378 pp, text figs, Routledge, London 1861 [=1860]. Later editions to 1901.

4094. —— *Common objects of the microscope.* Sm. 8vo, iv+132 pp, 12 col. pls, text figs, Routledge, London 1861. Many later editions up to [1938].

4095. —— *Natural history picture book for children. Mammalia. (Reptiles, Birds).* 8vo, 3 pts, text figs, Routledge, London 1861–63.

4096. —— *Routledge's Illustrated natural history.* Roy. 8vo, 3 vols, text figs, Routledge, London [1859–] 1861–63; ed. 1870. Two volumes, by J.G.W., *The natural history of man* are uniform.

4097. —— *The common shells of the sea shore.* Sm. 8vo, iv+132 pp, 12 col. pls, Warne, London 1865; 2ed. 1866; 3ed. 1869.

4098. —— *Homes without hands.* 8vo, xvi+iii+632 pp, 21 pls, text figs, Longman, London [1864–] 1865. Many later editions to

1892. See also abridgement No. 4103, and extracts Nos 4121, 4122, 4124 & 4125.

4099. —— *Routledge's Popular natural history.* Roy. 8vo, 1 pl., text figs, Routledge, London 1867. Later editions up to 1905, some with title *Natural History.*

4100. —— *The fresh and salt-water aquarium.* Sm. 8vo, vii+150 pp, 8 col.+3 plain pls, text figs, Routledge, London 1868. Later undated editions. Plates I, III–XII.

4101. —— *The common moths of England.* Sm. 8vo, 186 pp, 12 col. pls, text figs, Routledge, London [1870]. Later editions, some with title *Common British moths.*

4102. —— *Common British beetles.* Sm. 8vo, 175 pp, 12 col. pls, text figs, Routledge, London 1871. Later editions.

4103. —— *Strange dwellings . . . abridged from Homes without hands.* 8vo, xii+411 pp, 1 pl., text figs, Longman, London 1871. Later editions to 1903.

4104. —— *Insects at home.* Roy. 8vo, xx+670 pp, 1 col.+20 plain pls, text figs, Longman, London [1871–] 1872. Later editions to 1892.

4105. [——] *The bird picture-book.* 8vo, 64 pp, 10 col. pls, Routledge, London [1873]. Abridged from *The illustrated natural history.*

4106. —— *The calendar of the months; or the young naturalist abroad.* 16mo, text figs, Routledge, London 1873; 1878.

4107. —— *Insects abroad.* Roy. 8vo, xii+780 pp, 20 pls, text figs, Longman, London 1874. Later editions to 1892.

4108. —— *Man and beast here and hereafter.* Sm. 8vo, 2 vols, Daldy & Isbister, London 1874; 2ed. 1 vol, 1875.

4109. —— *Out of doors: a selection of original articles on practical natural history.* Sm. 8vo, vi+242 pp, 6 pls, Longman, London 1874. Later editions.

4110. —— *Trespassers, showing how the inhabitants of the earth, air and water are enabled to trespass on domains not their own.* 8vo, vi+351 pp, 6 pls, text figs, Seeley, London 1875.

4111. [——] *The animal picture-book.* 8vo, 64 pp, col. pls, Routledge, London [1876]. Abridged from *The illustrated natural history.*

4112. —— *Nature's teachings: human invention anticipated by nature.* 8vo, 1 pl., text figs, Daldy, London 1877 [=1876]. Later editions to 1907.

4113. —— *Lane and field.* Sm. 8vo, x+246 pp, text figs, SPCK, London 1879. Natural History Rambles Series.

4114. —— *Picture book of animals, fishes. (Insects).* 16mo, 2 vols, Routledge, London 1881.

4115. —— *Common British insects selected from the typical beetles, moths and butterflies.* 8vo, viii+284 pp, 1 pl., text figs, Longman, London 1882; 1883. Selected from *Insects at home.*

4116. —— *First (Fourth) natural history reader.* 8vo, 4 vols, 4 pls, text figs, Isbister, London 1882–84.

4117. —— *Half hours in field and forest.* Sm. 8vo, xi+371 pp, text figs, Isbister, London 1885.

4118. —— *Half hours with a naturalist. Rambles near the shore.* Sm. 8vo, xii+340 pp, 1 pl., text figs, Isbister, London 1885. Later editions to 1897, the latter unsigned.

4119. —— *The handy natural history.* 8vo, xvi+367 pp, pls, text figs, RTS, London 1886.

4120. —— *My back yard zoo.* Sm. 8vo, text figs, Isbister, London 1887; 1893.

4121. —— *Homes underground.* Sm. 8vo, 1 pl., text figs, Longman, London 1887. Extracted from *Homes without hands.*

4122. —— *Branch builders.* Sm. 8vo, x+159 pp, 1 pl., text figs, Longman, London 1887. Extracted from *Homes without hands.*

4123. —— *Romance of animal life. Chapters on natural history.* Sm. 8vo, Isbister, London [1887].

4124. —— *Wonderful nests.* Sm. 8vo, 1 pl., text figs, Longman, London 1887; 1892. Extracted from *Homes without hands.*

4125. —— *Social habitations and parasitic nests.* Sm. 8vo, 1 pl., text figs, Longman, London 1888. Extracted from *Homes without hands.*

4126. —— *Zoo. Series 1. (-Series 5).* 4to, 5 ser., col. pls, SPCK, London 1886–96. Series 3–5 by J.G.W. & Wood, Theodore.

4127. —— *The brook and its banks.* 4to, RTS, London 1889; [ ].

4128. —— *The dominion of man.* Roy. 8vo, xx+440 pp, 1 pl., text figs, Bentley, London 1889.

4129. —— & WOOD, Theodore *The field naturalist's handbook.* 8vo, vi+167 pp, Cassell, London [1880]; 3ed. 1883; 4ed. 1886; 5ed. 1893.

4130. —— & —— *Hughe's Illustrated anecdotal natural history.* Sm. 8vo, 194 pp, text figs, Hughes, London 1882.

—— & —— *see* No. 4126.

4131. WOOD,Mary Bristow *The entomological researcher.* 12mo, 190 pp, 2 pls, Painter, London 1849.

4132. —— *The botanical researcher.* 12mo, 190 pp, 2 pls, Painter, London 1849.

4133. WOOD, Neville *British song-birds.* 8vo, xii+408 pp, Parker, London 1836.

4134. —— *The ornithologist's text-book.* 12mo, 3+232 pp, Parker, London 1836.

4135. WOOD, Sara *The dwellers in our gardens, their lives and works.* Sm. 8vo, 184 pp, 6 col. pls, text figs, Groombridge, London 1875; 2ed. 1877; [ ].

4136. WOOD, Theodore *Practical lessons on insect life.* 12mo, 171 pp, text figs, Hughes, London 1882.

4137. —— *Legs and wings.* Sm. 8vo, Fletcher, London 1884. Naturalist's Library.

4138. —— *Our insect allies.* Sm. 8vo, 238 pp, text figs, SPCK. London 1884.

4139. —— *Our insect enemies.* SM. 8vo, text figs, SPCK, London 1885.

4140. —— *Our bird allies.* Sm. 8vo, text figs, SPCK, London 1886.

4141. —— *Nature and her servants; or sketches of the animal kingdom.* Sm. 8vo, text figs, SPCK, London 1896.

—— *see* Wood, J. G.

4142. [WOOD, Thomas William] *Curiosities of entomology.* 8vo, 64 pp, 10 col. pls, text figs, Groombridge, London [1871]. Reprinted from *The intellectual Observer*, 1861–. See also No. 4143.

4143. [——] *Curiosities of ornithology.* 8vo, 64 pp, 10 col. pls, Groombridge, London [1871]. By T.W.W. Reprinted from *The intellectual Observer*, 1861–. A combination volume with No. 4142 is also found.

4144. WOOD, William *Zoography; or the beauties of nature displayed.* Imp. or Roy. 8vo, 3 vols, 60 pls, text figs, Cadell & Davies, London 1807.

4145. —— *General conchology. Vol. 1.* 8vo, lxi+7+iv+246 pp, 60 col. pls, Booth, London 1815; issue 1835. All publ. Pls 1–59, 4*. 1835 issue omits 'Vol. 1.'

4146. —— *Index testaceologicus; or a catalogue of shells, British and foreign.* 8vo, viii+188 pp, 8 col. pls, for W. Wood, London 1818; ed. [1824–] 1825, 38 col. pls; suppl. iv+59 pp, 8 col. pls, 1828; ed. [1855–] 1856 title *An illustrated catalogue of British and foreign shells.* See also No. 4148.

4147. —— *Illustrations of the Linnaean genera of insects.* 12mo, 2 vols, 86 col. pls, by the author, London 1821.

4148. —— *A list of the plates of the Index testaceologicus with the Lamarckian names adapted to the figures in each plate.* 8vo, 34 pp, London [1829]. See No. 4146 for the *Index.*

4149. —— *Index entomologicus; or, a complete illustrated catalogue . . . of the lepidopterous insects of Great Britain.* Roy. or demy 8vo, xii+266 pp, 54 col. pls, Ward, London [1833–] 1839; suppl. 5 col. pls, 1852; ed. 1852; ed. 1854.

4150. —— *A complete illustration of the British fresh-water fishes.* 8vo, 3 pts, 12 col. pls with letterpress, London [?1842–43]. All publ.

4151. WOODFORDE, James *A catalogue of the indigenous phenogamic plants growing in the neighbourhood of Edinburgh.* 8vo, xi+86 pp, Edinburgh 1824.

*The woodland companions.* 1802. *see* Aikin, John

*Woodland gleanings; being an account of British forest trees.* 1837. *see* Tyas, R.

*Woodland, moor and stream, being notes of a naturalist.* 1889. *see* Jordan, D.

4152. WOODS, Henry *An introductory lecture on the study of zoology.* 8vo, vi+92 pp, Hunt for the Bath Literary and Scientific Institution, Bath 1825.

4153. WOODS, Joseph *The tourists' flora . . . of the British Islands, France, Germany, Switzerland, Italy, and the Italian Islands.* 8vo, lxxxii+610 pp, 1 pl., Reeve, London 1850.

4154. WOODWARD, Bernard Barham *The young collector's handbook of shells.* Sm. 8vo, 32 pp, text figs, Sonnenschein, London [1883]. Young Collector's Handbook Series.

—— *see* Gray, P.

4155. [WOODWARD, Bernard Bolingbroke] *The natural history of the year for children.* 8vo, Ward, London 1852. Many later editions.

4156. ——, WILKS, Theodore C. & LOCKHART, Charles *A general history of Hampshire.* Fol., 3 vols, 73 pls, 1 portr., Maps, Virtue, London [1861–69].

4157. [WOODWARD, John] *Brief instructions for making observations in all parts of the world: as also for collecting, preserving, and sending over natural things.* 4to, 20 pp, Wilkin, London 1696; ed. in *Fossils of all kinds*, 1728.

4158. WOODWARD, Samuel Peckworth *A manual of the Mollusca.* 12mo, 3 pts, xvi+485 pp, 25 pls, 1 map, text figs, Weale, London 1851–56; 2ed. 1866; appendix 1868; 3ed. 1875; 4ed. 1880; 1890. Rudimentary Series Nos 72–75.

4159. —— *Generic catalogue of recent and fossil shells.* 4to, 23 pp, Weymouth 1859.

4160. —— *Tabular view of the orders and families of the Mollusca.* 4to, 4 pls, SPCK, or Stanford, London [c.1860]. Diagrams of Natural History No. V.

Woolhope Naturalists' Field Club *see*
Bull, H. G.
Purchas, W. H. &.Ley, A.

4161. WOTTON, Edward *Edoardi Wottoni Oxoniensis De differentiis animalium libri decem.* Fol., 220 numbered leaves, Vascasanus, Paris 1552. Colophon dated 1551.

4162. WRAXALL, *Sir* Frederick Charles Lascelles *Bart. Life in the sea, or the nature and habits of marine animals.* 8vo, vii+312 pp, 1 pl., text figs, Houlston & Wright, London 1860.

4163. [WRIGHT, *Mrs* Anne] *The observing eye; or letters to children on the three lowest divisions of animal life. The radiated, articulated and molluscous.* 12mo, 3 pts, 264+vi+68 pp, text figs, Jarrold, London [1851–52]. Later editions. By the author of *Passover feasts and scripture sacrifices.*

4164. WRIGHT, Chauncey *Darwinism: being an examination of Mr. St George Mivart's Genesis of species.* 8vo, 46 pp, Murray, London 1871. Reprinted with additions from *The North American review*, July 1871.

4165. WRIGHT, Edward Perceval *Spicilegia biologica: or, papers on zoological and botanical subjects. Part 1.* 8vo, 84 pp, 5 pls, for the author, [?Dublin] 1870. All publ. 75 copies.

4166. —— *Animal life: being a series of descriptions of the various sub-kingdoms of the animal kingdom.* 8vo, text figs, Cassell, London [1879].

4167. WRIGHT, J. Hall *Ocean-work, or evenings on sea and land.* 12mo, xii+168 pp, 1 pl., Tegg, London 1845.

4168. WRIGHT, *Mrs* Julia McNair *Sea-side and way-side, No. 1.* 8vo, iv+87 pp, 1 pl., text figs, Educational Supply Association, London [1888]. U.S.A. text.

4169. WRIGHT, Lewis *A popular handbook to the microscope.* Sm. 8vo, 256 pp, 1 pl., text figs, RTS, London 1895; ed. [1922] title *The microscope.*

4170. WYATT, Claude Wilmott *British birds, being illustrations of the British species of passerine birds . . . also picarian birds, birds of prey and pigeons.* Roy. 4to, 2 vols, 67 col. pls, Wesley, London 1894, 99.

4171. WYATT, Mary *Algae Danmonienses, or dried specimens of marine plants, principally collected in Devonshire.* 4to, t.ps & 236 specimens, Cockrem printed, Torquay [1834–40]. 5 vols. Vol. 5 is supplement with title . . . *to which are added some rare species from the coast of Cornwall.*

4172. WYATT, W. & THORNTON, C. G. *Flora Repandunensis. The wild flowers of Repton, with a list of: I. The birds. II. The moths and butterflies.* 8vo, vi+98 pp, 1 pl., 1 map, Bemrose, London 1866; 2ed. 1881 title *Contributions to the flora and fauna of Repton.*

Wycliff Museum, Newcastle *see* Fox, G. T.

4173. WYBURD, Henry *Introduction to the Linnaean classification of plants.* 12mo, iv+94 pp, 21 pls, Darton, London 1810.

4174. WYLDE, James *editor The circle of the sciences.* 4to, 3 vols, 44 pls, 10 portrs, 6 maps, The London Printing & Publishing Co., London [1862–67]. With title pages I–III, but pagination and index in 2 vols. Biology is in Vol. II.

4175. —— *editor The Royal natural history.* Roy. 4to, ii+8+847 pp, 34 col. pls, text figs, The London printing & publishing Co., London [1881– ]; ed. [ ].

# Y

4176. YARRELL, William *A history of British fishes.* 4to, Imp., Roy. or demy 8vo, 2 vols, text figs, Van Voorst, London [1835–] 1836; suppl. 2 pts, 1839; 2ed. 1841; 3ed. 1859; 2nd. suppl. 1860. Natural History of the British Islands. See also Newman, E.

4177. —— *A history of British birds.* 4to, Imp., Roy. or demy 8vo, 3 vols, text figs, Van Voorst, London 1837–43; suppl. 72 pp, 1845; 2ed. 1845; 2nd suppl. x+71 pp, 1856; 3ed. 1856; 4ed. 4 vols, 1871–85. Natural History of the British Islands. See also Newman, E. and Saunders, H.

4178. —— *On the growth of salmon in fresh water.* Ob. 4to, 3 pp, 3 col. pls, Van Voorst, London 1839.

—— *see* Temminck, C. J.

4179. *A year of country life; or, the chronicle of the young naturalists.* 16mo, SPCK, London 1853.

*A year with the birds.* 1886. *see* Fowler, William Warde

Yeats, Thomas Pattinson *see* Linnaeus, C. 1773.

4180. [YONGE, *Miss* Charlotte Mary] *The herb of the field.* 8vo, viii+288 pp, Mozley, London 1853; 2ed. 1858; ed. 1887. Reprints of 'Chapters on flowers' in *The magazine for the young.* By the author of *The kings of England.*

4181. [——] *The instructive picture book: Vol. II, Lessons from the vegetable world.* Fol., 31 col. pls with letterpress, Edmonston & Douglas, Edinburgh 1857; ed. 1858; 5ed. London 1877; 6ed. 1882. By the author of *The heir of Redclyffe.* For Vol. I see White, A. & Stark, R. M.

4182. YORK, Frederick *Animals in the gardens of the Zoological Society, photographed from life.* 4to, London 1872.

4183. YOUMANS, *Miss* Eliza Ann *The first book of botany.* 8vo, 202 pp, text figs, Kegan Paul, London 1872. Later editions. First edition is New York 1870.

4184. *The young botanist.* 16mo, vi+184 pp, text figs, RTS, London 1875. By the author of *Flowers and their teachings.*

4185. *The young botanists.* 12mo, 112 pp, 1 pl., text figs, RTS, London [1847].

4186. *The young botanists; in thirteen dialogues.* 12mo, v+149 pp, 12 col. pls, Phillips, London 1810.

Young Collector Series (Sonnenschein) *see*

Bagnall, J. E. *Mosses.* 1886.
Bath, W. H. *Ants, bees, dragonflies* etc. 1885.
—— *British birds.* 1888.
Butler, E. A. *Pond life.* 1886.
Gray, P. & Woodward, B. B. *Seaweeds, shells, and fossils.* 1886.
Holmes, E. M. & Gray, P. *Fungi, lichens, and mosses.* 1886.
Hopley, *Miss* C. *Reptiles.* 1888.
Hutchinson, W. *Grasses.* 1895.
Kirby, W. F. *Butterflies, moths, and beetles.* 1885.
Lowe, E. J. *Ferns.* 1891.
Macpherson, H. A. *British birds.* 1891.
Skuse, F. A. *Crustaceans and spiders.* 1887.
Smithson, T. S. *Pond life.* 1890.
Williams, J. W. *Land and fresh-water shells.* 1888.
Wood, H. *Wild flowers.* 1897.

Young Collectors' Handbook Series (Sonnenschein) *see*

Britten, J. *Flowering plants.* [1883].
Kirby, W. F. *Beetles.* [1883].
—— *Butterflies and moths.* [1883].
—— *Orders of insects.* [1883].
Sharpe, R. B. *British birds.* [1883].
Woodward, B. B. *Shells.* [1883].

4187. *The young ladies book of botany.* 16mo, Tyas, London 1838; 1840.

4188. *The young lady's book. A manual of elegant recreations, exercises and pursuits.* 8vo, 504 pp, 3 pls, text figs, Vizetelly, London 1829; 2ed. 1829; ed. 1859.

4189. *The young naturalist, a new history of foreign and British beasts.* 12mo, 24 pp, text figs, Marsden, Chelmsford [c.1830].

4190. YOUNG, Edward *The ferns of Wales.* 4to, v+29 pp, 1 col. pl. & 28 ll of 34 specimens, Thomas, Neath 1856.

4191. YOUNG, Henry C. *A contribution towards a complete list of the fauna and flora of Clydesdale and the West of Scotland.* 8vo, Glasgow 1876.

4192. YOUNG, William Weston *Proposals for publishing . . . British ornithology, or history of birds.* 8vo, 2 pp, 1 col. pl., London [c.1810]. Prospectus only, never published, with 1 col. aquatint of bullfinch.

*The young travellers.* 2ed. 1816. *see Longlands Castle.*

# Z

Zeller, Philipp Christoph *see* Stainton, H. T.

4193. ZIMMERMANN, Albrecht Wilhelm Phillipp *Botanical microtechnique, a handbook of methods.* 8vo, xii+296 pp, text figs, Constable, London 1896. Translation of *Die botanische Microtechnik*, Tübingen 1892, by James Ellis Humphrey.

4194. *The Zoological Gardens.* 8vo, 32 pp, Clarke, London 1852; ed. 12mo, [1853]; ed. 16mo, [1853].

4195. *Zoological keepsake; or zoology, and the garden and museum of the Zoological Society for the year 1830.* 12mo, text figs, Marsh & Miller, London 1830.

4196. *The zoological library; or, natural history, made familiar by a pleasing account of all the most remarkable quadrupeds, birds, fishes, reptiles & insects.* 12mo, 1 pl., text figs, Sears, London [ ].

*Zoological notes and anecdotes.* 1852. *see* Cooper, W. W.

4197. ZOOLOGICAL SOCIETY OF LONDON *Catalogue of the Mammalia contained in the Museum.* 8vo, 18 pp, for the Society, London 1828; 2ed. 1838; suppl. 1839.

4198. —— *Catalogue of the animals preserved in the Museum.* 8vo, 40 pp, text figs, for the Society, London 1829.

4199. —— *Guide to the Gardens.* 8vo, for the Society, London 1829; to 58ed. 1903. Continued as *Official guide* . . . Eds 1 & 2 by Mitchell, D. W.; eds 3–5 by Sclater, P. L.

4200. —— *List of animals in the Gardens.* 8vo, for the Society, London 1829. Later editions up to 1844.

4201. —— *Report of the farm of the Society at Kingston Hill.* 8vo, for the Society, London 1832.

4202. —— *Hand-book for visitors.* 8vo, 114 pp, text figs, Tyas, London 1838.

4203. —— *Popular guide to the Gardens.* 8vo, for the Society, London 1852. By Mitchell, D. W.

4204. —— *Guide to the fish-house in the Gardens.* 8vo, 50 pp, text figs, for the Society, London 1860. By Holdsworth, E. W. H.

4205. —— *List of vertebrated animals living in the gardens.* 8vo, 100 pp, the Society, London 1862; 2ed. 1863; 3ed. 1865; 4ed. 1867; 5ed. 1872; suppl. 1875; 6ed. 1877; 7ed. 1879; suppl. 1880; 8ed. 1883; 9ed. 1896; centenary ed, 3 vols, 1929.

4206. —— *Guide to the insect house.* 8vo, for the Society, London 1881; ed. 1886.

—— *see*

Aflalo, F. G.

Fraser, L.

Morley, C. & Friederichs, H.

York, F.

# PART II
# LIST OF TITLES IN DATE ORDER
# from 1495–1800

# LIST OF TITLES IN DATE ORDER from 1495–1800

[?1495] Bartholomaeus Anglicus *De proprietatibus rerum.* [First English ed.]
[?1521] *The noble lyfe & nature of man, of bestes.*
1538 Turner, W. *Libellus de re herbaria novus.*
1544 Turner, W. *Avium praecipuarum.*
[1548] Turner, W. *The names of herbes.*
1551, 62 Turner, W. *A new herball.*
1551 Wotton, E. *De differentiis animalium libri decem.*
1567 Maplet, J. *A green forest.*
1570 Caius, J. *De canibus Britannicis* etc.
1570 Pena, P. & L'Obel, M. *Stirpium adversaria nova.*
1578 Dodoens, R. *A nieuwe herball.*
1586 Le Moyne de Morgues, J. *La clef des champs.*
1597 Gerard, J. *The herball.*
1601 Plinius Secundus *The historie of the world.* [First full translation.]
1607 Topsell, E. *A historie of foure-footed beasts.*
1608 Topsell, E. *The historie of serpents.*
1621 Widdowes, I. *Natural philosophy.*
1622–34 Simson, A. *Hieroglyphica animalium.*
1627 Bacon, F. *Sylva sylvarum.*
[1629] Johnson, T. *Iter plantarum investigationis.*
1632 Johnson, T. *Descriptio itineris plantarum investigationis.*
1634 Johnson, T. *Mercurius botanicus.*
1634 Moffett, T. *Insectorum sive minimorum animalium theatrum.*
1635 Swan, J. *Speculum mundi.*
1640 Parkinson, J. *Theatrum botanicum.*
1641 Johnson, T. *Mercurii botanici pars altera.*
1646 Browne, *Sir* Thomas *Pseudodoxia epidemica.*
[c.1650] Hoole, C. *A plain and easy primer for children.*

1650 How, W. *Phytologia Britannica.*
1651 Harvey, W. *De excercitationes de generatione animalium.*
1651 Highmore, N. *The history of generation.*
1652 Boate, G. *Ireland's natural history.*
1655 Moffett, T. *Healths improvement.*
1656 Tradescant, J. *Musaeum Tradescantianum.*
1657 Brown, T. *pseudonym. Nature's cabinet unlocked.*
1657 Coles, W. *Adam in Eden.*
1657 Johnston, J. *A history of the wonderful things of nature.*
1659 Comenius, J. A. *Orbis sensualium pictus.*
1659 Lovell, R. *Enchiridion botanicum.*
1660 Childrey, J. *Britannica Baconica.*
1660 [Ray, J.] *Catalogus plantarum circa Cantabrigium nascentium.*
1661 Digby, *Sir* Kenelm *A discourse concerning the vegetation of plants.*
1661 Lovell, R. *Panzoologicomineralogia.*
1662 Cowley, A. *Plantarum libri duo.*
1664 [Evelyn, J.] *Sylva.*
1664 Hubert, R. *A catalogue of those rarities collected.*
1664 Power, H. *Experimental philosophy.*
1665 Hooke, R. *Micrographia.*
1666 Merrett, C. *Pinax rerum naturalium Britannicarum.*
1668 Charleton, W. *Onomasticon zoicon.*
1668 Wilkins, J. *An essay towards a real character.*
1669 Malpighi, M. *Dissertatio epistolica de bombyce.*
1669 Morison, R. *Hortus Regius Blesensis.*
1669 [Vaughan, T.] *A brief history . . . with refutation of such vulgar errors as our modern authors have omitted.*
1670 Frantze, W. *The history of brutes.*
1670 Ray, J. *Catalogus plantarum Angliae.*
1671 Legrand, A. *Philosophia veterum e mente Renati Descartes.*
1672 Grew, N. *The anatomy of vegetables begun.*
1672 Morison, R. *Plantarum umbelliferarum distributio nova.*
1672 Willis, T. *De anima brutorum.*
1673 Grew, N. *An idea of a phytological history propounded.*
[?1673] Plot, R. *Quae's to be propounded to the most ingenious in each County.*
1674 Ray, J. *A collection of English words not generally used.*
1675 Grew, N. *The comparative anatomy of trunks.*
1675–79 Malpighi, M. *Anatome plantarum.*

1676 Willughby, F. *Ornithologiae libri tres.*
1677 [Plot, R.] *The natural history of Oxfordshire.*
1678 Johnston, J. *A description of the nature of four-footed beasts.*
1678 Lister, M. *Historiae animalium Angliae tres tractatus.*
1680, 99 Morison, R. *Plantarum historiae universalis Oxoniensis.*
1680 [Tyson, E.] *Phocaena, or the anatomy of a porpess.*
1681 Grew, N. *Musaeum Regalis Societatis.*
1681 Swammerdam, J. *Ephemeri vita.*
1682 Goedart, J. *Of insects.*
1682 Grew, N. *The anatomy of plants.*
1682 Mullen, A. *An anatomical account of the elephant accidentally burnt.*
1682 Ray, J. *Methodus plantarum nova.*
1683 Browne, *Sir* Thomas *Certain miscellany tracts.*
1683 Sibbald, *Sir* Robert *Nuncius Scoto-Britannus.*
[1683–] 1684 Sibbald, *Sir* Robert *Scotia illustrata.*
1685 Collins, S. *A system of anatomy.*
1685 Lister, M. *De cochleis.*
1685 [–1697] Lister, M. *Historia conchyliorum.*
1686 Plot, R. *The natural history of Staffordshire.*
1686–1704 Ray, J. *Historia plantarum.*
1686 Willughby, F. *De historia piscium libri quatuor.*
1688 Perrault, C. *The natural history of animals.*
1688 Ray, J. *Fasciculus stirpium Britannicarum.*
1689 Malpighi, M. *Opera omnia.*
?1689 Newton, J. *Enchiridion universale plantarum.*
1690 Ray, J. *Synopsis methodica stirpium Britannicarum.*
1691–92 Plukenet, L. *Phytographia.*
1691 Ray, J. *The wisdom of God.*
1692 Boyle, *Hon.* Robert *General heads for a natural history of a country.*
1692 Sibbald, *Sir* Robert *Phalainologia nova.*
1693 Blount, T. P. *A natural history.*
1693 Wallace, J. *A description of the Isles of Orkney.*
1694 Lister, M. *Exercitatio anatomica.*
1695 Lister, M. *Exercitatio anatomica altera.*
1696 Lister, M. *Conchyliorum bivalvium . . . exercitatio anatomica tertia.*
1696 Pitcairne, A. *Dissertatio de legibus historiae naturalis.*
1696 Plukenet, L. *Almagestum botanicum.*
1696 Ray, J. *De variis plantarum methodis dissertatio brevis.*

1696 Ray, J. *Synopsis methodica animalium quadrupedum et serpentini generis.*
1696 Robinson, T. *New observations of the natural history of this world of matter.*
1696 [Woodward, J.] *Brief instructions for making observations.*
1697 Malpighi, M. *Opera posthuma.*
1697 Sibbald, *Sir* Robert *Auctarium Musaei Balfourniani.*
1698 Martin, M. *A late voyage to St Kilda.*
1700 Cox, T. *A topographical . . . and natural history of Great Britain.*
1700 Leigh, C. *The natural history of Lancashire, Cheshire and the Peak.*
1700 Plukenet, L. *Almagesti botanici mantissa.*
1703 Martin, M. *A description of the Western Islands of Scotland.*
1705 Plukenet, L. *Amaltheum botanicum.*
1705 Ray, J. *Methodus insectorum.*
[?1709] Petiver, J. *Brief directions for . . . making . . . collections.*
1709 Robinson, T. *An essay towards a natural history of Westmorland and Cumberland.*
1709 Sibbald, *Sir* Robert *A letter from Sir R—— S—— to Dr. A. Pitcairn.*
c.1710 Greenwood, J. *The London vocabulary.*
1710 Ray, J. *Historia insectorum.*
1710 *A short and easie method to give children an idea . . . of most things.*
1710 Sibbald, *Sir* Robert *A history . . . of the Sheriffdoms of Fife and Kinross.*
1710 Sibbald, *Sir* Robert *Vindiciae Scotiae illustratae.*
1711 Allen, B. *The natural history of the mineral waters of Great Britain.*
1712 Morton, J. *The natural history of Northampton-shire.*
1713 Derham, W. *Physico-theology.*
1713 Petiver, J. *Herbarii Britannici . . . Raii catalogus.*
1713 Petiver, J. *Botanicum Anglicum, or, the English herbal.*
1713 Ray, J. *Synopsis methodica avium et piscium.*
?1715 Culpepper, E. *The description and use of a set of portable microscopes.*
[1716] Petiver, J. *Graminum . . . Britannicarum concordia.*
1716 Petiver, J. *Papilionum Britanniae icones.*
[1716–] 1719–30 Tournefort, J. P. de *The compleat herbal.*

1718–19 Nieuwentijdt, B. *The religious philosopher.*
1718 Ray, J. *Philosophical letters.*
1719, 18 Aubrey, J. *The natural history . . . of the County of Surrey.*
1720 Albin, E. *The natural history of British insects.*
1720 Blair, P. *Botanick essays.*
1721 Bradley, R. *A philosophical account of the works of nature.*
1724 Bradley, R. *New experiments . . . relating to the generation of plants.*
1725 Douglas, J. *Lilium Sarniense.*
[1726] Casteels, P. *Icones avium.*
1726 Threlkeld, C. *Synopsis stirpium Hibernicarum.*
1727 Buchan, A. *A description of St Kilda.*
1727 Hales, S. *Vegetable staticks.*
1727 Martyn, J. *Methodus plantarum circum Cantabrigiam nascentium.*
1727 Stephens, W. *Botanical elements.*
1728 Martyn, J. *Historia plantarum rariorum.*
1728 [Woodward, J.] *Brief instructions for making observations.*
1729 Martyn, J. *The first lecture of a course of botany.*
1729 [Salter, J.] *A catalogue of rarities to be seen at Don Saltero's coffee-house.*
1730 [Boreman, T.] *A description of three hundred animals.*
1730 Denne, J. *The wisdom of God in the vegatable creation.*
1730 Southall, J. *A treatise of buggs.*
1730 Taylor, S. *The history and antiquities of Harwich.*
1731–38 Albin, E. *A natural history of birds.*
1731–33 Hales, S. *Statical essays.*
1732 Coker, J. *A survey of Dorsetshire.*
1732 Tournefort, J. P. de. *History of plants growing about Paris.*
1733 Cheselden, W. *Osteographia.*
1733 Denne, J. *The wisdom and goodness of God in the vegetable creation.*
1733–48 Pluche, N. A. de la *Nature display'd.*
1735 Keogh, J. *Botanologia universalis Hibernica.*
1735 Martin, B. *Grammatica philosophica.*
1736 Albin, E. *A natural history of spiders.*
1736 [Boreman, T.] *A supplement to A description of three hundred animals.*
1736 Collins, C. *Icones avium.*

1736 Salmon, W. *Antiquities of Surrey.*
1737 Albin, E. *A natural history of English song-birds.*
1737 Blackstone, J. *Fasciculus plantarum circa Harefield sponte nascentium.*
1738 Deering, G. C. *Catalogus stirpium . . . about Nottingham.*
[1738] Martin, B. *The description and use of a . . . pocket reflecting microscope.*
1739 [Boreman, T.] *A description of . . . creatures omitted in The description of three hundred animals.*
1739 Keogh, J. *Zoologia medicinalis Hibernica.*
1740 Alston, C. *Index plantarum . . . in Horto Medico Edinburgensi: (C. Linnaei Fundamenta botanica).*
1740 Williamson, J. *The British angler.*
1741 Dillenius, J. J. *Historia muscorum.*
1741 Petiver, J. *Opera.*
1741–42 Wilkes, B. *Twelve new designs of English butterflies.*
1742 Baker, H. *The microscope made easy.*
1742 Martin, B. *Micrographia nova.*
1742 Owen, C. *An essay towards a natural history of serpents.*
1743–46 Adams, G. *the father. Micrographia illustrata.*
1743 Baker, H. *An attempt towards the natural history of the polype.*
1743–51 Edwards, G. *The natural history of some uncommon birds.*
1743 Mortimer, C. *Some papers lately read before the Royal Society.*
1743 *Ornithologia nova.*
1744 [Monro, A. *first*] *An essay on comparative anatomy.*
1744 Wilson, J. *A synopsis of British plants in Mr Ray's method.*
1745 [Needham, J. T.] *An account of some new microscopical discoveries.*
1745 Parsons, J. *Microscopic theatre of seeds.*
1746 Blackstone, J. *Specimen botanicum . . . rariorum Angliae.*
1746 Coetlogon, D. de *A tour through the animal world.*
1747 Gould, W. *An account of English ants.*
1747–60 [Wilkes, B.] *The English moths and butterflies.*
1748[–1749] Dutfield, J. *A new and complete natural history of English moths and butterflies.*
1748–52 Hill, J. *A general natural history.*
1749 Needham, T. *Observations upon generation.*
1750 Heath, R. *A natural and historical account of the Islands of Scilly.*

1750 Maillet, B. de *Telliamid*.
1750 *The wonders of nature and art.*
1751 Hill, J. *A review of the works of the Royal Society of London.*
1752 Dodd, J. S. *An essay towards a natural history of the herring.*
1752 Hill, J. *Essays in natural history and philosophy*.
1752 Newton, J. *A complete herbal*.
1752 Parsons, J. *Philosophical observations on the analogy between the propagation of animals and that of vegetables*.
1753 Alston, C. *Tirocinium botanicum Edinburgense*.
1753 Baker, H. *Employement for the microscope*.
1753 *A pretty book of pictures.*
1753 Watkins, W. *A treatise on forest trees*.
1754 Linnaeus, C. *Flora Anglica.*
1754 *Queries proposed . . . in hope of obtaining . . . a better knowledge of . . . antiquities and natural history*.
1754 Watson, F. *The animal world display'd.*
1755 Ellis, J. *An essay towards the natural history of the corallines*.
1756 Hill, J. *The British herbal.*
1757 Botanista, Theophilus *pseudonym Rural beauties; or, the natural history of . . . Cornwall* etc.
1757 Hill, J. *The sleep of plants.*
1757 Hill, J. *A system of botany.*
1758 Borlase, W. *The natural history of Cornwall.*
1758–64 Edwards, G. *Gleanings in natural history*.
1758 Hill, J. *An account of a stone . . . which on being watered produces excellent mushrooms.*
1758–62 Hill, J. *Botanical tracts*.
1758 Hill, J. *Outlines of a system of vegetable generation*.
1758 Swammerdam, J. *The book of nature*.
1759 Hill, J. *Flora Britanica*.
1759 Hill, J. *The origin and production of proliferous flowers*.
1759–75 Hill, J. *The vegetable system.*
1759–63 Martin, B. *The natural history of England*.
1759–60 Miller, J. *Collection of plants and insects*.
1759 Stillingfleet, B. *Miscellaneous tracts relating to natural history*.
1760 Lee, J. *An introduction to botany*.
1760 Miller, P. *A short introduction to the knowledge of the science of botany*.

[c.1760] Roesel von Rosenhof, A. J. *A collection of curious insects.*
1761–66 [Pennant, T.] *British zoology.*
1761 Stillingfleet, B. *The calendar of flora, Swedish and English.*
1762 Hudson, W. *Flora Anglica.*
1763 Brookes, R. *A new . . . system of natural history.*
1763 Lyons, I. *Fasciculus plantarum circa Cantabrigium nascentium.*
1763 Martyn, Thomas F. R. S. *Plantae Cantabrigienses.*
1763 Wesley, J. *A survey of the wisdom of God in the creation.*
[1763] Wheeler, J. *The botanist's and gardener's new dictionary.*
1764 *England illustrated.*
1764 *Flora Anglica.*
1764 Macauley, K. *The history of St Kilda.*
1764 Martyn, Thomas F. R. S. *Heads of a course of lectures in botany.*
1765 [Gregory, J.] *A comparative view of the state and faculties of man with those of the animal world.*
1766 Bonnet, C. *Contemplation of nature.*
1766 Harris, M. *The aurelian.*
[1767] Harris, M. *An essay preceding a supplement to The aurelian.*
[1768–]1770 *The British herbal.*
1768 [Bruckner, J.] *A philosophical survey of animal creation.*
1768 Wildman, T. *Treatise on the natural history of bees . . . the natural history of wasps and hornets.*
1769–70 *A description of England and Wales.*
1769 Duttens, L. *An inquiry into the origins of the discoveries attributed to the moderns.*
1769–70 Hill, J. *Herbarium Britannicum.*
1769 Spallanzani, L. *An essay on animal reproductions.*
1769 Telltruth, T. *pseudonym Natural history of four-footed beasts.*
1769 Wallis, J. *The natural history and antiquities of Northumberland.*
1770 Edwards, G. *Essays upon natural history.*
1770 Forster, J. R. *A catalogue of British insects.*
1770 Hill, J. *The construction of timber from its earliest growth.*
1770 Linnaeus, C. *Termini botanici.*
[?1770] Martin, B. *Optical essays.*
[1770–]1777 Miller, J. *Illustratio systematis sexualis Linnaei.*
1770 Milne, C. *A botanical dictionary.*

1770 Telltruth, T. *pseudonym Natural history of birds.*
1771 [Curtis, W.] *Instructions for collecting and preserving insects.*
[1771–]1775 Hayes, C. *The portraits of British birds.*
1771 Linnaeus, C. *Institutes of botany.*
1771 Linnaeus, C. *Genera animalium.*
1771 Linnaeus, C. *Genera plantarum.*
1771 Martin, B. *The description and use of the graphical perspective and microscope.*
[1771] [Mendes da Costa, E.] *Conchology.*
1771 Pennant, T. *Synopsis of quadrupeds.*
1771 Pennant, T. *A tour in Scotland, 1769.*
1771 [Tunstall, M.] *Ornithologia Britannica.*
1771 Warner, R. *Plantae Woodfordienses.*
1772 Bladh, A. J. *Fundamenta entomologiae.*
1772 [Lettsom, J. C.] *The naturalist's and traveller's companion.*
1772 Rutty, J. *An essay towards a natural history of the County of Dublin.*
1773 Hill, J. *A decade of curious insects.*
1773 Linnaeus, C. *Institutions of entomology*
1773 Pennant, T. *Genera of birds.*
1774 [Curtis, W?] *A catalogue of the plants growing wild in the environs of London.*
1774 Dicquemare, J. F. *An essay towards elucidating the history of the sea-anemones.*
1774 Goldsmith, O. *A history of the earth and animated nature.*
1774 Hutchins, J. *The history of the County of Dorset.*
1774 Martin, B. *Typographia naturalis.*
1774 Martin, B. *The description and use of an opake solar microscope.*
1775–76 Buffon, G. L. Leclerc *Conte de Natural history.* [*First translation.*[
[1775–]1777–98 Curtis, W. *Flora Londinensis.*
1775 Harris, M. *The English Lepidoptera.*
1775 Linnaeus, C. *The elements of botany.*
1775 Linnaeus, C. *A generic and specific description of British plants.*
1775 Martyn, T. *F.R.S. Elements of natural history.*
1775 Person, C. *Elements of anatomy and animal oeconomy.*
1775–76 Ward, S. *and others A modern system of natural history.*
1776 Brown, P. *New illustrations in zoology.*

1776 Edwards, G. *Some memoirs of the life and works of George Edwards.*
1776 Ellis, J. *On the nature of the Gorgonia.*
1776[–1780] Harris, M. *An exposition of English insects.*
1776 Mendes da Costa, E. *Elements of conchology.*
[1776–92] Miller, J. F. *Icones animalium et plantarum.*
1776 Withering, W. *A botanical arrangement of all the vegetables naturally growing in Great Britain.*
1777 *Beauties of natural history.*
1777 Curtis, W. *Linnaeus system of botany.*
1777 Jacob, E. *Plantae Favershamienses.*
1777 Lightfoot, J. *Flora Scotica.*
1777 Robson, S. *The British flora.*
[1778] Delany, *Mrs* M. *A catalogue of plants copyed from nature in paper mosaick.*
1778 Rymsdyk, J. van & Rymsdyk, A. van *Musaeum Britannicum.*
1778 Walcott, J. *Flora Britannica indigena.*
1779 Giseke, P. D. *Index Linneanus in Leonardi Plukenetii . . . opera botanica.*
1779 Ingen-Heusz, J. *Experiments upon vegetables.*
1779 Mendes da Costa, E. *Historia naturalis testaceorum Britanniae.*
[?1780] J., R. *Juvenile rambles through the paths of nature.*
1780 [Legg, J.] *A discourse on the emigration of British birds.*
1780 Linnaeus, C. *Genera plantarum; ex editone decima tertia.*
1780 [Stuart, J. *Earl of* Bute] *A tabular distribution of British plants.*
1781 Barbut, J. *The genera insectorum of Linnaeus.*
1781 Barrington, *Hon.* Daines *Miscellanies.*
1781 Lindsay, A. *Dissertatio . . . de plantarum incrementi causis.*
1781–85 Latham, J. *A general synopsis of birds.*
1781 Linnaeus, C. *Select dissertations.*
1781 Pulteney, R. *A general view of the writings of Linnaeus.*
1782 [Broughton, A.] *Enchiridion botanicum.*
[1782] Curtis, W. *A catalogue of certain plants . . . in the environs of Settle.*
1782 Curtis, W. *A short history of the brown-tail moth.*
1782, 83 Linnaeus, C. *A system of vegetables.*
1782 Martyn, T. *F.R.S. Heads of a course of lectures in natural history.*

1783 Barbut, J. *The genera vermium.*
1784 [Aikin, J.] *The calendar of nature.*
[1784] Boys, W. & Walker, G. *Testacea minuta rariora.*
1784 Jones, W. *The religious use of botanical philosophy.*
1784 Linnaeus, C. *Characteres generum plantarum.*
1784–87 Martyn, T. *draughtsman The universal conchologist.*
1784–87 Pennant, T. *Arctic zoology.*
1784 Spallanzani, L. *Dissertations relative to the natural history of animals and vegetables.*
1785, 90 Bolton, J. *Filices Britannicae.*
1785–1801 Dickson, J. *Fasciculus primus [-quartus] plantarum cryptogamicarum Britanniae.*
1785 Linnaeus, C. *Reflections on the study of nature.*
1785 Martin, M. *The aurelian's vade mecum.*
1785 Martyn, W. F. *A new dictionary of natural history.*
[1785] Millar, G. H. *A new . . . body . . . of natural history*
1785 Monro, A. *second The structure and physiology of fishes.*
1785 Relhan, R. *Flora Cantabrigiensis.*
1785 Rousseau, J. T. *Letters on the elements of botany.*
[?1785] Stuart, J. *Earl of* Bute *Botanical tables containing the different families of British plants.*
1786 *The description of a new universal microscope.*
1786 Ellis, J. & Solander, D. C. *The natural history of many curious and uncommon zoophytes.*
1786–91 [Galton, S. J.] *The natural history of birds.*
1786 Hunter, J. *Observations on certain parts of animal economy.*
1786 Johnson, R. *Juvenile rambles through the paths of nature.*
1786 Linnaeus, C. *A dissertation on the sexes of plants.*
1786–87 Martin, M. *Observations on marine vermes, insects.*
1786 Pennant, T. *Histoire naturelle des oiseaux . . . systematically disposed.*
1787 Adams, G. *the son Essays on the microscope.*
1787 [Curtis, W.] *General observations on . . . the seeds of our best grasses.*
1787 [Darwin, R. E.] *Principia botanica.*
1787 [Fyfe, A. & Rotherham, J.] *A system of anatomy and physiology.*
1787 Hope, T. C. *Tentamen . . . de plantarum motibus et vita complectens.*
1787 Kentish, R. *An essay on the method of studying natural history.*

1787 Linnaeus, C. *The families of plants.*
1787 Relhan, R. *Heads of a course of lectures on botany.*
1787[–1789] [Taylor, C.] *Surveys of nature.*
1788 *Animals drawn from nature.*
1788–89 Bolton, J. *An history of the fungusses growing about Halifax.*
1788, 95 Cordiner, C. *Remarkable ruins.*
1788 Curtis, W. *Companion to the Botanical magazine.*
1788 Martyn, T. *F.R.S. Thirty-eight plates . . . to illustrate Linnaeus's System of vegetables.*
[1788] Sowerby, J. *the elder An easy introduction to drawing flowers.*
1788 Sturm, C. C. *Reflections on the works of God.*
1789 Darwin, E. *The loves of the plants.*
[1789–90] [Donovan, E.] *An essay on the minuter parts of plants in general.*
1789 Ferryman, R. *A catalogue of the . . . Museum of Rev. R. Ferryman.*
1789[–1794] Lewin, W. *The birds of Great Britain with their eggs.*
1789 *A new moral system of natural history.*
1789 Pilkington, J. *A view of the present state of Derbyshire.*
[1789–]1790–1813 Shaw, G., Nodder, F. P. & Nodder, R. P. *The naturalist's miscellany.*
1789 Walcott, J. *Synopsis of British birds.*
1789 [White, G.] *The natural history and antiquities of Selborne.*
1790 [Bewick, T.] *A general history of quadrupeds.*
1790 Latham, J. *Index ornithologicus.*
[c.1790] *Lilliputian Spectacle de la nature.*
1790 Martinet, J. F. *Catechism of nature.*
1790 Pole, T. *The anatomical instructor.*
1790 Pulteney, R. *Historical and biographical sketches of the progress of botany in England.*
1790 Riley, G. *The beauties of creation.*
1790 Rooke, H. . . . *Oaks in the park at Welbeck.*
1790 Rotherham, J. *The sexes of plants.*
1790 Shaw, G. *Speculum Linnaeanum.*
1790, 99 Smellie, W. *The philosophy of natural history.*
1790–1814 Smith, *Sir* James E. *English botany.*
1790 Swayne, G. *Gramina pascua.*
1791, 90 Darwin, E. *The botanic garden.*
1791 Gilpin, W. *Remarks on forest scenery.*

1791 *The history of birds.*
1791[–1795] Lord, T. *Entire new system of ornithology.*
1791–92 *A new system of natural history of quadrupeds, birds, fishes and insects.*
1791 Townley, R. *A journal kept in the Isle of Man.*
1792 Boys, W. *Collections for an history of Sandwich in Kent.*
1792 Curtis, W. *Proposals for a course of herbarizing excursions.*
1792–1813 Donovan, E. *The natural history of British insects.*
[1792] Forster, B. M. *Peziza cuticulosa.*
1792 Heron, R. *Elegant extracts of natural history.*
1792 Linnaeus, C. *The animal kingdom.*
1792 Martyn, T. *draughtsman The English entomologist.*
1792–94 Martyn, T. *F.R.S. Flora rustica.*
1792 *The natural history of insects.*
1792 Saunders, S. *A short and easy introduction to scientific and philosophical botany.*
1792, 96 Shaw, G. *Musei Leveriani explicatio.*
1792 Walker, J. *Institutes of natural history.*
1793 Falconer, W. *Miscellaneous tracts.*
1793 [Hope, J.] *Characteres genera plantarum.*
1793 [Jones, S.] *The natural history of birds.*
1793 [Jones, S.] *The natural history of fishes.*
1793 Martyn, T. *draughtsman Aranei, or the natural history of spiders.*
1793 Martyn, T. *F.R.S. The language of botany.*
1793 Milne, C. & Gordon, A. *Indigenous botany.*
1793 Ure, D. *The history of Rutherglen and East Kilbride.*
1794–96 Bolton, J. *Harmonia ruralis.*
1794–96 Darwin, E. *Zoonomia.*
1794 Donovan, E. *Instructions for collecting and preserving various subjects of natural history.*
1794[–1819] Donovan, E. *The natural history of British birds.*
1794 Forsyth, W. *A botanical nomenclature.*
1794, 99 Hayes, C. *Portraits of rare and curious birds . . . of Osterley Park.*
1794 *A history of esculent fish.*
1794 [Peacock, L.] *The visit for a week.*
[1794–1804] Silby, E. *An universal system of natural history.*
1794 Sibthorp, J. *Flora Oxoniensis.*
1794 Wade, W. *Catalogus systematicus plantarum . . . Dublinensi.*

1794 Wakefield, *Mrs* P. *Mental improvement.*
1795 Anderson, J. *Recreations in agriculture, natural history.*
[1795–]1805 Church, J. *A cabinet of quadrupeds.*
1795 Ferryman, R. *A catalogue of the . . . British Zoological Museum.*
1795 Herdman, J. *Essay on the causes and phenomena of animal life.*
1795 *History of fishes in miniature.*
1795 Lewin, W. *British birds' eggs.*
1795 Lewin, W. *Insects of Great Britain.*
[?1795] *A natural history of sea anemones.*
1795 Smith, *Sir* James E. *Syllabus of a course of lectures on botany.*
[1795–]1797–1809 Sowerby, James *the elder. Coloured figures of English mushrooms.*
[1795–]1801 Stackhouse, J. *Nereis Britannica.*
1795 Velley, T. *Coloured figures of marine plants.*
1795 White, G. *The naturalist's calendar.*
1796 Coyte, W. B. *Hortus Britannicus Gippovicensis.*
1796 Fabricius, J. C. *Nomenclator entomologicus.*
1796 Gregory, G. *The economy of nature explained.*
1796 Raff, G. C. *A system of natural history adapted for the instruction of youth.*
1796 Saint-Pierre, B. *Studies of nature.*
1796 Wakefield, *Mrs* P. *An introduction to botany.*
1797, 1804 [Bewick, T.] *History of British birds.*
1797 Covolo, G. dal *A discourse concerning the irritability of some flowers.*
[1797] Fenn, *Lady* Elleanor *A short history of insects.*
1797, 1809 Freeman, S. *Select specimens of British plants.*
1797 Hopper, R. *Observations on the structure and economy of plants.*
1797 [Jackson, *Mrs* M. E.] *Botanical dialogues.*
[1797] Martyn, T. *draughtsman. Psyche. Figures of nondescript lepidopterous insects.*
1797 Maton, W. G. *Observations . . . on the western Counties of England.*
1797 Rooke, H. *A description of the great oak in Salcey forest.*
1797 Stead, J. *Description and natural history of English song birds.*
1797 Strickland, *Miss* C. & Strickland, *Miss* J. S. *Select specimens of British plants.*

1798 Abbot, C. *Flora Bedfordiensis*.
1798 Aikin, J. *The natural history of the year*.
1798 Brown, T. *Observations on the Zoonomia of Erasmus Darwin*.
1798 *The children's cabinet*.
1798–1807 Leeuwenhoek, A. van *The select works*.
1798 Pratt, S. J. *Pity's gift*.
1798 Smith, *Sir* James E. *Tracts relating to natural history*.
1798 Sole, W. *Menthae Britanniae*.
1798 Symons, J. *Synopsis plantarum Insulis Britannicis indigenerum*.
1798 Thompson, J. *Botany displayed*.
1799 Bullock, W. *A companion to Mr. Bullock's Museum*.
[1799–]1800–04 Donovan, E. *The natural history of British shells*.
1799 Faujas de Saint-Fond, B. *Travels in England, Scotland, and the Hebrides*.
1799 Fothergill, C. *Ornithologia Britannica*.
1799 Hull, J. *The British flora*.
1799 Lesser, F. C. *Insecto-theology*.
1799–1802[–1803] *The naturalist's pocket magazine*.
1799 Pulteney, R. *Catalogue of the birds, shells . . . plants of Dorsetshire*.
1799 Rooke, H. *Ancient and present state of Sherwood Forest*.
1799 Spallanzani, L. *Tracts on the nature of animals and vegetables*.
[1799–]1807 Thornton, J. R. *New illustrations of the sexual system of C. von Linnaeus*.
1799 Townson, R. *Tracts and observations in natural history*.
1799 Trimmer, *Mrs* S. *An easy introduction to the knowledge of nature*.
1799 White, C. *An account of the regular gradation in man, and in different animals and vegetables*.
1800 Bewick, T. *Figures of British land birds*.
1800 Binglay, W. *A tour round North Wales*.
1800 *A concise abridgement of natural history*.
1800 Evans, J. *A tour through part of North Wales*.
1800 Hull, J. *Elements of botany*.
[c.1800] *The infant's library. Book 5* [Quadrupeds].
1800–04 Linnaeus, C. *A general system of nature*.
1800 Lobb, R. *The contemplative philosopher*.

1800 Mavor, W. F. *The lady's and gentleman's botanical pocket book*.
1800 Mavor, W. F. *Natural history for the use of schools*.
1800 *The natural history of birds and beasts*
[c.1800] *The new book of natural history of various British birds.*
1800–26 Shaw, G. *General zoology*.
1800 Smith, *Sir* James E. *Compendium florae Britannicae*.
1800–04 Smith, *Sir* James E. *Flora Britannica*.

# SUBJECT INDEX

# SUBJECT INDEX

Numbers in italics indicate those works which, in the opinion of the author, are the most important contributions to the subject of the entry.

## A

## B

**C**

## D

**E**

## F

**G**



**H**

## I

## J

## K



## L

## M

## N

## O

## P

**Q**



**R**

## S

## T

**W**

## Y



## Z